The footprint of polar tourism:
Tourist behaviour at cultural heritage sites
in Antarctica and Svalbard

Circumpolar Studies

Volume 7

Circumpolar Studies is a series on Dutch research in the Polar regions by the University of Groningen, Arctic Centre, the Netherlands.

For the other publications in Circumpolar Studies, see the website of the publisher at www.barkhuis.nl

This research was funded by the University of Groningen, Groningen Institute of Archaeology, as part of the University's Ubbo Emmius Scholarship Programme.

Photos cover:

1) Airship *Italia* taking off from Ny-Ålesund, May 1928. Photo © Norwegian Polar Institute. Reproduced with permission
2, 3) The airship mooring mast at Ny-Ålesund, August 2007. Photos by Ricardo Roura

Cover design: Nynke Tiekstra, Coltsfootmedia, Noordwolde
Book design: Hannie Steegstra, Drachten
Printing: Ten Brink, Meppel

ISBN: 9789077922873

RIJKSUNIVERSITEIT GRONINGEN

THE FOOTPRINT OF POLAR TOURISM:

TOURIST BEHAVIOUR AT CULTURAL HERITAGE SITES IN ANTARCTICA AND SVALBARD

Proefschrift

ter verkrijging van het doctoraat in de
Letteren
aan de Rijksuniversiteit Groningen
op gezag van de
Rector Magnificus, dr. E. Sterken,
in het openbaar te verdedigen op
donderdag 28 april 2011
om 16:15 uur

door

Ricardo Mariano Roura

geboren op 20 januari 1961
te Buenos Aires, Argentinië

Promotor:	Prof. dr. L. Hacquebord
Copromotor:	Dr. P. Groote
Beoordelingscommissie:	Dr. S. Barr Prof. dr. C.J. Bastmeijer Prof. dr. D.C.M. Raemaekers Prof. dr. F. Vanclay

Table of Contents

List of illustrations

List of Tables

Dedicated to my immigrant ancestors for the opportunities they gave us
and for the cultural heritage we inherited from them;
to my parents and brothers;
and to dear Janet, Isabel, and Matías.

Preface

> *I saw the hut...it was exactly as I had imagined it, only instead of being imaginary, it was real. E. Arthur (1994: 499).*[1]

On 25 January 1990 the Greenpeace ship MV *Gondwana* approached Home Beach at Cape Evans, Ross Island, to resupply the organisation's Antarctic station, "World Park Base", and swap crews for the incoming year. I was one of the replacement crew of four. A low cloud hung over the dark basalt of Cape Evans and gave a monochromatic, bi-dimensional feel to the landscape. Mt. Erebus was invisible at the time, and since this was my first visit to that area I could not even imagine its dominant presence. That day Cape Evans did not appear to be a very welcoming place.

This was not my first Antarctic trip but it was going to be the longer, and I would have normally been curious to get to know my future home. Yet as Home Beach came into sight what attracted me was not the bright green Greenpeace building on the port side, but rather a dull grey wooden building to starboard, which I knew was RF Scott's *Terra Nova* hut. Of course, the story of Scott's last expedition is well known. Nationalism and imperial nostalgia abound in some narratives about this expedition, which tend to turn Scott and his companions into stilted heroic characters rather than real people. These narratives influence the way the hut itself is represented. However, there was something touching about seeing such a building in that remote location. Rather than a shrine to heroic deeds the hut looked vulnerable, somehow out of place in the Antarctic wilderness, and I thought about the people who had built that place and had lived there for a while.

The building and its ghosts would be my neighbours for a year. Many of the building's contents, large and small, lie spilled on the landscape around the hut and beyond. These too would become familiar and tell stories of wind and weathering as they changed in minute but noticeable ways through the year. The memorial cross on Wind Vane Hill, which commemorates three men who disappeared on the sea ice, was an everyday reminder that this could be a dangerous place.

At first we were completely isolated. We only saw people when we ventured on the sea ice to McMurdo Station and Scott Base some 40 kilometres away. After about six months, as the winter darkness began to recede, a trickle of people would make the journey from those bases and visit Cape Evans and its famous hut. As the summer approached, more and more people came: carpenters, an archaeologist, VIPs, and a writer who wanted to absorb the ambiance of the hut for her novel in progress. As did the early occupants, we all left traces at Cape Evans, some ephemeral, some longer lasting, superimposed on the natural landscape and on each other. I never saw a tourist ship there at the time or on several visits afterwards, and I was surprised when I learnt that every summer a few hundred paying tourists made their way to this place, which is literally located on one of the most southernmost beaches on Earth. Change

---

1 E. Arthur (1994): *Antarctic Navigation*. New York: Alfred A. Knopf.

did not stop after the hut had last been occupied in 1917. Despite its timeless appearance, the *Terra Nova* hut is far from being frozen in time, but rather is the locus of substantive contemporary activity.

I was interested in the ongoing process of writing and re-writing the landscape as different actors take the scene. In particular, I found the juxtaposition of contemporary tourism and historic sites interesting. Historic sites represent continuity, as symbolised by the lasting remains of past activities and events whereas polar tourism illustrates the rapid change affecting the Polar Regions. What happens at this interface? How are historic sites changing? What is the contribution of tourism to this process? How can this process be managed? Management frameworks seemed technically adequate to assess, manage and monitor impacts, but lacked the depth to understand and conceptualise the interactions between tourism and its destinations.

Consequently, for this research I have chosen behavioural archaeology as the main theoretical framework and as a guide for field methods. "Archaeology" brings to mind past societies and excavations at places such as Pompeii and Çatalhöyük rather than tourism at polar historic sites. However, behavioural archaeology was conceived as an approach to investigating the interactions of people and artefacts in all times and places (Reid, 1995; Reid *et al.*, 1974). It addresses the fundamental issue of interactions between people, their artefacts, and the external environment, in all possible combinations. Michael B. Schiffer defines behavioural archaeology thus:

> *As a theoretical program, it privileges behavior – what people actually do, recognizing that all behavior (defined as activities) consists of people-artifact interactions. Also implied is an interest in the life histories of artifacts, whether one represents these as flow models or activities in behavioral chains. Behavioral archaeology also strives to understand and take into account, in archaeological inference, all relevant cultural and noncultural formation processes of the archaeological record. (n.d).* [2]

At its most basic, polar tourism is a form of interaction between people (and their artefacts) and the natural and cultural environment of the Polar Regions. Multiple transformations may result from this interaction. Some of the consequences of tourism on the environment would be regarded, in management terms, as tourism impacts. The term "impact", however, is not unambiguous, and relies on value judgements that can be influenced one way or another by different agendas. Judging an impact as not significant means that it is subsequently ignored, and so the action that caused the impact in the first place. Behavioural archaeology perspectives enable one to examine tourism impacts from a more detached point of view by which the value-laden term "impacts" becomes the more neutral tourism "artefacts" (i.e. made or modified by effect of tourism).

Although I have simplified behavioural archaeology to suit my research needs, I have found that it is a useful conceptual scheme to decipher how polar tourism works and how human behaviour transforms polar landscapes. The behavioural perspective avoids *a priori* value judgements about the significance of particular tour-

2 http://www.u.arizona.edu/~schiffer/bio.html, accessed 31 January 2010.

ism effects. Rather, all forms of behaviour and all consequences of behaviour are of interest to the understanding of tourism, which – as a second step – helps to identify management needs.

I am not formally trained as an archaeologist so looking into archaeological theory and methods did not initially seem the most obvious choice of research perspective. Cultural heritage, however, can be usefully examined from archaeological perspectives. In addition, by the time I began this research I had been professionally involved with the Antarctic for over two decades, looking at landscapes and at human presence in the polar environment first through the eyes of a geomorphologist and then as a social scientist: how landscapes formed and evolved; how human behaviour modified the natural environment; and how environmental processes affected cultural remains (and often, again, the environment). Behavioural archaeology described a familiar territory, albeit using a new theoretical and methodological language. In addition, behavioural archaeology formed a bridge between the different perspectives I had used so far to examine natural and cultural processes. Since beginning this research I have learnt some more about this framework and how it can be applied to polar tourism. Other investigations on polar tourism (or indeed on other contemporary polar activities) inspired by behavioural archaeology may follow from this preliminary step.

Ricardo M. Roura
Amsterdam, March 2011

And the remarkable thing about the hut was not just that it seemed as if its inhabitants had just stepped out a few moments, a few years, before...It was also that the building itself was moving in slow time....It had missed the twentieth century, remained almost literally in the century's second decade. E. Arthur (1994: 504-505).[1]

Cruise to the very heart of Antarctica – gateway to the South Pole and guardian of some of the richest Antarctic treasures. These expeditions include visits to the four huts of the historic period – those of Borchgrevink, Scott and Shackleton – plus the American, New Zealand and Italian scientific bases. Some great wildlife, including good opportunities for Emperor Penguins. Heritage Expeditions (2010).[2]

1 E. Arthur (1994): *Antarctic Navigation*. New York: Alfred A. Knopf.

2 Publicity for cruises to the Ross Sea area of Antarctica retrieved from http://heritage-expeditions.com/destination/the-ross-sea. Accessed 12 October 2010.

The footprint of polar tourism: tourist behaviour at cultural heritage sites in Antarctica and Svalbard

R. M. Roura

ABSTRACT

1 INTRODUCTION

This thesis aims to describe and interpret the effects of tourism on historic sites in Antarctica and Svalbard (also known as Spitsbergen), and to assess the implications for management. Explorers, whalers, seal hunters, scientists and others have left many material remains in the Polar Regions that are significant because they tell the history of the exploration and exploitation of these regions. Contemporary polar tourism represents a new phase in this "exploration" and exploitation of the Polar Regions. The potential for the transformation of historic sites has increased following the substantial expansion of polar tourism in recent decades. Key cultural heritage sites are regularly included in standard tourist itineraries and are also the subject of specialized tourism. In this context, the central research question of this thesis is:

> *What are the effects of tourism on polar historic sites, and what are the implications of this for the management of tourism and these historic sites?*

2 CONCEPTUAL FRAMEWORK

In this study tourism is examined from the perspective of behavioural archaeology (Schiffer, 2010 and refs.). "Archaeology" may suggest excavating the remains of past cultures. However, behavioural archaeology was conceived as an approach in order to study people-material interactions in all times and places, including aspects of landscape formation processes. Through behaviour, people (tourists) and their artefacts (e.g., boots, walking sticks, and photographic cameras) interact acoustically, chemically, mechanically, thermally, and visually with the external environment (both natural and cultural). The transformations of the environment that result from these interactions can also be regarded as tourism "artefacts". In management terms, these tourism "artefacts" could be regarded as "impacts", depending on a value judgement about their significance.

This main conceptual framework is complemented by two theories. The creation of tourism attractions is analysed on the basis of Leiper (1990). The designation of material remains of past activity as "heritage" is discussed by means of Thompson's Rubbish Theory (1979).

3 RESEARCH METHODS

This thesis examines case studies from either Antarctica or Svalbard, or from both regions. Four field seasons were carried out between 2005 and 2008 in Svalbard (two trips) and Antarctica (two trips). Earlier field work in Antarctica was also included as part of this research. Research methods included site surveys, ethnographic observations, and repeat photography. Complementary research methods were semi-structured interviews; participant observation; the examination of tourist weblogs; and analysis of tourism statistics. The data obtained using the various methods were integrated to produce assessments for the various case studies.

4 MAIN THEMES

This research has explored several interconnected themes regarding tourism and historic sites in the Polar Regions, while recognising that there are other relevant themes that have not been addressed here. The themes explored in this thesis are:

Tourism as behaviour: Tourism operates in a hierarchy of behavioural scales – from tourism as a whole, to individual visitors. Key processes in these various scales include the recruitment and consolidation of particular sites as tourism destinations; the patterns of behaviour displayed by visitors, both in groups and individually, during shore visits; and discrete interactions between visitors and site features of various kinds. The effects of tourism on historic sites result from interactions at all these levels.

Tourism behaviour consists largely of a basic repertoire of actions that may result in interactions between tourists, their artefacts, and historic sites, and result in the creation of tourism "artefacts". Visual and acoustic interactions (instructions, narratives) may lead tourists towards (or away from) particular site features. In behavioural terms, a pattern of visitor activity may result in a corresponding "archaeology" of direct and cumulative traces of visitation, which define a tourist activity area superimposed over a historic activity area. These two areas may overlap but are not necessarily the same. Mechanical interactions such as trampling and handling may change the condition and integrity of historic site components and damage the archaeological record on the ground surface and subsurface.

Tourism as a site formation process, in the context of other (natural and cultural) formation processes: Polar historic sites are not static but subject to transformation and decay resulting from both cultural and non-cultural (natural) processes. Tourism may contribute to this process. Recent time-serial changes were observed in several case studies of site features. No feature degraded substantially during the monitoring period; rather, several features were restored in different ways. Changes were interpreted as resulting from a range of cultural processes (including conservation, research, and tourism) and natural processes (mainly wind action). Local changes take place in the context of broader regional developments in Antarctica and Svalbard.

The designation of past activity sites as "cultural heritage sites": Assessing whether or not a change to something is an impact relies on a judgement as to the value of what is being impacted. In both Antarctica and Svalbard the criteria and processes used to categorise the sites and remains of past activity ("historic sites") as "heritage" are defined by law. Designation of past activity sites as heritage has environmental and geopolitical implications.

The use of cultural heritage as a tourist attraction: Tourism narratives play a central role in the process of constructing meaning and assigning value to historic remains, which helps to create consumable tourist products and tourism attractions. Information about the sites influences tourist behaviour and therefore the degree of exposure or protection of the sites. Tourism narratives enable the different elements of the tourism attraction system to "click" together into a coherent whole, particularly at sites that because of their original function or present condition reveal little of their history.

Research methods to study the interaction of tourism with historic sites: The methods selected for this research allowed the production of a rapid and reasonably informed appraisal of the status, recent changes, tourism modalities, and management needs of the sites investigated. A rapid, low technology approach is appropriate to the Polar Regions, where site visits may be infrequent, opportunistic, and brief. Naturally, more comprehensive assessment methods may be required for different (or more specific) research or management needs

The management of tourism and historic sites: Tourism tends to focus on important site features so that it may have a greater impact on elements that have a greater relative value than the rest of a site. Tourism impacts are generally localized within sites, which can be reflected in localised forms of management. Cumulative impacts are potentially a cause of concern. Some dedicated management tools may be required to regulate the movement of people within historic sites; minimal distances from sensitive features; the use of established tracks; and limits in visitor numbers.

5 CONCLUSIONS

The tourism "footprint" consists of an assemblage of behavioural traces, which develop on different temporal, spatial, and behavioural scales.

Tourism does not takes place at unchanging sites that are "frozen in time", but rather at sites that are undergoing decay.

Tourism contributes to this process. Tourism traces may modify traces of previous activities, without necessarily removing them entirely. Tourism has a minor effect on historic sites in relation to that of natural processes, but tends to focus on significant site features that may be vulnerable to visitation.

Tourism impacts may detract from the archaeological value of historic sites. These impacts should be managed despite the ongoing decay of historic sites using, inter alia, dedicated environmental impact assessment and monitoring; qualified guides with enhanced responsibility to enforce regulations; and precautionary action.

Reference

Schiffer, M.B. (2010): Behavioural archaeology: principles and practice. London: Equinox.

1 General introduction

1.1 TOURISM AND HISTORIC SITES IN ANTARCTICA AND SVALBARD

Many historic sites in the Polar Regions look abandoned, seemingly unchanged since their original occupants left them. This impression is only broken by the increasingly common sight of tourists walking around, looking at the sights and documenting them in photography and film. When historic sites become regular tourism destinations they may experience changes to their condition, appearance, and assigned values. In this context, the objective of this thesis is to describe and interpret the effects of tourism on historic sites in Antarctica and Svalbard, and to assess the implications for management.

These issues are topical because polar historic sites have always attracted visitors and are valuable resources for the global tourism industry. The potential for tourism-related transformation of historic sites and other types of destinations has increased following the substantial growth, expansion and diversification of polar tourism over the past several decades. These developments take place in the context of comprehensive environmental protection regimes for both Antarctica and Svalbard, which – in different ways – also protect the cultural heritage of each region. However, the impacts of tourism on polar cultural heritage have in comparison been less thoroughly studied than those on the natural environment. The outcome of this study will enable a better understanding of the practice of polar tourism, and how it affects or might affect historic features. This information can be used to inform the management of both tourism and historic sites.

The Polar Regions (Fig. 1.1) are vast areas of wilderness where human presence is comparatively recent relative to more temperate regions. People have populated the Arctic for some 10,000 years, yet there are parts of this region, such as the high Arctic archipelago of Spitsbergen (Svalbard),[1] and the whole of the Antarctic continent, that have remained inaccessible up until comparatively recent times. This relative isolation had significant consequences for later developments, three of which will be highlighted here. First, indigenous peoples did not populate Spitsbergen or Antarctica. Rather, people originating at lower latitudes discovered those lands, and over time endeavoured to explore, research, and occupy them, and to exploit their natural resources. Dotted through the apparently pristine polar landscapes of Antarctica and Svalbard are the sites and material remains of these earlier activities. Some sites have experienced

1 Spitsbergen (or Spitzbergen) is the Dutch name for the archipelago as given by Wilhem Barentz. Svalbard is the modern Norwegian name of Spitsbergen, adopted after the signature of the 1920 Treaty Concerning the Archipelago of Spitsbergen. In modern Norwegian usage Spitsbergen is the name given to the larger island of the archipelago, formerly known as "West Spitsbergen". In this thesis the name Spitsbergen will be used to refer to the whole archipelago in a historic context, whereas Svalbard will be used to refer to the modern times.

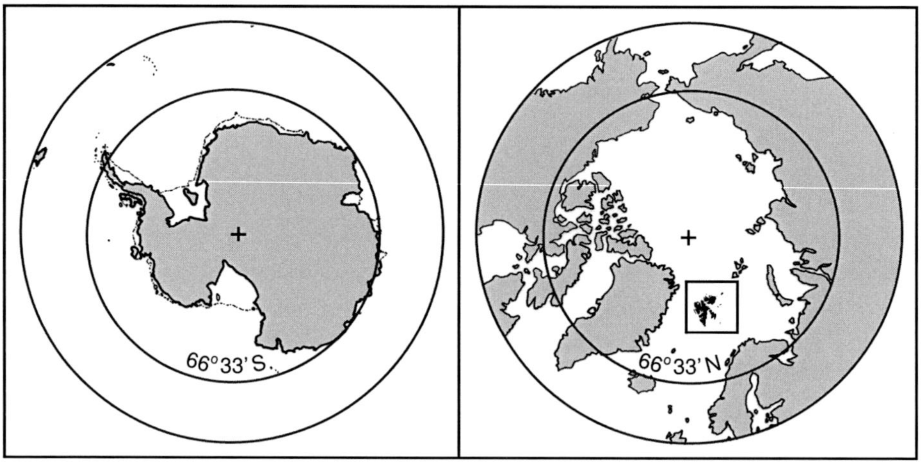

Fig. 1.1 Antarctica and Svalbard. Sketch by author.

a more or less constant occupation since their establishment, while others were abandoned for many years and subject to degradation before being rediscovered and in some cases assigned a new status as "cultural heritage", or simply "heritage".

Second, as a result of their isolation and lack of indigenous population, both Antarctica and Spitsbergen have been regarded as terra nullius or "no-man's" land. This status encouraged territorial ambitions in several states that were at some point active in these regions.[2] Resulting tensions over the control and ownership of territory were in both instances resolved through international treaties. The 1920 Treaty Concerning the Archipelago of Spitsbergen grants the "full and absolute" sovereignty of Spitsbergen to Norway[3] while maintaining equal rights for all contracting parties,[4] subject to some

2 "*Terra nullius* is no man's land, while *res communis* refers to the common areas of the world, such as the high seas and outer space. The legal difference between the two concepts is that sovereignty over *terra nullius* may be acquired by occupation, whereas *res communis* cannot be the object of occupation, except under general acquiescence from other states." (Ulfstein, 1995: 37 and refs.).

3 Spitsbergen Treaty Art. 1.

4 Spitsbergen Treaty Arts. 2 and 3.

Norwegian laws.[5] The 1959 Antarctic Treaty "freezes" the status quo regarding territorial claims for the duration of the Treaty.[6] These international treaties, which came into force in 1925 and 1961 respectively, form the basis for the present day governance of Antarctica and Svalbard (e.g. Mathisen, 1954; Beck, 1986; Triggs, 1987; Ulfstein, 1995). Despite these arrangements, geopolitical considerations – the assertion or denial of territorial claims in Antarctica, and the protection of Norwegian strategic interests and sovereignty in Svalbard – have been and remain a fundamental factor underlying international relations and national practices concerning these regions (e.g. Triggs, 1987; Jabour and Weber, 2008; Pedersen, 2009; Hemmings and Stephens, 2010; Timothy, 2010; Avango *et al.*, 2011; Dodds, forthcoming). There are however significant differences between the two cases.

Third, despite centuries of occupation the regions have remained sparsely populated (albeit to different degrees), remain relatively unspoiled, and are now protected as wilderness areas. Historic remains in the midst of these natural reserves are a reminder that the exploitation of the Polar Regions started decades or centuries ago. Some historic remains, despite their protected status, look very much like waste materials. In fact, some protected polar artefacts *are* waste and have been waste since their primary deposition. As a consequence, there is a basic tension at the interface between the original non-human environment and the remains of more recent human activities.

There is a growing literature, particularly in the wake of the International Polar Year 2007-2009, comparing and contrasting different aspects of the Arctic and Antarctic regions, covering topics as diverse as, *inter alia*, legal regimes (Jabour and Weber, 2008; Leary, 2008), environmental protection and management (Poland *et al.*, 2003; Bastmeijer and Koivurova, 2005); tourism (Snyder, 2007); and climate change (Turner and Overland, 2009). Some comparisons can seem contrived because of the vast differences that exist in the basic characteristics of the environment, cultural geography, geopolitics, and legal regimes of the Arctic and Antarctic regions. Antarctica and Spitsbergen are also substantially dissimilar in their sheer spatial scale, the temporal scale of human occupation, endemic fauna and flora, and sovereignty status (Table 1.1). However, fundamental similarities between Antarctica and Spitsbergen, which result from their remoteness, isolation, lack of indigenous people, historic *terra nullius* status, status as natural reserves, and their being subject to international treaties, make comparative analysis meaningful and relevant for research, governance and management.

---

5 The Treaty Concerning the Archipelago of Spitsbergen grants Norway "absolute and full sovereignty" (§1), but it also contains important restrictions on the exercise of Norwegian sovereign rights (see Ulfstein, 1995: 472-475). § 2 of the Spitsbergen Treaty sets out a principle of non-discrimination towards all 44 treaty members, granting equal rights of fishing and hunting in the treaty area. § 2 also deals with the preservation of Svalbard's natural environment: "Norway shall be free to maintain, take or decree suitable measures to ensure the preservation and, if necessary, the re-constitution of the fauna and flora of the said regions, and their territorial waters". According to Ulfstein (2001: 10) it is up to the Norwegian government to decide what "suitable measures" are. These have included, in Norwegian interpretation, the protection of the cultural heritage.

6 Antarctic Treaty, Art. IV.

Fig. 1.2 – A tourist landing at Whalers Bay, Deception Island, Antarctica, with the remains of a whaling factory and a research station in the background. Photo by author 2002.

Furthermore, as will be discussed in later chapters, many aspects of tourism in these regions are very similar.

The exploration and exploitation of the Polar Regions can be conceived as a series of *waves of activity* (Sugden, 1982; White, 1994; Avango *et al.*, 2011). This should be understood as fundamentally the exploitation of natural resources, interlinked with exploration/science and the geopolitical use of territory. Developments in Antarctica and Spitsbergen have followed a broadly comparable sequence. Subsequent to the discovery of Spitsbergen in 1596, and of Antarctica no later than 1820 (Headland, 1989: 26), both regions have experienced broadly analogous waves of activity leading to the establishment of regimes of governance based on international treaties. However, the timing and tempo of these developments differed, and key events in Spitsbergen's history – hunting and trapping, mining, and active warfare during World War II – have limited or no Antarctic equivalent.

Following those earlier developments, contemporary polar tourism represents an additional wave of activity in the regions (Sugden, 1982; White, 1994) (Fig. 1.2). Conceptually, there is an apparent inevitability to the trajectory of polar tourism: it builds on the long-established use of the Polar Regions as resource-rich areas in the periphery available to the centres of economic power (Sugden, 1982); and also follows from the intrinsic expansive tendency of the industry, which tends to spread over new areas (Cohen, 1984). As noted by Spennemann:

> *Humanity has an affinity for adventure and discovery. History has shown that in places previously deemed out of reach, intrepid pioneers first visit, followed by an audacious few, and then finally by the segment of general tourist population that can afford it. Thus, increasingly remote heritage places have come under threat from visitation* (2007: 898).

Figs.1.3 a-b – Publicity for cultural heritage tourism in Antarctica. Advertisement in a Canberra newspaper based on an advertisement supposedly placed by E. Shackleton while looking for crew members[7] (left); and promotion for a centennial cruise commemorative of the "Nimrod" Antarctic Expedition, 1907-1909 in the historic port of Lyttelton, New Zealand (right). Left photo by A.D. Hemmings 2006, right photo by author 2007.

Polar tourism is not a recent activity (e.g., Palazzi, 1993; Headland, 1994; Kolltveit, 2006), but the rising pace of tourism developments since the late 1980s coincides with the increased integration of the Polar Regions into the global economic system (AHDR, 2004; Hemmings, 2007). Features of tourism interest include the pristine nature of the Polar Regions; archetypical elements of their flora and fauna, such as penguins and polar bears; the remains of past activities of exploration and exploitation; and contemporary settlements, such as research stations. It is apparent that wildlife and nature tend to generate the most interest among contemporary tourists (e.g., Bauer, 2001; Maher, 2010; Basberg, forthcoming), but polar history is part of mainstream tourism as well as a niche market for specialised tourism (Figs.1.3 a-b).

The attraction of tourism to polar historic sites is a function not only of the historic nature of the sites, but also of practical factors that led to the original occupation of the sites (e.g., Kaltenborn and Emmelin, 1993). These included logistic convenience (such as sheltered bays, flat ground surfaces, the availability of water or drift wood, and location relative to other sites), natural resources (which provide food, commercially valuable raw materials, and research subjects), and/or the strategic interest of specific sites (at a local and/or regional scale). Many of these qualities are also of interest to the

7 The original advertisement presumably read: "Men wanted for hazardous journey. Low wages, bitter cold, long hours of complete darkness. Safe return doubtful. Honour and recognition in event of success." However, the date and name of the newspaper in which this add appeared have not yet been found, despite a modest recompense being on offer for nearly a decade, so the advertisement is believed to be apocryphal (http: //www.antarctic-circle.org/advert.htm, accessed 12 October 2010).

tourism industry. The presence of material cultural remains and stories of past events add tourism interest to a site, and over the past few decades many historic sites in Antarctica and Svalbard, particularly in coastal areas, have become "destinations" where tourism landings are a regular seasonal feature (e.g., Kaltenborn and Emmelin, 1993; Hughes, 1994; Hughes and Davis, 1995; Geitz, 2004; Lynch *et al.*, 2009).

Representations of polar historic sites in Antarctica and Svalbard as reflected in, for instance, tourism brochures, the popular media, and contemporary debates about tourism management or heritage conservation, describe some of these sites as veritable "time capsules". The 1989 Oxford English Dictionary defines a time capsule as "a container used to store for posterity a selection of objects thought to be representative of life at a particular time". "Time capsule" representations describe historic sites in the Polar Regions as isolated, abandoned, and virtually unchanged since they were abandoned. This perception is reflected, for instance, in narratives about huts from the early period of Antarctic exploration, or "Heroic era": "Visitors to the huts are fascinated by finding the legendary explorers' belongings preserved *as though they had just walked out*" (NHM, n.d., original emphasis). Cold is the invisible but ever present factor in these narratives, freezing for posterity cultural remains. Permafrost, snow and ice play a role comparable to that of volcanic lava at Pompeii, leaving things as they were at a particular moment in time.

The "time capsule" concept is marketable because it highlights both the authenticity of the remains and the exclusivity of the experience – only a few people have ever been to these sites, which remain in an "authentic" state.[8] However, polar historic sites are affected by natural processes and, despite their remoteness, are the loci of a range of contemporary activities. Some sites may have experienced a hiatus after their original occupation, but others have been occupied continuously. In addition, while the polar environment protects many sites from the type of decay encountered in temperate regions, many sites have experienced gradual but significant deterioration since abandonment (Blanchette *et al.*, 2002; Blanchette *et al.* 2004; Barr, 2008). In fact, the "time capsule" appearance is sometimes the result of careful work to replicate a "frozen moment" (Carman, 2002; AHT-NZ, 2004; McKay, 2005).

Polar tourism has its own narrative (e.g. Nutall, 2010), primarily as an environmentally friendly industry (e.g. Splettstoesser, 1999, 2000). Tourism activities both in Antarctica and in the Arctic are broadly guided by the adage "Take only photographs, leave only footprints" (Stonehouse, 1990; Mason, 1994).[9] This adage serves as a visitor code of conduct,[10] and sends a message to the regulatory authorities that tourism "treads lightly" on its destinations. In addition, it may serve to reassure consumers that the impact of their visits can be minimal. In the case of Antarctica, this concept is incorporated in the bylaws of the International Association of Antarctica Tour Operators (IAATO), which – paraphrasing the 1991 Protocol (Art. 8) – commit members not to

8 For a discussion on object authenticity in tourism, see Reisinger and Steiner (2006).

9 "Enjoy yourself and remember: Take nothing but photographs. Kill nothing but time. Leave nothing but footprints." (Mason, 1994: 97).

10 http: //www.abercrombiekent.com.au/luxury-travel/antarctica/index.cfm; http: //www.oceanwide-expeditions.com/Frequently-Asked-Questions#General. Accessed January 2011.

have "more than a minimum or transitory impact" (IAATO, 2010). Along similar lines – but focusing on process rather than output – the objective of IAATO's Arctic counterpart, the Association of Arctic Expedition Cruise Operators (AECO), is:

> *To ensure that expedition cruises and tourism in the Arctic is carried out with the utmost consideration for the vulnerable, natural environment, local cultures and cultural remains, as well as the challenging safety hazards at sea and on land.* (AECO, n.d.).

The objective, and also the argument underlying these various statements, is that responsible tourism has no discernible or at least no significant effects on its destinations: "Tourism can be projected to be a sustainable activity because of how it operates" (Splettstoesser *et al.*, 2004: 35; see also Headland, 1994; Splettstoesser, 1999, 2000; Splettstoesser and Folks, 1994; Splettstoesser *et al.*, 2004). These assessments reflect the views of regular lecturers on tourism cruises, which suggest hands-on expertise in polar tourism but also raise some questions about their objectivity. Nevertheless, these arguments find a degree of support with government officials and environmental managers.

When representations of historic sites and tourism are combined they depict a frozen world in which tourists are able to step in and out in unchanging "time capsules" without causing any lasting disturbance, while enjoying the privilege of being among the privileged few to visit a site years or decades after its abandonment by the original occupants. However, there is an apparent paradox between these representations and the recent trajectory of tourism in Antarctica and Svalbard (Hall and Johnston, 1995; Geitz, 2004; Hall and Saarinen, 2010; see also Chapters 3 and 8). Over the past few decades the industry has grown, expanded spatially, and diversified, increasing accordingly the potential for tourism to modify its destinations. Not only are polar historic sites undergoing constant change, but at the same time tourism is potentially a cause and a consequence of change.

The authorities that are responsible for the governance of Antarctica and Svalbard are generally supportive of responsible tourism, and some go as far as to promote it, but they are also concerned about the impacts that may be associated with this activity. For instance, in 2009 several Antarctic Treaty Parties stated that, "when properly managed, tourism in Antarctica should be welcomed", while also noting the need to minimise environmental impact, maximise the safety of operations, and avoid interference with scientific activities (ATCPs, 2009, paragraph 182). To this statement other Parties retorted that "... the absolute priority of the Treaty is Antarctic scientific research and environmental protection and that tourism should not be encouraged but rather strictly regulated." (ATCPs, 2009, paragraph 183). Consequently, the growth of tourism has triggered discussions and initiatives on the regulation and active management of this activity. According to a high-ranking official who represents an influential Antarctic Treaty state "...we all know if 20,000 people visit a place they may be doing something

– it has got to be bad!" (Anonymous, personal communication, 13 March, 2008).[11] This fundamental concern – whether or not justified by events on the ground – forms the basis of debates and initiatives concerning tourism management and regulation both in Antarctica[12] and Svalbard.[13] In this context, the objective of this thesis is to critically examine, by means of field research, the effects of tourism on polar historic sites, and to ascertain whether and how tourist behaviour is a factor of transformation of the historic environment of Antarctica and Svalbard. However, it should be emphasized that the issue of the contemporary use of polar historic sites is not only about tourism, and that polar tourism does not solely consist of visiting historic sites.

Before going further it is necessary to define some key concepts.

1.2 KEY CONCEPTS

1.2.1 Heritage, historic sites, and the historic environment

The notion of heritage includes three key characteristics: it belongs to a past; it can be tangible or intangible; and it is susceptible to being used in the present for a range of purposes, including research, outreach, and enjoyment, rather than solely being preserved as a legacy to the future (Pérez-Juez Gil, 2006: 32). Accordingly, cultural remains from past activities can be categorised according to their tangible material characteristics (i.e. "historic sites", "artefacts", or simply "things") or more broadly according to their recognised values (here referred to as "cultural heritage" or the broader term "heritage"). When applied to specific sites, ICOMOS (2008) defines a Cultural Heritage Site as "...a place, locality, natural landscape, settlement area, architectural complex, archaeological site, or standing structure that is recognized and often legally protected as a place of historical and cultural significance".

In this thesis a "historic site" is defined as "any location that demonstrates past human activity...evidenced by the presence of artefacts, ecofacts, features, structures,

---

11 This comment has been kept anonymous as the official is serving at the time of writing, and the comment was made as part of a professional exchange rather than in a research context.

12 One of the earlier instruments addressing Antarctic tourism, Recommendation ATCM VI-7 (Tokyo, 1970) *Effects of tourists and non-governmental expeditions to the Antarctic Treaty Area,* states:

> The representatives...Noting the increase in recent years in the number of tourists and also in the number of visitors who are not sponsored by the Consultative Parties to the Antarctic Treaty area; considering that the activities of such visitors can have lasting and harmful effects on scientific programmes, on the Antarctic environment, particularly in Specially Protected Areas, and on historic monuments... Recommend to their Governments that...visitors must not enter Specially Protected Areas and must respect designated historic monuments."

13 Recently some sites in Eastern Svalbard have been closed off to tourism traffic on precautionary grounds (Governor of Svalbard, 2009).

or other material remains" (Kipfer, 2007: 50). This definition of historic site is broadly consistent with the legal definition used in Svalbard, and with the notion of "historic site" as used in practice in Antarctica.[14]

While tangible heritage is the focus of this thesis, the polar cultural heritage includes both tangible and intangible elements, which may in fact be different dimensions of the same heritage (e.g. Blake, 2000; Ahmed, 2006; Henare *et al.*, 2007). Overall, defined succinctly, "heritage denotes everything we suppose has been handed down to us from the past" (Lowenthal, 2005 – see also Blake, 2000; Howard, 2003, 6; Timothy and Boyd, 2003, 2). Legal protection implies that remains of past activity have been valued and categorised as a material legacy of the past, or "heritage" (Carman, 2002).

In some conceptions of heritage there is not only a sense of passively receiving something from the past, but also of actively using it for contemporary purposes: "Heritage is that part of the past which we select in the present for contemporary purposes, whether they be economic or cultural (including political and social factors), and choose to bequeath to a future." (Ashworth and Graham, 2005: 7). Under these and similar definitions (e.g. Graham *et al.*, 2000; Tunbridge and Ashworth, 1996), "...heritage is not simply the past, but the modern-day use of elements of the past" (Timothy and Boyd, 2003: 5). Graham *et al.* (2000) further note that heritage has several key characteristics:

- It is essentially a geographic concept, defined by magnitudes of scale, distribution, and location;
- It concerns issues of power relations, defined by questions, such as: Who decides what is heritage? Who owns it?; and
- It is used as an economic instrument, defined by questions such as: Who owns? Who uses? Who manages? Who benefits? And who pays? .

The cultural heritage protection criteria used in Antarctica and Svalbard are legally defined. Consequently, the approach that will be used in this thesis is that legal designation is what determines what is regarded and managed as heritage (see e.g. Pérez-Juez Gil, 2006: 32). Legal considerations are also important for the regulation and management of tourism and of its interface with historic sites.

The 2001 Svalbard Environmental Protection Act (the Svalbard Act, or the Act) defines "cultural heritage" (*kulturminner*) as "traces of past human activity, including structures and sites of all kinds, on or below the surface of the land, in the sea, on the seabed and in watercourses."[15] The Act differentiates between "structures and sites"

---

14 The historic sites of Antarctica and Svalbard originate from periods documented by historic sources, and therefore are not archaeological sites in that sense. However, archaeological approaches can be used to study these sites – including those of historic archaeology, industrial archaeology, and archaeological approaches to anthropology.

15 2001 Svalbard Environmental Protection Act, § 3 (definitions), (f) (Official English translation - http: //www.regjeringen.no/en/doc/Laws/Acts/Svalbard-Environmental-Protection-Act.html?id=173945).

and "movable historic objects".[16] Thus, the Act focuses on the tangible aspects of cultural heritage, which it broadly equates with "historic sites" in the definition used here (Tables 1.2 and 1.3). The legal designation of historic features in the Antarctic Treaty System is "Historic Sites and Monuments" (HSMs). The 1991 Protocol of Environmental Protection to the Antarctic Treaty (the Protocol) does not define "historic site" or "cultural heritage", but refers to "sites or monuments of recognised historic value"[17] and more broadly to "historic values" [18] and to "areas...of historic significance."[19] Additional ATCM instruments provide detail as to what these values might entail by establishing non-mandatory criteria to list new sites[20] and to de-list existing sites.[21] Some of these criteria emphasise the material aspects of historic sites, while others emphasise the values assigned to them (through e.g. association). HSMs can be classified in three partly overlapping categories: the material remains of past activities; monuments and memorials of different kinds dedicated years or decades after a past activity;[22] and monuments and memorials that are contemporary to past activities, some of which might also be regarded as "sites" themselves (Table 1.4).[23] To further complicate matters, any given HSM may contain elements from more than one of these categories.

The distinction between "historic site" and "heritage" is relevant to this research in a number of ways. First, under the definitions used here "historic sites" and "heritage" are not always one and the same thing. Not all sites that contain remains of past activities (i.e. "historic sites) might be legally protected because of their significance (and therefore categorised as "heritage"). Rather, they might be regarded as inactive or abandoned sites, or simply as waste. This is particularly the case in Antarctica where the listing of "historic sites" is primarily done on a case by case basis, rather than using an all-encompassing criterion of age, as in Svalbard. However, any of such sites may potentially be listed for future protection. This is exemplified by the continued expansion of the listing of Antarctic historic sites and monuments since the first designations

16 Art 3a defines "structures and sites" (*fast kulturminne*) as "any element of the cultural heritage that is physically attached to the ground or associated with a specific locality"; Art 3 b. defines "movable historical objects" (*løst kulturminne)* as "any element of the cultural heritage that is not defined as a structure or site." (Official translations - http: //www.regjeringen.no/en/doc/Laws/Acts/Svalbard-Environmental-Protection-Act.html?id=173945).

17 Protocol Annex V, Art. 3(2) h, Art. 4 (2)(b), Art. 8 (1), and Art. 8(2).

18 Protocol Art. 3(b)(2)(vi); and Annex III, Art. 3(2)

19 Protocol Annex V, Arts.3 (1).

20 (ATCM) Resolution 8 (1995) *New Historic Sites and Monuments: Suggested guidelines for the designation of historic sites*, and (ATCM) Resolution 3 (2009) *Guidelines for the designation and protection of Historic Sites and Monuments.*

21 (ATCM) Resolution 4 (2001) *Review of Historic Sites and Monuments.*

22 These are usually national and religious symbols of different kinds, as embodied in plaques, busts, religious statues, crosses, etc.

23 This might be, for instance, memorial crosses, tombs, and rock cairns.

in 1972. In fact, some of the newly listed historic sites post-date the 1950s, and some of these are considerably more recent.

Second, the condition "heritage" does not imply that a site might be used for tourism. What makes a historic site a tourism destination is not legal protection *per se*, but the site's inherent attractions as conceived from a tourism perspective, which includes a suitable combination of sights, stories, and access. Some past activity sites used as tourism destinations are not legally protected, and conversely some legally protected sites or monuments are not tourist destinations. This underscores that tourism has its own logic when it comes to assigning tourism value to cultural remains.

Finally, elements of the past may be subject of a range of contemporary uses and narratives. In their seminal 1996 book, Tunbridge and Ashworth discuss the concept of heritage dissonance, which they describe as "discordance or lack of agreement and consistency as to the meaning of heritage". The concept of heritage dissonance is relevant to historic sites and tourism, both separately and in their interface. The same historic site may be represented in different ways by those connected to that heritage (a raft of actors that include nations, societies and individuals, past and present), and also by the narratives that represent that heritage for the benefit of tourists: "the object remains, but its messages change" (Graham *et al.*, 2000: 25).

Historic sites often reflect territorial ambitions, whether at their original establishment or subsequently through designation as heritage. Whether in their original form or as historic sites/heritage, they help to maintain a presence over a discrete swath of territory, and to refer to actual or supposed events from the past. Contemporary narratives and uses of historic sites may also reflect tensions between hegemonic and other groups. This may be, for instance, between Norwegian and non-Norwegian heritage in Svalbard or, in the case of Antarctica, the multiple heritages of early arrivals (early explorers and the original signatories of the 1959 Antarctic Treaty) against those of later arrivals. As noted above, the sheer presence of historic remains in a wilderness area, when regarded from contemporary environmental and cultural heritage perspectives, may give place to dissonant versions of the value of those remains – somebody's "heritage" may be somebody else's "rubbish". The scientific uses and value of the environment, which is common currency in Antarctica and Svalbard, are sometimes in tension with other uses and values of the same environment.

All these factors come to the fore when historic sites are used for tourism purposes. Conflicting views as to the meaning of heritage are reflected through contemporary tourism narratives in the present; through the ways in which visitors may regard historic remains (particularly those of industrial origin – e.g. Basberg, forthcoming); through conflicts of use of particular sites; and through the commodification of heritage at frequently visited sites, in particular when this is used to legitimise claims of sovereignty in contested territories (e.g. Timothy, 2010).

To summarise, this thesis focuses primarily on those sites that contain the material remains of past activities, or historic sites that have been designated as "heritage" according to legislation. These sites may be regarded, and are sometimes referred to, in a broad sense, as the *historic environment* of the study regions.[24]

1.2.2 Tourism, polar tourism, and polar tourists

There is no universally accepted definition of tourism or tourists (e.g. France, 1979; Leiper, 1979; Hunter and Green, 1995), and ambiguities about these concepts also exist in a polar context (Hall and Johnson, 1995; Stewart *et al.*, 2005; Hall and Saarinen, 2010). A dictionary definition of tourism is "Organised touring; operations of tours as business; provisions of things and services that attract tourists", while a tourist is a "Person who makes a tour, traveller, esp. for recreation".[25] Hunter and Green (1995) use the term *tourist* to describe somebody who travels for the primary *purpose* of commerce, personal business, visiting friends or relatives and pleasure, and *tourism industry* in a broad sense to represent the amalgam of services and resources used by tourists. Leiper (2004: 25) suggests that being a tourist has three attributes: it requires travelling, visiting, and having leisure experiences – leisure being the antithesis of work (France, 1997). There is no greater clarity as to what constitutes *polar* tourism, other than it takes place in the Polar Regions. Hall and Johnston (1995: 8) define polar tourism as "all travel for pleasure and adventure within Polar Regions, exclusive of travel for primarily government, commercial, subsistence, military or scientific purposes". This definition is broadly applicable and in common usage in Svalbard and Antarctica, however there are some differences as to how the term is used in these regions, discussed below.

Svalbard's *Regulations relating to tourism and other travel in Svalbard*[26] do not define the term "tourist" but define "tour operator" as meaning "...anyone who for payment organises journeys with appurtenant services, including transportation and accommodation. Payments to meet actual expenses and where no profit is calculated are also regarded as payment." Implicitly, tourists are the customers of tour operators, or those sharing the costs of travelling. This is reinforced by the definition of "individual travellers", who are defined as "...visiting tourists, persons permanently resident on Svalbard, scientists, and other persons who travel outside settled areas and who are not taking part in journeys organised by a tour operator." This definition excludes "... persons carrying out official duties in Svalbard ".

There are no official definitions of tourist or tourism in the context of the Antarctic Treaty System. The earlier Antarctic tourism instrument (the 1966 Recommendation IV-27, *Effects of Antarctic tourism*), recommends governments to exchange information about any "tourist and other non-scientific expedition" organised from their countries. In later ATCM instruments, the term "tourism" is usually linked to "non-governmental

24 Worthing and Bond (2008: 188-189) discuss the similarity between the concepts of "historic environment" and "cultural landscape". For simplicity, in this thesis only the first of these concepts will be used.

25 *The Concise Oxford Dictionary of current English*. Sixth edition, 1976. Oxford: Oxford University Press.

26 Issued by (Norwegian) Royal Decree of 18 October 1991.

activities".[27] Confusingly, the term "visitors" as currently used in the Antarctic Treaty System specifically " ... does not include scientists conducting research within such sites, or individuals engaged in official governmental activities."[28]

Debating Antarctic tourism, Stonehouse (1992) defines a tourist as "one who travels for pleasure or culture, visiting a number of places for their objects of interest, scenery or the like", which is quoted from the *Shorter Oxford English Dictionary 1983.* Enzenbacher (1992: 17) defines Antarctic tourists as "Visitors who are not affiliated in an official capacity with an established National Antarctic Program. They include both fare-paying passengers...and private expedition members and adventurers [sic] aboard sea or airborne vessels...". Tracey (2001) takes a similar approach, reflecting a common approach in the Antarctic Treaty System. Other authors include the *purpose* of the activity (Bauer, 2001; Murray and Jabour, 2004; Molenaar, 2005; Lamers, 2009), which may include pleasure, education, and adventure.

IUCN makes a distinction between commercial and private tourism activities in Antarctica, noting that the former involve more people, have greater potential for environmental impact or disruption of activities, and generally demand greater management effort (Ceballos-Lascuraín, 1992). Along similar lines, Hemmings *et al.* (1991) make a distinction between *commercial* and *non-commercial* ventures. For White (1994: 246) "... [Antarctic] tourists are largely defined as the pleasure seekers who are not part of a scientific program and have usually paid for their trips from their own funds."

Overall, in the Antarctic context "tourism" and "tourist" tend to be defined implicitly by exclusion from government activities, and always in connection with non-governmental activities, whatever these might be. The problem with this approach is twofold: there are historic and current links between some National Antarctic Programs and tourism (Palazzi, 1993; Headland, 1994; Bastmeijer and Roura, 2008); and some non-governmental activities in Antarctica have little in common with commercial tourism in terms of purpose, pattern, or *modus operandi.* The Antarctic Treaty Parties' unwillingness or inability to agree on an official definition of tourism underscores the political

27 These include, *inter alia*, Recommendation XVIII-1 *Tourism and Non-Governmental activities*, (ATCM) Measure 4 (2004) *Insurance and Contingency Planning for Tourism and Non-Governmental Activities in the Antarctic Treaty Area*, (ATCM) Resolution 4 (2004) *Guidelines on Contingency Planning, Insurance and Other Matters for Tourist and Other Non-Governmental Activities in the Antarctic Treaty Area*, (ATCM) Resolution 4 (2007) *Ship-based Tourism in the Antarctic Treaty Area*, (ATCM) Resolution 5 (2007) *Tourism in the Antarctic Treaty Area*, and (ATCM) Resolution 7 (2009) *General Principles of Antarctic Tourism.*

28 (ATCM) Resolution 4 (2009).

nature of the issue and has resulted in some instances in tourism regulation being put in place while deliberately avoiding providing definitions of what is being regulated.[29]

To cut the Gordian knot of defining what tourism is, a heuristic definition of polar visitors, tourists, and tourism will be used here. Based on Leiper (1979, 2004), tourism is conceived as a system consisting of five elements: tourists, tourism industries, generating regions, transit regions, and destination regions. In polar tourism the Polar Regions are the destination regions, and the tourism industries facilitate the transport of tourists to those regions. Visitors to a historic site in Antarctica and Svalbard are those persons who are present at that site. [30] This broad definition includes anybody spending time at a historic site, regardless of factors such as purpose or means of travel, institutional affiliation, or the purpose or duration of their visit. Visitors' activities at historic sites may be entirely recreational, entirely professional, or combine both. Tourists – a subset of the broad group of visitors – are the paying passengers travelling in organised visits with commercial tour operators.

Using these definitions, visitors to historic sites may also include those spending time there for other reasons, such as conservators or scientists; residents from nearby research stations or settlements; and VIPs, students, and private non-commercial expedition members. Certain activities of visitors who are not paying tourists are very similar to those of organised tourists, while other activities would be fundamentally different. The basic patterns of visitation (the number of visitors, the frequency of visits, the parts of a site visited, etc.) may differ substantially between paying tourists and other kinds of visitors.

Overall, the definitions of tourism, tourist and visitor, used here are broadly compatible with the concept of tourism as used in Svalbard but somewhat different to the understanding of tourism in the Antarctic context. However, there are considerable similarities between tourism in both regions (Table 1.5).

1.2.3 Activity area and footprints, palimpsests and impacts

A family of terms describe the transformation caused by effect of human activity on the. Some of these terms are used primarily in archaeology and others in environmental management, and sometimes describe similar or identical concepts. Since historic sites combine material culture and landscape modifications, and tourism is potentially a factor of transformation of those sites, it is appropriate to define these terms.

---

29 See for instance XXXII ATCM Final Report, paragraph 197 "Germany noted that definitions are not needed for this Measure [ATCM Measure 15 (2009) on *Landing of persons from passenger vessels in the Antarctic Treaty Area*] and may lead to confusion with regard to the existing instruments. They should therefore be deleted." The text of the Measure, which bans landings from vessels carrying more than 500 passengers, uses terms such as "tourism activity" and "tourism industry" but do not define them because it is apparent that only commercial vessels have such passenger capacity.

30 The term "visitor" is preferred over that of "tourist" as more appropriate for conservation purposes as it encompasses the entire range of actors that may be active at a particular site (Davis, 1995; Maher, 2007).

From an archaeological perspective, activity area is defined as "...the place an activity occurs, whether it is repeated many times or takes place once." (Rathje and Schiffer, 1982: 45). This research concerns two (or more) overlapping areas of activity: a historic area of activity (or activities), presently defined by material remains (i.e. "historic site"); and the activity area of contemporary tourism at that historic site. In the Antarctic context, "footprint" usually refers to the spatial extent of disturbance (Jabour, 2009; Australia, 2010; New Zealand, 2010) including, for instance, the area around station buildings, including service ducts, tracks, masts and other infrastructure, and other visible or detectible disturbance such as fuel spills and emissions plumes.[31] Footprint normally refers to physical disturbance but may also refer to other forms of human presence e.g. visual, acoustic, or chemical (Australia, 2010; New Zealand, 2010). As the term "impact", to which it is related, "footprint" has management connotations and it is often used in juxtaposition to "wilderness" (that is, the footprint of human activities encroaches upon unmodified wilderness). When applied specifically to past activity sites, "footprint" is identical to the definition of "historic site" used here. In addition, footprint can be used as a synonym of "behavioural signature" – or traces of activity – in a tourism activity area, including the most literal connotation of the imprint of a foot.

The concept of palimpsest (originally a term referring to pieces of a scroll or a book in which the original text has been scraped off so that the scroll or book could be reused) describes "...a superimposition of successive activities, the material traces of which are partially destroyed or reworked because of the process of superimposition" (Bailey, 2007: 203-205). This term is appropriate to describe the sequential transformation of many historic sites in Antarctica and Svalbard effected by later cultural processes, including tourism. The concept of palimpsest may be explored further to identify several different types relevant to this research, two of which will now be highlighted. At one end, in "true" palimpsests all but the most recent traces of activity have been erased; at the other end, in "cumulative" palimpsests earlier activity traces have been only partially transformed and modified by more recent activities, resulting in traces that reflect a succession of activities, each of which cannot necessarily be distinguished from the material record (Bailey, 2007). In both Antarctica and Svalbard evidence of the earliest forms of occupation remains – such as historic sites – which sometimes represent the only occupation a site has experienced until being rediscovered as a tourism site. Tourism represents a temporary but often regular form of occupation. The extent to which tourism contributes to the modification of historic sites is the subject of this thesis, but the assumption can be made that tourism at polar historic sites results in cumulative palimpsests that may modify in some degree, but do not entirely erase, previous traces of activity.

To the extent that tourism leaves traces, these may potentially be regarded as "impacts". Impact is a management term – as applied for instance in Environmental Impact Assessment (EIA) processes – that usually includes an implicit value judgment assessing the intensity and duration of the impact , and establishing "...what is

31 The term "footprint" as used here should not be confused with other uses of the word such as "carbon footprint" or "ecological footprint" meaning the area of land per capita which is needed to meet the resources consumed by a nation.

important, desirable or acceptable." (Lawrence, 2007: 757). The Antarctic EIA guidelines define (impact) significance as "...a value judgment about the severity and importance of a change in a given environment or environmental value" (ATCPs, 2005). Changes that are minor in spatial extent, or ephemeral, are not usually regarded as impacts in practice, and result in no management actions. For the purposes of this research, however, both minor and major changes are of interest, since both may reflect the footprint of tourism, and ultimately the transformational effect of tourism and other processes of change to historic sites.

The term *impact* is extensively used both in the 1991 Protocol and the 2001 Svalbard Act, but it is not defined in these instruments. The International Association for Impact Assessment defines the impact of an actual or proposed action as "...the difference between what would happen with the action and what would happen without it" (IAIA, 2009: 1). The 2005 Antarctic EIA guidelines define an "impact" (or "effect") as "[a] change in the values or resources attributable to a human activity. An impact is the consequence of an agent of change, not the agent itself " (ATCM, 1999, 2005).[32] Impacts can be direct, indirect or cumulative. A direct impact is a change that results from direct cause-effect consequences of interaction between the exposed environment and outputs of a particular activity. An indirect impact is a change in environmental components that results from interactions between the environment and other direct or indirect impacts. Finally, cumulative impacts are "...the combined impact of past, present, and reasonably foreseeable activities. These activities may occur over time and space and can be additive or interactive/synergistic" (ATCPs, 2005). Potentially tourism can have all these types of impacts,[33] however the main concerns about tourism – aside from those resulting from major accidents – have been its possible cumulative impacts (De Poorter and Dalziell, 1996; Hofman and Jatko, 2000; Geitz, 2004).

Overall, these terms describe the different aspects of the transformation of the landscape through the effects of successive activities. Each of these activities may have a certain activity area and a footprint, and may sequentially contribute to a palimpsest. In management terms, some of these changes (such as changes caused to historic sites by later activities) may be regarded as impacts and merit active management.

1.3 OBJECTIVES AND RESEARCH QUESTIONS

According to Carman (2002: 22) the basic principles of heritage management are that the heritage is finite and non-renewable; it is a matter of public concern; it is regulated by legislation; and requires an assessment of value, because not all tangible heritage

32 (ATCM) Resolution 1(1999), *Guidelines for Environmental Impact Assessment in Antarctica*; and (ATCM) Resolution 4 (2005), *Updating of Guidelines for EIA in Antarctica*.

33 For instance, Recommendation ATCM VI-7 (Tokyo, 1970) *Regulation of Antarctic Tourism* notes that the activities of visitors "....who are not sponsored by the Consultative Parties to the Antarctic Treaty...can have lasting and harmful effects on scientific programmes, on the Antarctic environment, particularly in Specially Protected Areas, and on historic monuments."

can be preserved. However, in view of the regular and increased visitation to which many sites are subject it is valid to subject tourism to a critical examination and to ask if it contributes to the degradation of destination sites, and if so, how this happens, and whether or not it matters.

Most dedicated research on the impacts of polar tourism has focused on the natural environment. Perspectives have included behavioural studies of various species (Culik *et al.*, 1990; Wilson *et al.*, 1991; Giese, 1996; Burger and Gochfeld, 2007; Bricher *et al.*, 2008; van Polanen Petel *et al.*, 2008); visitation levels (Thomson, 1977; Cobley and Shears, 1999; Trathan *et al.*, 2008) and an assessment of cumulative impacts (De Poorter and Dalziell, 1996; Hofman and Jatko, 2000). Aspects of the impacts of tourism on the polar historic environment have been discussed by Hacquebord (1992), Hall (1993), Hughes (1994), Hughes and Davies (1995), Hofman and Jatko (2000), Geitz (2004); and Vistad *et al.* (2008). Tracey (2001: 112) notes that knowledge about many environmental impacts of tourism is largely assumption-based on account of few research programs having " ... rigorous experimental design, limited monitoring, and few empirical data sets." The same assessment is largely applicable to the impacts of tourism on the historic environments of Antarctica and Svalbard.

Many historic sites are inherently fragile because they have been preserved for decades (or centuries) even though their intended use was only temporary. Polar conditions have sometimes facilitated preservation, and sometimes contributed to deterioration. Tourism finds a new use for the sites and contributes to perpetuate them and the stories they embody. Tourism may in the process transform the sites in obvious or subtle ways, whether directly, indirectly, or cumulatively. Repeated tourism visits can result in the deterioration of historic features in the Polar Regions (e.g. Governor of Svalbard, 1999a: 22; Hofman and Jatko, 2000; Geitz, 2004: 47; Vistad *et al.*, 2008). These effects may further be magnified as a result of climate change (Barr, 2008).

As outlined earlier, an apparent paradox exists in the argument that tourism causes no discernible effects at its destinations, even after several decades of regular use of some sites, sustained growth, and diversification of tourism activities. Furthermore, if any effects of visitation were discernible at all, these have generally been regarded as both minor and ephemeral, and therefore negligible. This has in part resulted from uncritical discussions of tourism (e.g. Headland, 1994; Splettstoesser and Folks, 1994; Splettstoesser, 1999, 2000; Splettstoesser *et al.*, 2004). In Antarctica, negative effects of visitation have been assigned to visitors such as crews of private yachts or off-duty personnel of national research programs rather than to mainstream tourists (Riffenburgh, 1994; Chaplin, 2008; Lamers, 2009; UK *et al.*, 2010) even thought the latter constitute the bulk of visitors at many sites.

In this context, the objective of this thesis is to undertake a critical examination of tourism as a behavioural system, with an emphasis on the interactions between people and the historic environment, and to determine the effects (or footprint) of tourism on that environment.

The central research question of this thesis is:

- What are the effects of tourism on polar historic sites, and what are the implications of this for the management of tourism and historic sites?

This research recognises the multifaceted nature of tourism and historic sites, both individually and in their interface. According to Geitz (2004) whether or not visitors contribute to the degradation of polar historic sites depends on several factors, including the inherent sensitivity of the site in the context of ongoing degradation processes, such as erosion; the overall volume of visitation; tourist behaviour; visit duration; the expertise level of guides; and the enforcement of rules. When these dynamic factors are intertwined, as they are empirically, the assessment of tourism impacts on historic sites becomes rather complex. In particular, it is difficult to ascertain which changes are brought about by tourism, if any, which by other processes affecting the sites, and whether any tourism-related change matters.

This thesis has explored several interconnected themes, while also recognising that this listing is necessarily incomplete and that there are other relevant themes. These themes constitute the building stones for various chapters, which collectively form the structure of this thesis. The main themes are:

1. Tourism as a behavioural system and a site formation process, in the context of other (natural and cultural) formation processes.
2. The designation of historic sites as heritage.
3. The use of cultural heritage as tourism attractions.
4. The management of tourism and historic sites.
5. Research into the interaction of tourism with historic sites.

1.4 RESEARCH STRATEGY: BEHAVIOURAL ARCHAEOLOGY

Assessing the impact of tourism on a historic site requires determining how the activity affects the values of the site or, more specifically, how it affects what makes the site valuable. To start this assessment is important to recognise that impacts are a consequence of behaviour. In this context, an activity-output-exposure-impact framework, usually applied to the natural environment (ATCPs 2005 ((EIA guidelines)) is useful to differentiate conceptually the activity itself from the consequences of the activity, or impacts. Each activity has specific outputs that have an effect on particular components of the natural and cultural environments. Exposure of a certain environmental component to a particular output may or may not result in an impact. Overall, this framework links specific forms of behaviour with specific consequences. In different ways, this framework forms the basis of environmental management in Antarctica and in Svalbard.

Impact assessment relies on value judgements to assess the significance of particular changes (Lawrence, 2007; Asko *et al.*, 2010). This means that in the process of impact assessment some changes will be judged to be an impact and others will not.

Box 1.1 – Research themes and questions

Tourism as a behavioural system and a site formation process, in the context of other (natural and cultural) formation processes:

- What are the changes experienced by historic sites in the Polar Regions in the recent past? What are the causes of these changes? And what do these changes tell us about broader contemporary developments in the Polar Regions? (Chapter 4)
- How does tourism compare, with respect to the use of territory, with earlier activities? And what are the processes of site formation affecting polar historic sites? (Chapter 7)
- What are the patterns of tourism and historic sites, both regionally and locally? (Chapter 8)
- What are the effects of tourism on historic sites? (Chapter 8)
- How do visitors behave at historic sites, and what are the traces (or impacts) of their behaviour? (Chapter 8)

The designation of past activity sites as "cultural heritage sites":

- How are heritage sites designated for protection, and what is the interplay of cultural and environmental values in this process? (Chapter 5)

Historic sites as tourism attractions:

- How does cultural heritage in the Polar Regions operate as a tourist attraction? (Chapter 6)
- What is the role of tourism narratives in creating a tourism attraction? (Chapter 6)

Research into the interaction of tourism with historic sites:

- What (opportunistic, low technology) research methods can be used to assess the status and change of polar historic sites and the effects of tourism at those sites? (Chapter 7)

The management of tourism and historic sites:

- How can cumulative impacts at sites frequently visited by tourists be assessed and prevented? (Chapter 3)
- Should the current tourism management measures be characterised as conditions or restrictions? (Chapter 3)
- Is the precautionary principle being applied in the absence of scientific information about tourism impacts? (Chapter 3)
- What are the implications of tourism behaviour and impacts for site management? (Chapter 7)
- How effective is the protection of historic sites in the context of existing tourism policies? (Chapter 8)

Some forms of behaviour (and some types of changes related to that behaviour) will be dismissed on the grounds that these are not significant even though they may be inherent to the conduct of certain activities. In addition, changes that are individually negligible could, cumulatively, result in significant impacts (De Poorter and Dalziell, 1996; Hofman and Jatko, 2000). Furthermore, judging the significance of impacts is bound to be influenced by vested interests so that, for instance, particular impacts may be understated (or exaggerated) to suit particular agendas. As whole the use of the value-loaded term "impact" seems an unsatisfactory basis to conduct an enquiry into the effects of tourism on historic sites, which requires the ability to assess subtle differences in behaviour and behavioural consequences, even though it is an useful concept in other management contexts.

As a strategy of enquiry for this research, tourism has been considered from the perspectives of behavioural archaeology (Schiffer, 1987, 1995, 2010). "Archaeology" may suggest excavating the remains of past social systems; however, behavioural archaeology was conceived as an approach to study people-material interactions in all times and places (Reid *et al.*, 1974; Reid *et al.*, 1975) including, among others, aspects of landscape formation processes (Heilen *et al.,* 2010; Hollenback, 2010). The basis of this discipline are outlined in a 1974 article titled "Expanding archaeology" by Reid, Rathje and Schiffer, who proposed that "...archaeologists can apply their method and theory to the study of material culture in *modern* industrial societies for the purpose of deriving explanatory statements of *modern* human behaviour" (1974: 125; original emphasis). Subsequently, Reid *et al.* (1975) described four strategies for the study of relationships between various combinations of past/present human behaviour and material culture, which constitute the framework of behavioural archaeology (Box 1.2).

Box 1.2: This research in the context of the four strategies
(after Reid *et al.*, 1975; Schiffer, 1995: 69; Gamble, 2008: 70)

		MATERIAL CULTURE	
		Past	**Present**
BEHAVIOUR	**Past**	**Strategy 1** "Traditional" archaeology; study of particular past cultural systems.	**Strategy 2** Ethnoarchaeology and experimental archaeology used in general archaeological interpretation.
	Present	**Strategy 3** Archaeological research to understand general aspects of past and present behaviour. Long term cultural change.	**Strategy 4** Study of particular present cultural systems. This research.

Strategy 4 is the analysis of present material objects in order to answer questions regarding present human actions and behaviour (1975: 866; see also Rathje, 2001;

Buchli and Lucas, 2001a, 2001b). This research takes Strategy 4 as a starting point to investigate problems concerned with the effects of tourism at historic sites in Antarctica and Svalbard. A simplified account of the main tenets of behavioural archaeology, as they apply to the topic of this research, will be described below.

In later writings, behavioural archaeologists reviewed the dichotomy between "behaviour" and "material culture" to suggest human behaviour extends beyond the body itself to include artefacts, and that the "empirical reality" of human life is a constant interaction between different kinds of artefacts. The basic premise is that "human life consists of ceaseless and varied interactions between people and myriad kinds of artifacts" (Schiffer, 2010: 13). An artefact is "any material, in contradiction to spiritual or cognitive, phenomenon that exhibit one or more formal properties produced directly or indirectly through human activity" (Schiffer, 2010: 13). Human behaviour involves interactions between people and their artefacts with external environmental phenomena (or *externs* in the terminology of Schiffer, 2010), such as, in the Polar Regions, wind, snow, polar bears and penguins. Through human agency, some *externs* become themselves artefacts. People, artefacts and externs are – various combinations of discrete interactions – people-people; people-artefact; people-extern; artefact-artefact; and artefact-extern – the *interactors* in activities (Schiffer, 2010: 13). Behaviour links people and elements of the material world with which they interact – mechanically, visually, and in other ways. Such a system is comprised of activities – which are also patterned behaviours that process matter, energy, and information (La Motta and Schiffer, 2001).

This framework is applicable to polar tourism as follows. Tourism may be regarded as a behavioural system, defined as "a set of patterned behaviours that articulate a human group with the physical world and with other behavioural systems" (La Motta and Schiffer, 2001: 27). Tourist behaviour, as defined above, consists of interactions with artefacts and *externs*. Through behaviour, people (tourists) and their artefacts (e.g., boots, walking sticks, and photographic cameras) interact acoustically, chemically, mechanically, thermally, and visually with the environment. Different forms of interaction may have a range effects – from a footprint on the ground to a photograph recorded on the memory card of a digital camera. It should be noted that contemporary polar tourism usually involves some types of activities and behaviour to the exclusion of others that are discouraged for safety or environmental reasons.

Strategy 4, as defined above, concerns forms of behaviour that – for the purposes of this research – involve artefacts of three kinds:

- The historic sites themselves – which reflect a history of occupation and abandonment – and their constitutive artefacts, whether small or large, movable or unmovable;
- The artefacts (or equipment) used by tourists; and
- The artefacts that result from the modification of the cultural or natural environment as a result of discrete interactions with tourists (and their equipment) – such as trampling or handling.

In management terms, these latter artefacts would be regarded as "impacts". The challenge is to identify tourism impacts (or artefacts) in the context of other site formation processes – another of the basic tenets of behavioural archaeology.

Schiffer (1987 and refs.) developed the concept of site formation processes to identify the factors that create the archaeological (unwritten) record following the abandonment of a site by its original occupants. In cultural formation processes the agency of transformation is human behaviour. In non-cultural, or natural, formation processes natural environmental processes provide the agency of transformation. Thus, any cultural or natural process acting on a historic site contributes to the transformation of that site.

Polar historic sites are not static but subject to transformations resulting from both cultural and non-cultural (natural) processes. Natural and cultural processes contribute to transform sites of past cultural activity, which were established originally for the purposes of, *inter alia*, exploration or exploitation of natural resources, and subsequently abandoned. These same sites are today's historic sites (in the sense of abandoned bases and worksites, or sites of past activity). Site formation processes of different kinds may have acted uninterruptedly or intermittently since the sites were initially abandoned. These processes create their own specific patterns that may modify, obscure or preserve earlier patterns. An assessment of the effects of tourism on historic sites must be made with an awareness of the existence of other transformation processes with their own signatures.

The abandonment of sites by their original occupants marks a transition from the "systemic context" – the state of a site as part of an adaptive system – to an "archaeological context" – the state of a site in the archaeological record (Rathje and Schiffer, 1982). Abandoned sites become then sites of past activities (or historic sites in the definition used here). Site establishment, abandonment, rediscovery and/or reoccupation have usually characterized the evolution of historic sites in Antarctica and Svalbard. The materials or artefacts contained at the sites may have been subject to reuse at the site itself or at a different location. Later occupants may reuse the sites themselves, albeit not necessarily with a continuity of purpose in relation to the first occupants. Over time some sites have been afforded legal protection, which may have resulted in further changes through, for instance, preservation, active management, or adaptation to tourism, or a combination of different new uses. In contrast, other sites have been partly or fully cleaned up for environmental or economic reasons.

The examination of landscapes from a behavioural perspective requires the explicit consideration of different spatial, temporal, and behavioural scales (Heilen *et al.*, 2008; Hollenback, 2010). The polar tourism experience is composed by a myriad of interactions at different spatial, temporal, and behavioural scales: from specific landing sites to sub-regions of Antarctica; from the duration of a single landing to an entire season and multiple seasons; and from discrete interactions between one person and particular features, to the entire behavioural system of tourism. In this research, the analysis focuses primarily on the interactions between individual tourists and groups of tourists with historic polar sites and their components, over periods of time ranging from discrete interactions lasting a few seconds; visits lasting a few hours; repeated visits to landing sites; and the sustained use of some locations of periods of a decade or more. Interactions between tourism and historic sites at these various spatial, temporal and behavioural scales define the life history of particular sites with respect to their use as tourism destinations. This includes the disturbances potentially caused by tourism on the various elements that compose historic sites – artefacts and ecofacts, deposits, buildings, features, and the surrounding landscape.

1.5 COMPLEMENTARY THEORIES

1.5.1 Tourism attraction systems

The use of historic sites as tourism destinations and attractions shape the life history of particular sites. In this research, a *destination* is a site where tourism has become consolidated, so that these are "...sites that will continue to be visited repeatedly by tourists on a seasonal basis for the foreseeable future and that will therefore be exposed, in a sense, to permanent human activity" (Bastmeijer and Roura, 2004: 770; see also Chapter 3). Some historic sites are not exposed to tourism impacts as they are not tourism destinations; conversely, some are tourism destinations and frequently visited. The tourism use of particular sites is based around the concept of *attraction* – which is a site-specific feature that makes a particular site worth visiting. The use of material culture in the creation of tourism attractions will be analysed on the basis of the concept of tourism attraction systems developed by N. Leiper (1990) (Chapter 6).

MacCannell (1976) suggests that tourist attractions are signs i.e. they represent something to someone. Based on McCannell (1976) and Gunn (1972), Leiper (1990: 371) defined a tourism attraction as a system comprising three elements: "a tourist or human element; a nucleus or central element; and a marker or informative element. A tourist attraction comes into existence when the three elements are connected." Thus, the creation of a tourism attraction is part of an empirical process by which these three elements come together. Conversely, if one of these elements is remains unconnected then there would not be an attraction. N. Leiper (1990: 382) suggests that each of the elements is an important part of the attraction system; e.g., a place with no visitors would not be regarded as a tourist attraction.

Although not part of Leiper's original formulation, tourism attraction systems can be examined from the perspective of behavioural archaeology as a form of interaction between people and artefacts (which serve as "sights" in Leiper's terminology. The interactions between tourists and sights consist of markers or information about the site – which are primarily visual or acoustic (through narratives) but also mechanical (tactile) or chemical (through the smell of particular sites). Other forms of information about the sights (e.g. interpretation signs, plaques, brochures, etc.) are also markers. At sites that are established as tourism destinations, these types of interactions occur frequently on seasonal bases. As it will be discussed through this thesis, the location and attributes of particular historic features that serve as sights influence tourist behaviour.

1.5.2 The value of historic sites: Rubbish Theory

Analysing the effects of tourism on historic sites by means of behavioural archaeology – with its emphasis on human-artefact interactions and site formation processes – is not to suggest an absence of value systems. Indeed, value systems differentiate acceptable and unacceptable forms of behaviour, and have an influence on what the consequences of particular forms of behaviour are or may be. Concerns about the effects of tourism on historic sites are dependent on the notion that the sites have some significance for which they merit protection, as is often the case. As noted above, the significance of historic sites in Antarctica and Svalbard is characterised in legal terms. Otherwise, the sites of past activities would normally be regarded as waste and removed or just left abandoned to their fate, neither protected nor preserved. The designation of material remains of past activity into legally protected historic sites will be discussed by means of Thompson's Rubbish Theory (1979).

In the Rubbish Theory, Thompson (1979) identified three categories of objects: the "durable" – an object that increases in value over time and theoretically has an infinite lifespan ; the "transient"– an object that decreases in value and has a finite lifespan; and "rubbish" – that which has "zero and unchanging value" and usually does not disappear but "continues to exist in a timeless and valueless limbo" where at some later day it may be discovered and revalued (Thompson, 1979: 7-10; Carman, 2002; Howard, 2003: 47). The basic process is thus:

Transient →Rubbish →Durable

This is a categorization process (Carman, 2002) by which past material culture is defined as either "artefact" or "rubbish". Those remains that are classified as "artefact" are regarded as having some significance according to a range of legal criteria, and are designated as "heritage", that is, something to be handed over to future generations (e.g. Tunbridge and Ashworth, 1996; Graham *et al.*, 2000; Lowenthal, 2005). Those remains that are regarded as "rubbish" have no value and would need to be dealt with accordingly. The categorization of historic sites can in itself be regarded as a management concept, in that by legally categorising a past activity site as protected or otherwise it becomes a subject of management (even if management consists in allowing it to decay). It should be noted that some historic remains may experience iterations of this process, and be categorised and managed in several different ways through their life history.

According to Carman (2002: 169) Thompson's categorization process is identical to Schiffer's transformation of a site (or artefact) from adaptive use in the systemic context, through to abandonment in the archaeological context, and later to rediscovery as heritage.

Systemic context →Archaeological context →Heritage

The concept of *palimpsest* was introduced above to describe the effect of successive activities on particular locations. On a similar vein, the concept of *palimpsest of mean-*

ing may be applied to characterise the changing values and meanings assigned to historic remains. Palimpsest of meaning has been described as:

> *... the succession of meanings acquired by a particular object, or group of objects, as a result of the different uses, contexts of use and associations to which they have been exposed from the original moment of manufacture to their current resting place, whether in the ground, a museum, a textbook, an intellectual discourse, or indeed as objects still in circulation and use. (Bailey, 2006: 208)*

1.6 TRANSLATING BETWEEN DIFFERENT TERMINOLOGIES AND FRAMEWORKS

Behavioural archaeology provides a framework to analyse polar tourism that underscores the interactions between tourists and historic sites in a polar setting, and the effects that result from these interactions, without making *a priori* value judgments about the significance of tourism-related effects. Behavioural archaeology perspectives also reflect awareness about other site formation processes affecting tourism destinations. This perspective is a departure with the management-oriented focus that characterises much of polar tourism research (e.g. Hall and Johnston, 1995; Kaltenborn, 2000; Tracey, 2001; Lamers, 2009). The findings of this research, however, may have management implications by the way of providing new perspectives on the patterns of polar tourism as it applies to historic sites. Consequently, a first stage of this thesis involved *translating* a management language to a behavioural archaeology language to describe and interpret tourist behaviour, and to make inferences about the changes brought about by visitants to historic sites. A second stage was to translate the findings back to a management language. This information could then be used to identify recommended management actions. Box 1.3 contains an equivalence of terms

Box 1.3: A comparison of key terms of behavioural archaeology and management

Behavioural archaeology	Management
Interaction, activity	Action, activity
By-product	Output
Interaction	Exposure
Behaviour traces; use wear	Impact
Behavioural signature	Footprint
Palimpsest; life-history	Cumulative impact
Site formation processes	Pressures, threats

from these two different conceptual frameworks, while noting that some terms are not absolutely equivalent.

1.7 CONTRASTING ANTARCTICA AND SVALBARD: TIMELINE OF HISTORIC EVENTS AND PROTECTED CULTURAL HERITAGE

As noted earlier, the basic similarities between Antarctica and Svalbard include their lack of indigenous people; the fact that their governance results from agreements made as an effect of international treaties; and their current status as protected natural areas. In addition, both regions have been seen as natural resource frontiers (e.g. Hacquebord and Avango, 2009) and have been the subject of distinct but fairly comparable waves of economic and geopolitical activity since their discovery. Beside these basic facts, the regions differ considerably in just about every other respect – such as scale, flora and fauna, climate, proximity to populated parts of the world, human geography, and regimes of governance (Table 1.1). A comparative approach is used here because issues concerning cultural heritage and tourism are similar in both regions (Tables 1.2-1.5). A summary outline of the history of these regions is offered below (see also Fig. 1.4).[34]

The Dutch pilot Wilhem Barentz discovered the High Arctic archipelago of Spitsbergen on 17 June 1596.[35] The archipelago quickly became the focus of whaling activities by Dutch, English, Basque and Dano-Norwegian operators based on land-based outposts during summer. Whaling declined around 1670 along with the depletion of whales, and all but disappeared towards 1750. Russian (Pomor) hunters and trappers originating in the White Sea region of Russia operated in the territory year round hunting for walrus, reindeer, fox, and seal. The early dates of Pomor activity are the subject of some controversy (see e.g. Arlov, 2005). Pomor activities peaked towards the second half of the 18th century, and diminished through the 19th to come to a halt towards 1850. Norwegian hunting and trapping in the winter began in 1795-96, and peaked towards the end of the 19th century.

34 For more detailed historic accounts readers are referred to the following publications: Sugden (1982) for the Polar Regions as a whole, Conway (1906), Arlov (1996) and Driveness and Jølle (2006) for Svalbard, and Fogg (1992) for Antarctica. Headland (1989, 2009) lists chronologically Antarctic expeditions and related historical events. Palazzi (1993) outlines the basic factors of Antarctic geography and history that influence geopolitics in the region. For political context see Auburn (1982), Fogg (1992), Beck (1986), Triggs (1987); and Triggs and Riddell (2007) for Antarctica; and Ulfstein (1995) and Pedersen (2009) for Svalbard. Firsthand accounts by the original actors are readily available for much of the modern period e.g. Scott (1913), Cherry-Garrard (1922), Amundsen and Ellesworth (1927), and Nobile (1961).

35 Marstrander (1999: 119-120) lists three theories about the earlier discovery of Spitsbergen by Stone Age peoples, Vikings, or Russian Pomor hunters, none of which is widely accepted at present.

Research expeditions in Spitsbergen began in the 18th century, sometimes in combination with other activities such as recreational hunting, adventure, and, later, the assertion of territorial interests. The islands, which are relatively accessible by sea, were used as a starting point for a range of expeditions by air to the North Pole. The existence of mineral resources in Spitsbergen has been known since the 17th century, but the excavation of coal began in 1899. This was followed by a major period of development of coal and other minerals such as chalk and marble prior to World War I. Many of these developments were based on an optimistic assessment of the mineral resource potential of the archipelago, and eventually failed. However, coal mining continues to this day. Following the coming into force of the 1920 Spitsbergen Treaty in 1925 Norway took over the governance of the region. During World War II the Allies evacuated all civilians from Spitsbergen in 1941; several German weather stations were established in parts of the archipelago, and there was active warfare. All remains pre-dating 1 January 1946 are now legally protected and constitute the bulk of the cultural heritage in Svalbard.

Human activity in the Antarctic region began in the late 18th century with the seasonal exploitation of marine living resources, mainly seals and whales.[36] A period of early exploration took place between the late 19th century and the early part of the 20th century. These were goal oriented activities that for the most did not result in a permanent presence, barring historic remains and one active research station.[37] The assertion of territorial claims by a number of states in the first half of the 20th century resulted in the strategic establishment of research stations through claimed territory. However, the greatest expansion of base facilities took place only in the period following World War II and in particular the International Geophysical Year (IGY) of 1957-1958. Whaling and sealing declined from the 1960s, while in the 1970s distant water fishing fleets began to operate in the Southern Ocean harvesting krill and finfish. Initially, the Antarctic region was essentially isolated from the rest of the world, with the Southern Ocean acting as an all but impassable "moat". However, the arrival of global commercial interests and activities to the Antarctic region has been a key factor (albeit not the only one) in marking the end of Antarctic isolation (Hemmings, 2007).

Tourism developments in Antarctica and Svalbard took place at substantially different times, but in both instances these happened after industries based on the exploitation of living resources had become consolidated or run their course. The remains of these activities provided some of the early tourist attractions in these regions, in addition to the regions' natural attractions and contemporary settlements. The first tourism cruise to Spitsbergen was organized in 1871 (Conway, 1906: 302) while organised tourism in the Antarctic Treaty area began in the 1950s, first organised by Argentina and Chile, and subsequently by private operators (Palazzi, 1993; Headland, 1994:

---

36 Human activities in the Antarctic region north of what is today the Antarctic Treaty Area (i.e. south of 60° south) began earlier than below this parallel.

37 The stone hut built by the Scottish National Expedition to Laurie Island, South Orkney Islands, which was occupied between 1902 and 1904, was transferred by its owners to the Argentine Government in 1904. The site has been occupied uninterruptedly since then under the name "Orcadas". It is the longest occupied station in Antarctica.

275). Tourism in Svalbard was intermittent during War World II, but Antarctic tourism continued more or less interruptedly at a relatively constant low level since it began (Enzenbacher, 1992; Viken and Jørgensen, 1998; Kolltveit, 2006).

The 1990s marked an inflection point in which tourism in these regions began to increase significantly and eventually became a major presence (Viken and Jørgensen, 1998; Kolltveit, 2006; see also Chapter 3). The end of the Cold War facilitated the expansion of Antarctic tourism through the availability of ice class vessels and skilled crews, formerly used for scientific research and surveillance in the Soviet Arctic (Hemmings, 2007; see also Chapter 3). Some of these ships also became active in Svalbard, and a Norwegian policy change in 1990 encouraging tourism resulted in an increase of this activity (Viken and Jørgensen, 1998).

Historic details for specific case study sites are listed in Chapters 3-8. The main periods in the history of Antarctica and Svalbard and their associated historic sites/ cultural heritage are briefly outlined in Tables 1.3 and 1.4. Different aspects of the polar cultural heritage are described in Barr and Chaplin (2004, 2008) (for both polar regions), Marstrander (1999), Governor of Svalbard (2000), and Jørgensen (2005) for Svalbard, and Harrowfield (2005), Bizzarri (2006), and Guyomard (2008) for Antarctica.

The 2001 Svalbard Act automatically protects structures, sites, and movable historical objects predating 1946, as well as cultural features such as human graves and certain animal slaughtering sites, irrespective of their age. In addition, cultural remains postdating 1946 that are of particular historic or cultural value may also be protected by a decision of Norway's Directorate of Cultural Heritage. Several international conventions are also applicable to the preservation of Svalbard's cultural heritage: the 1992 Malta Convention, which protects the archaeological heritage of Europe, and the 1985 Granada Convention, which protects Europe's architectural heritage (Governor of Svalbard, 2000; Geitz, 2004).

In Antarctica the Antarctic Treaty Consultative Meeting assesses each proposed Historic Site and Monument on case-by-case bases, following the initiative of a Party or Parties, using a set of criteria agreed by ATCPs.[38] In practise, these criteria are applied loosely and the Antarctic Treaty Consultative Meeting approves most proposals without much ado. [39] However, uncommon proposals may raise some questions.[40] A few past activity sites in Antarctica are protected by the domestic legislation of Antarctic Treaty states but not by instruments of the Antarctic Treaty System (e.g. Evans, 2007), so that under international law they are not protected. Antarctic historic sites are

38 (ATCM) Resolution 3 (2009) on *Guidelines for the designation and protection of Historic Sites and Monuments.*

39 Personal observation at several Antarctic Treaty Consultative Meetings, 1997-2010.

40 For instance, at the XXXIII ATCM Chile proposed changes in the description of HSM 37 (a bust of Chile's independence hero Bernardo O'Higgins) (Chile, 2010). The proposed change asked for the addition of several facilities pre-dating the IGY, including one building that no longer exists located a considerable distance away. A proposal was postponed for the following year to allow time for consultation with experts (CEP, 2010: paragraphs 204-205), which is unusual.

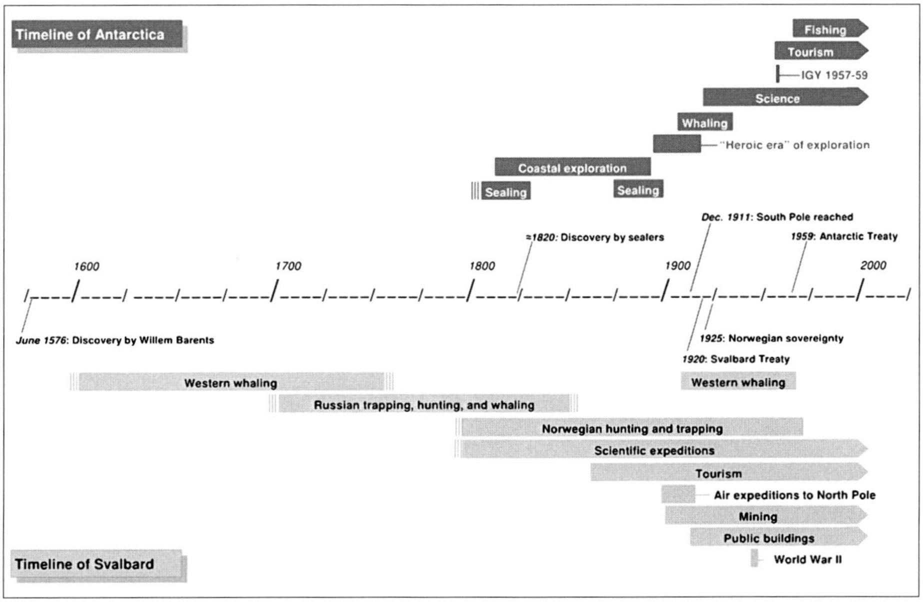

Fig. 1.4 – Simplified time lines for Antarctica and Svalbard, with an emphasis on activities whose material remains constitute today's historic sites and tourism attractions. Graph by author.

not located in the territory of sovereign states and therefore do not meet the criteria for nomination under the 1972 World Heritage Convention.[41]

1.8 THE GEOGRAPHY OF CULTURAL HERITAGE IN ANTARCTICA AND SVALBARD

As noted by Graham *et al.* (2000: 4), heritage is inherently a spatial phenomenon, defined by its location, distribution, and scale. These factors will be examined briefly here with respect to the cultural heritage of Antarctica and Svalbard.

The *location* of heritage triggers two basic geographic questions – Where is it? Why is there? (Graham *et al.*, 2000: 4). Historic sites in Antarctica and Svalbard are located where they are found for a range of reasons including logistic convenience and access to resources. While there are historic sites in remote locations – for instance, somewhere in the Polar Plateau lie the buried remains of the tent left at the South Pole by Roald Amundsen and his colleagues in 1911 – many historic sites are located in ice-free coastal areas. The original location of these historic features and the ability to

41 "Only countries that have signed the World Heritage Convention, pledging to protect their natural and cultural heritage, can submit nomination proposals for properties on their territory to be considered for inclusion in UNESCO's World Heritage List." http: //whc.unesco.org/en/nominations (Retrieved October 2010).

visit them in the present underscores the relative accessibility of many polar historic sites, which is relevant to their development as tourism destinations.

The cold conditions prevalent in the Polar Regions influenced the establishment of the original sites and similarly influence their contemporary use as tourism destinations. Cold has also enhanced the preservation of historic features after abandonment. Polar historic sites and the elements they contain would be preserved on or below ice-free (but often frozen) ground,[42] on or below snow or ice, or underwater in cold waters. Conversely, some characteristics of the polar environment would have contributed to the decay of historic remains in quite unique ways. Gravity-fed katabatic winds, the accumulation and melting of snow, and glacier ablation by calving would all have affected not only the original activities but also have contributed to the destruction of the remains of those activities. For instance, Amundsen's base camp at Bay of Whales on the Ross Ice Shelf has long calved into the sea, whereas Scott's *Terra Nova* hut at nearby Cape Evans survived several decades without any maintenance. Furthermore, Scott's hut contains a vast collection of artefacts including textiles, leather, furs, and food. While decaying, these artefacts are far better preserved than they would have been were the hut not located in a cold, dry environment.

The *distribution* of the polar heritage reflects patterns of occupation, which in turn reflect logistic convenience, access to resources, scientific interest, and territorial claims. As noted by Graham *et al.* (2000: 5) "some places will have more or less, or a different heritage". This is apparent in the type of remains that result from different periods in the history of Antarctica and Svalbard (Tables 1.3 and 1.4), and in the specific distribution of some remains. For instance, Norwegian hunting and trapping in Svalbard was carried out from outposts consisting of one main centre with a network of minor outposts about a day's walk away, which cover most of the accessible part of Spitsbergen. In contrast, most whaling and mining sites are located along the large fjords in the W of Svalbard, close to the resources and to shipping.

Heritage, as places, exist within a hierarchy of spatial scales (Graham *et al.*, 2000: 4). These range from the local to the regional and the intercontinental/international. Plainly the cultural heritage of Antarctica and Svalbard is local in that it is based in specific locations, but it originates outside the regions, given the absence of indigenous people. At the same time it retains historic associations with other places within and outside the regions. For instance, the hut built by the 1902-1903 Swedish Antarctic Expedition at Snow Hill Island is not just an isolated feature, but it is associated with expedition remains found elsewhere. These include emergency stone huts built by two separate groups of expedition members at Paulet Island and at Hope Bay; and a signpost erected at Seymour (Marambio) Island to alert potential rescuers. The Snow Hill Island hut is also associated with other sites including the location where the expedition ship *Antarctic* sank, some 40 kilometres E from Paulet Island; with the Argentine vessel *Uruguay* that rescued expedition members, now a museum in Buenos Aires; and with specific places in the home countries from where the expedition members originated (Sweden, Norway, and Argentina).

42 Depending on the ground moisture content historic materials would be subject to more or less churning effected by freeze and thaw, which would influence their basic properties.

Heritage elements can have different scales (Graham *et al.*, 2000: 4) illustrated by the sites examined as part of this research, which included:

- Industrial sites consisting of mining or whaling remains, including standing buildings, standing ruins or ruins and industrial remains of various kinds, including artefacts and waste, extending over substantive areas (about 1 km^2);
- Research stations consisting of several buildings and associated structures (about 0.1 km^2);
- Expeditions huts consisting of a main hut and associated structures and artefact scatters (about 0.01 km^2);
- Trappers' huts consisting of a small hut and some associated remains (a few square metres each);
- Sealers' shelters usually consisting of single dry-wall enclosures (a few square metres each);
- A wooden oar located on a ridge about 300 metres above the sea level, some kilometres away from a whaling factory.

Conceptually, it would be possible to provide further examples of an increasingly smaller record of past activities in Antarctica and Svalbard down to sites of past activity where nothing remains other than perhaps the archaeological record that is invisible to the naked eye (e.g. Weiner, 2010).

1.9 RESEARCH APPROACH: CREATING RESEARCH OPPORTUNITIES

1.9.1 Dedicated vs. opportunistic research

The Polar Regions in the early 21st century are more accessible than ever before, yet they remain remote places where conducting research is expensive and potentially dangerous. Even if one manages to journey to Antarctica or Svalbard, accessing sites of interest may be difficult. Visits may be brief, and factors such as weather conditions or snow cover may limit or prevent fieldwork. In this research, practical circumstances like funding, transport, living arrangements, and safety placed rather considerable limits to the conduct of fieldwork. In addition, much of the actual fieldwork was conducted single-handedly, which is unusual in the Polar Regions. Nevertheless, fieldwork was an essential component of research, and limiting circumstances were a motivation to create fieldwork opportunities, for instance by looking for jobs or commissions in the study regions. Case study sites were selected in the proximity of relatively accessible places, and research methods were adapted to limiting factors. Research in Svalbard was dedicated – in the sense that I was able to travel there specifically for research purposes – while Antarctic research was conducted opportunistically in the margins of other professional activities (e.g. Roura and Pérez-Muñoz, 2002; Bastmeijer and Roura, 2008; Roura *et al.*, 2008).

The ideal case study for this research required being at a historic site when tourists were visiting it, in order to conduct observations. In practice , this research unfolded through a range of sites and circumstances, including ideal and less than ideal cases studies – such as sites with no visitors, visitors who were not tourists, or sites with no historic remains. Less than ideal cases, however, served as controls for the ideal cases, in that they allowed the identification of the effects of site formation processes (other than tourism) on historic sites or other kinds of sites. These are further described in Chapter 2.

Working solo was not a reason to dispense with safety concerns. Most of the work was conducted with relatively easy access to, and fluid communications with, local support centres. Previous Antarctic fieldwork experience gave me prior knowledge of many sites, which greatly facilitated research. Barring a brief visit in the late 1990s, Svalbard was largely unknown to me prior to this research. Some aspects of conducting research in Svalbard are comparatively easier than in the Antarctic. However, polar bears (*Ursus maritimus*) are predators that pose risks to humans, even though the species is increasingly threatened. This factor alone has a significant influence on fieldwork in Svalbard. A course on polar bear awareness, weapon handling and shooting practice provided some confidence and enabled independent fieldwork beyond populated centres.[43] Fieldwork in Svalbard was facilitated by access to the Netherlands Arctic Station at Ny-Ålesund, which served as a centre of operations (there is no equivalent facility for Dutch researchers in the Antarctic.) A discounted berth on a trip to NW Svalbard was negotiated with Oceanwide Expeditions, a Dutch tour operator. Environmental protection is a matter of utmost importance both in Antarctica and in Svalbard, and applicable legislation was adhered to.[44]

---

43 As noted by a colleague, carrying a rifle is very useful to build confidence as long as one does not come across a polar bear – at which point one is bound to go weak at the knees (M. Loonen, personal communication July 2007)..

44 Environmental Impact Assessments (EIAs) are required for all activities in Antarctica. The EIA of the various expeditions in which he was involved, from which opportunistic research was carried out, covered the author's activities in Antarctica. Operational standards were according to those used in the various Antarctic or Arctic programs that provided logistic support to the expeditions.

1.9.2 Case study approach

There are 80 listed sites and monuments in Antarctica[45] and approximately 920 cultural monuments in Svalbard.[46] Since many of these are tourism destinations, selectivity was of the essence. A case study approach to research may be defined as "...an intensive study of a single unit or a small number of units (the cases), for the purpose of understanding a larger class of similar units (a population of cases)" (Guerring, 2007: 37). Multiple case studies (Yin, 2003; Guerring, 2007) enabled a deeper understanding of how formation processes affect comparable historic sites that are managed under the different regimes in place in Antarctica and Svalbard.

Case studies were selected amongst the most significant possible to include historic sites representative of different types and functions with tourism or other forms of visitation at those sites acting as a common theme (Table 1.6). In line with the objectives set at the outset of this research, fieldwork was the basis for most of the case studies. The challenge, in this type of work, is connecting the findings (Yin, 2004) – or finding the "red line" that links all case studies.

Given the spatial extent of the area and logistic constraints field work has focused on some particular coastal areas. The *foci* of work in Svalbard were whaling and mining sites in the NW of the archipelago. The *foci* of Antarctic work were sealing, whaling and exploration sites in the South Shetland Islands and Ross Island in the Atlantic and Pacific sectors of the Southern Ocean, respectively. The Antarctic Peninsula has a closer similarity to Svalbard than other parts of the continent. This includes the distance to populated centres outside the region; the location and distribution of historic sites along ice-free coastal areas; and the broad patterns of tourism. To a lesser extent this also applies to the Ross Sea region. Historically, these three distinct areas have been focal points of exploration activities and of the exploitation of natural resources, and in the case of Antarctica also the starting point for exploration into the Polar Plateau. Specific sites are described in Chapters 4-8.

1.9.3 Unit of analysis

The unit of analysis for this research are sites that contain remains of past activities, or historic sites. The sites studied here have different degrees of complexity – from sites composed by a single artefact, to complex sites composed by remains from multiple periods of occupation. The oldest site visited was a 17th century Dutch whaling factory in Svalbard; the most recent was a Chilean monument in the Antarctic Peninsula dedicated in 2001, and designated as a "historic monument" in 2007. Case study sites ranged from no more than 1 m^2 to approximately 1 km^2 in surface area, and

45 By ATCM XXXIII (2010) there were 80 listed historic sites and monuments numbered 1-85 (some have been de-listed or subsumed into other HSMs).

46 Marstrander (1999: 134) lists 1,872 cultural monuments in Svalbard including 972 individual graves. These graves are located in 20 gravesites; taking the gravesites, rather than individual graves, as cultural monuments there is a total of 920 cultural monuments (PK Reymert, pers. comm. 10 July 2007), which is the figure used here.

involved ruins, standing ruins, and buildings and other facilities in varying conditions and degrees of active management. Monuments of various kinds were examined as well. All the sites were located in ice-free coastal areas and many were regular tourism destinations and attractions.

Focusing on individual sites results in lesser coverage of the area *in between* sites (O'Connor and Evans, 1999: 137). The area between sites was examined on a micro-regional scale at Fildes Peninsula in King George Island and at Deception Island. This approach gave a better picture of the extent of historic remains in an area than otherwise, although dating some remains may be problematic. From the perspective of tourism research historic sites were of greatest interest, since most organised tourism landings take place at substantive historic sites and not in between them.

1.9.4 The boundaries of research

Temporal boundaries for this research, in a historic sense, are defined by the age of the earliest material cultural remains in the study regions i.e. the 17th century for Svalbard and the early 19th century for Antarctica. However, greater emphasis has been placed on the time period since the current regimes of governance for Antarctica and Svalbard began to develop (1920 in Svalbard and 1959 in Antarctica), and particularly on the past three decades (i.e. the 1980s onwards). This has been a period of substantive change in Antarctica and Svalbard including the expansion of polar tourism, the establishment of comprehensive environmental protection regimes, and emerging evidence of climate change, which have helped both to the exposure and to the preservation of historic remains. The most recent time frame concerns the period in which tourism observations were conducted (2002-2008).

In this research the **spatial boundaries** of Antarctica and Svalbard are based on legal criteria. These differ from those defined using ecological, climatic, or other criteria.[47] The rationale is that the concept of cultural heritage as used in this research is equated to historic sites that are legally protected. The spatial boundaries of Antarctica correspond to the area south of 60° S or the Antarctic Treaty Area. Svalbard's boundaries are defined by the Treaty Concerning the Archipelago of Spitsbergen, which comprises all the islands situated between 74° to 81° N and 10° to 35° E. The legislation applicable to the designation and management of historic sites and the regulation of tourism is different on either side of these legal boundaries. However, the cultural heritage of these regions has significant historic and contemporary associations with locations outside those boundaries – for instance in the Sub-Antarctic islands.

---

47 The Antarctic region encompasses the Antarctic continent, surrounding islands, and the Southern Ocean south of the Antarctic convergence. The Antarctic convergence is a dynamic zone that marks the encounter between the warm waters of the northern seas and the cool and less salty waters of the Southern Ocean. At this boundary there is a sudden change in sea water temperature. The latitude of the Antarctic convergence varies between 45° S and 60° S and the location of this boundary is not static. The Arctic is variously defined by physical markers such as the limit of the tree line; the 10C July isotherm line; the Arctic circle at 66° 33' N; and 60° N.

The ground surface (combined with ice and/or snow) marks the **vertical boundary** (or third dimension) of this research. Historic features may be buried, but historic remains, as tourism attractions, are largely those remains that are visible above the ground surface. Visitor interactions with historic remains are also largely on the surface, even though human locomotion may affect artefacts below the ground surface (e.g. Gifford-González *et al.*, 1985).

The **scale boundaries** are set by the macroscopic record, that is, whatever is visible to the naked eye.

Heritage and tourism, both individually and taken together, are linked to many other issues. Just as it is important to define what this research is about, it is also important to define what this is *not* about. Rapid climate change is apparent in the Arctic and in parts of the Antarctic (SCAR, 2010) with consequential effects on historic sites (ICOMOS, 2008; Barr, 2008). Generally, the effects of climate change in the Polar Regions may affect both the landscape and the historic sites it contains. Effects may include the erosion of coastal sites, increased forms of biological decay, and increased tourism penetration (Barr, 2008). The specific linkages between cultural heritage and climate change merit a detailed, stand-alone treatment and are not discussed here. The regulation of polar tourism has been discussed *ad nauseam* from a myriad of perspectives (e.g. Davis, 1995; Hall and Johnston, 1995; Johnston, 1997, 1998; Enzenbacher, 1992, 1995; Hemmings and Roura, 2003; Haase *et al.*, 2007; Bastmeijer *et al.* 2008; Lamers, 2009). Tourism regulation will be discussed in Chapter 3 and elsewhere as a context but not as a central theme of this research.

1.10 STRUCTURE OF THIS THESIS

This thesis is presented in ten chapters covering different theoretical and practical aspects of the interaction between tourism and historic sites in Antarctica and Svalbard. A choice has been made to publish this research as a collection of articles in scholarly publications. Consequently, Chapters 3-8 are self-contained research papers. This was a tactical decision to speed up publication based on the fact that the polar tourism field is very competitive, particularly in the context of the International Polar Year 2007-2009, and the polar heritage field is not far behind. There are redundant sections in some of the chapters in sections concerning the definition of basic concepts, heritage protection regimes, and research methods. This has disadvantages for those attempting to read the thesis from cover to cover, but it facilitates reading it out of order.

Chapter 2 describes the research methods. As noted earlier, finding out research methods suitable to research the interactions between tourism and historic sites required some research in itself. Therefore, this chapter also addresses the issue of how to study the effects of polar tourism on historic sites, and describes the "tool box" developed for the purposes of this research.

Chapter 3 sets the scene by examining key issues of polar tourism, taking as a case study Antarctic tourism observed from a regional legal perspective. The chapter discusses the regulation of tourism, the assessment and prevention of cumulative impacts, and the application of the precautionary principle. These concepts are

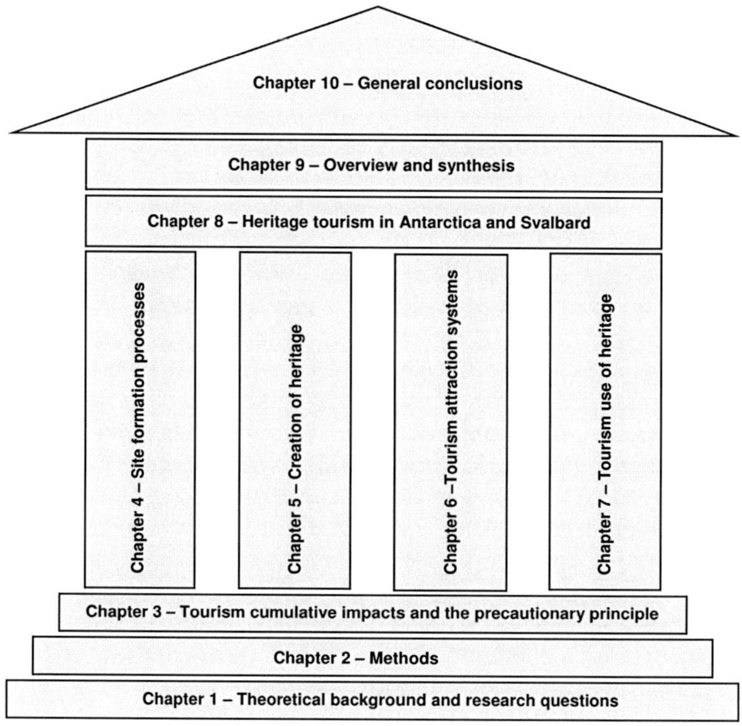

Fig. 1.5 – A "Roman temple" overview of this thesis.

applicable to polar tourism as a whole. In this regard this chapter serves as a foundation for later chapters.

Chapter 4 focuses on site formation processes. It analyses the recent transformation of historic features in Antarctica and Svalbard as assessed using repeat photography. Heritage, in this context, is multi-scalar and hierarchical, and local changes take place in the context of broader regional developments in the Antarctica and Svalbard.

Chapter 5 examines the creation of the polar cultural heritage, based on Thompson (1989), from the evolving and at times contrasting perspectives of the protection of environmental and cultural values. The focus is on cross case studies of a specific type of historic site – Antarctic scientific bases – however, the questions it addresses are applicable to other polar cultural heritage.

Chapter 6 focuses on the use of polar historic sites as tourism attractions. It explores the use of narratives in the transformation of historic sites in the Polar Regions into attractions and consumable tourism products, based on the theory developed by Leiper (1990) on tourism attraction systems. The analysis is based on a case study of organised visits to the airship mooring mast at Ny-Ålesund, Svalbard, which was used for the transpolar flights of the airships *Norge* and *Italia* in 1926-1928.

Chapter 7 discusses the use of polar historic sites for tourism. Based on the case study of visitation to a historic marble quarry in Ny-London, Svalbard, this chapter explores the application of practical and theoretical concepts discussed in this thesis.

Chapter 8 reviews key concepts of tourism and historic sites in the study regions. It compares and contrasts organised cruise tourism at protected historic sites in Antarctica and Svalbard, and identifies the patterns and impacts of organised tourism and historic sites in the context of existing tourism policies for each region.

Chapter 9 provides an overview and a synthesis of tourism as a behavioural system and a site formation process.

Chapter 10 is the concluding section and covers the pure academic questions addressed in this research, complemented by an evaluation of research methods and a discussion of key management implications.

Some chapters (Chapters 3, 5, 6 and 7) focus on case studies from either Antarctica or Svalbard. Although the legal and governance frameworks are different in each region, the basic issues are conceptually applicable to both. Table 1.6 illustrates how each of the central chapters (Chapters 3-8) relates to specific case studies.

As noted above, the basic narrative of this thesis concerns several interlinked themes. These themes are important to sustain the narrative of each chapter, and each chapter is important to sustain the narrative of the thesis to its conclusion. Fig. 1.5 illustrates the basic architecture of the thesis and how the different chapters build a coherent whole.

TABLES CHAPTER 1

Table 1.1 – Overview of Antarctica and Svalbard

Topic	Antarctica	Svalbard
Discovery	No later than 1820.	1596 by Wilhem Barentz.
Governance	Antarctic Treaty System.	Norway.
Surface area	13,829,430 km² (including ice shelves and islands within ice); 12 272 800 (excluding ice shelves) (BAS, 2004).	61,022 km², of which 36,502 km² (59.82%) are glaciers (Statistics Norway, 2010).[1]
Ice free surface area	44,890 km² (0.32%) (BAS, 2004).	24,520 km² (40.18%) (Statistics Norway, 2010).[2]
International treaties	1959 Antarctic Treaty. Entered into force 1961.	1920 Treaty Concerning the Archipelago of Spitsbergen. Entered into force 1925.
Treaty Parties	12 original signatories. 49 parties (2009) – 28 Consultative Parties and 18 Non-Consultative Parties.	14 original signatories (including the United Kingdom and five overseas Dominions). 39 parties (2009).
Sovereignty	Art. IV of the Antarctic Treaty "freezes" territorial claims while the Treaty is in force.	Assigned to Norway in the Treaty of Svalbard (Art. 1) subject to "equal opportunity" to all contracting Parties (Arts. 2 and 3.)
Key institutions	29 National Antarctic Programs operating in Antarctica. Antarctic institutions located outside the continent include the ATCM, Antarctic Treaty Secretariat, CEP, CCAMLR, CCAS, COMNAP, and SCAR.	*Norway:* Governor of Svalbard; Kings Bay AS; Svalbard Museum; The University Center in Svalbard (UNIS); Store Norske Spitsbergen Coal Company (SNSG); the Norwegian Polar Institute; Svalbard Tourism; Svalbard Science Forum; Svalbard Global Seed Vault. *Poland:* the Polish Polar Station (Hornsund). *Russia*: Trust Arktikugol (Barentsburg).
Indigenous people	None. A few people have been born in the continent[3]	None. Some people have been born in the islands.[4]
Population	111 active sites and facilities through the continent, including 40 stations operating year-round. Winter average population at least 1,094. Peak population excluding tourists at least 4,460 (COMNAP, 2010).[5]	Five main settlements. 2,085 people registered in the Norwegian settlements at Svalbard (1 January 2009). Total population including non-Norwegian settlements, around 2,570 people (Statistics Norway, 2010).[6]
Main industries	Fisheries, tourism, polar services, biological prospecting.	Mining and quarrying, construction, education, public administration, hotels and restaurants, financial (Statistics Norway 2010).[7]

Notes to table 1.1:

1 http://www.ssb.no/english/subjects/00/00/20/nos_svalbard_en/nos_d330_en/tab/012.html Accessed 29 January 2010
2 Calculated from data in http://www.ssb.no/english/subjects/00/00/20/nos_svalbard_en/nos_d330_en/tab/012.htmlAccessed 29 January 2010.
3 Palazzi (1993: 164-165) lists ten births at Antarctic bases between 1978 and 1985.
4 Currently no children are born in Svalbard, but may be registered there if the mother is a Svalbard resident http://www.ssb.no/en/befsvalbard/arkiv/art-2009-03-05-01-en.html Accessed 29 January 2010.
5 https://www.comnap.aq/facilities, accessed 29 January 2010.
6 http://www.ssb.no/en/befsvalbard/arkiv/art-2009-03-05-01-en.html. Accessed 29 January 2010.
7 http://www.ssb.no/english/subjects/00/00/20/nos_svalbard_en/nos_d330_en/tab/127.html. Accessed 29 January 2010.

Table 1.2 – Overview of tourism in Antarctica and Svalbard

Topic	Antarctica	Svalbard
Tourism modalities	Mostly ship-borne, with and without landings; also yachts and limited air-supported and land-based tourism (few refuges and semi-permanent camps); overflights. "Adventure" tourism.	Ship-borne (including day cruises from local operators, and longer cruises from local/foreign operators); yachts; and air-based and land-based (hotels, refuges, semi-permanent camps). "Adventure" tourism.
Tourism industry associations profile (~ scale of the industry)	IAATO (20 Affiliate, 5 Associate B1, 35 Associate B2, 50 Members); 42 vessels (January 2010)	AECO (12 Members and one Provisional Member); 23 vessels (January 2010)
Tourism attractions and activities	Mostly ship-based visits to research stations, historic sites, penguin rookeries, cruising in scenic areas, and reaching the "furthest south". Activities such as extended walks, kayaking, diving, camping, skiing, and climbing. Lectures and other on-board activities. Limited but growing facilities such as souvenir shops, post offices, and museums.	Mostly ship-based visits to settlements, historic sites, walrus hauling sites, bird cliffs, cruising in scenic areas, and reaching the "furthest north". Activities such as extended walks, kayaking, diving, camping, skiing, and climbing. From a Longyearbyen base, excursions on speedboats, dog sledging, snow mobiles. Tourist facilities such as replica of Behuiden Huise,[8] Svalbard Museum, Spitsbergen Airship Museum.
Operating season	October-April	May-September; some operators also year round. Longyearbyen open for tourism year round.
Number of tourists	More than 35,000 ship borne tourists visited Antarctica in the austral summer of 2006-2007, of which more than 28,000 landed ashore (IAATO 2008).	28,000 tourists arrived in Svalbard in overseas cruise ships, and more than 12,000 tourists boarded 'expedition' cruises and day cruises (Governor of Svalbard 2006: 6-8).
Most visited sub-regions	Antarctic Peninsula, Ross Sea	NW Spitsbergen, Ijsfjorden
Number of tourism landing sites	About 200 sites (2006) but many are difficult to access. Tourism concentrates at some 35 sites.	About 200 sites (2006), with no natural access restrictions. A few sites in NW Spitsbergen receive many visitors, but tourism is otherwise dispersed.

Note to table 1.2

8 "Behuiden Huise" ("'t Behouden Huys") is the name of the hut built by Wilhem Barentz in Novaya Zembla (now Russia).

Table 1.3 – Cultural heritage (legally protected historic sites) in Antarctica and Svalbard

Topic	Antarctica	Svalbard
Number of historic sites	79 listed sites and monuments numbered 1-84 (2010)	Approximately 920 cultural monuments (Maarstrander, 1999)
Main protection instruments	1991 Protocol of Environmental Protection to the Antarctic Treaty and instruments approved by the Antarctic Treaty Consultative Meeting.	2001 Svalbard Environmental Protection Law and Government of Svalbard Regulations.
Most common cultural heritage by type	Monuments and cairns, and buildings	Ruins, buildings, and graves
Most common cultural heritage by function	Expedition remains from prior to 1917	Russian hunting, Norwegian hunting, and Western whaling
Earliest cultural heritage	Early 19th century sealers' shelters; late 19th century to early 20th century "heroic era" buildings and ruins.	17th century Dutch and English whaling activities
Modern cultural heritage	Monuments and memorials of various kinds from the 1970s-2000s; plaque commemorating a US nuclear power plant from the 1950s; German fibreglass hut from the 1970s; British base and hut from the 1950s and 1970s.	The Swedish-Finnish-Swiss International Geophysical Year station Kinnvika (1957-1958); the aerial transport system for coal from the mines to the transport centre in Longyearbyen.

Table 1.4 – Overview of cultural remains on Svalbard based on main periods of history (Based on Marstrander, 1999; [Norwegian] Government White Paper N° 22, 1994-1995)

Period	Activity	Remains	Fieldwork locations (this thesis) (*)
Prehistory - approximately 900-1300	Hypothesis about Spitsbergen's discovery by stone age cultures, Vikings, and Russian Pomors.	None.[9]	--
1576	Discovery by Wilhem Barentz	None (geographic names).	--
Approximately 1600-1750	Whaling	Whaling stations with blubber ovens and dwelling sites. Burial places.	Whaling station at Smeerenburg
Approximately 1700-1850	Russian hunting, trapping and whaling	Hunting, trapping and whaling stations with dwelling sites. Two remaining buildings. Shipwreck remains ("lodjer"). Burial places. Votive crosses.	--
Approximately 1800-ongoing	Norwegian hunting, trapping and whaling	Hunting, trapping and whaling stations with huts and dwelling sites. Permanent animal traps: fox traps, polar bear traps. Whaling boats. Burial places.	Two huts in NW Spitsbergen.
19th and 20th centuries	Scientific expeditions;	Winter-over bases with buildings and dwelling sites. Remains of tools and technical equipment.	--
1871- ongoing	Tourism	Site of first hotel; hotel.	--
1896-1928	Air expeditions to the High Arctic, North Pole.	Dwelling sites. Remains of balloon and airship hangars. Mooring mast for airships and other equipment.	Airship mooring mast at Ny-Ålesund
1900 – ongoing	Exploitation of minerals	Buildings and dwelling sites. Mines. Technical equipment for digging and transport. Burial sites.	Mining remains at Ny-Ålesund, Ny-London
1911 – ongoing	Public buildings and equipment	Buildings. Coastal lights and beacons. Technical equipment.	--

Table 1.4 continued

Period	Activity	Remains	Fieldwork locations (this thesis) (*)
1940-1945	Second World War	Cabins and dwelling sites. Gun-posts. Weather forecasting stations with technical equipment and escape depots.	--
21st century	Norwegian and Russian settlements; research stations.	Aircraft remains. Cultural environment, with mining works, transport facilities and other technical developments. Buildings with various purposes	Tourism landings at Pyramiden, Barentsburg, Ny-Ålesund.

(*) *Sites where substantive fieldwork has been carried out. The listing of sites visited is more extensive.*

Note to table 1.4:

9 There is a substantive discussion about the chronology of Russian Pomor activity, particularly the absolute dating of earlier remains (e.g. Arlov, 2005: 16-17).

Table 1.5 – Overview of cultural remains in the Antarctic Treaty Area based on main periods of history (*)

Period	Activity	Remains[10]	Fieldwork locations (this thesis) (*)
1500-1818	Early notions of Antarctica; first crossings of the Antarctic Circle; sealing and whaling in Sub Antarctic Islands.	None.	--
1819-1894	Mapping the coast-line, sealing and exploration.	Sealers' shelters and associated artefacts and ecofacts. Ship-wrecks. Landing site.[11] Monuments commemorating 1820s activities.[12]	Sealers' shelters, shipwreck at Fildes Peninsula, King George Island, South Shetland Islands.
1895-1917	First winter over in Antarctica and first landing on the main-land Early exploration of the continent; expeditions to the South Pole, early scientific research, early whaling.	Wooden and stone huts and shelters, expedition remains, depots, message posts, rock cairns, memorial crosses, research base; whaling factory remains, plaques and monuments. Cemeteries, graves and memorials to the dead.	Expedition huts and associated remains at Ross Island.
Pre IGY period 1922-1957	Scientific expeditions, occupation of territory, Operation Tabarin.	Rock cairns, lighthouse, plaques, memorials, magnetic observatory, abandoned installations, commemorative shelters, busts, cemeteries, crosses, research bases (active, inactive, and abandoned) including buildings, artefacts and waste of scientific expeditions. Cemeteries, graves and memorials to the dead.	Scientific bases, memorials and monuments at Deception Island and Ross Island.
Post IGY period 1958-present	Scientific expeditions, occupation of territory, tourism.	Stations, camps, refuges, and depots (active, inactive, and abandoned); and associated features (roads, airstrips, waste dumps, spill sites); accident sites; cemeteries, graves and memorials to the dead; ship-wrecks, etc.	Stations and past activity sites in King George Island, Deception Island, and Ross Island.

(*) Sites where substantive fieldwork has been carried out. The listing of sites visited is more extensive.

Notes to table 1.5

10 Note that some historic/past activity sites are protected but many are not. Some monuments are contemporary to the events, while others were designated decades after.

11 The rock where French Admiral Dumont D'Urville is believed to have landed on 21 January 1840 has been designated as a protected site (HSM 81), following a proposal by France. There are however no on-site cultural remains from that event.

12 Two monuments (HSM 57 and HSM 58), which were dedicated in the second half of the 20th century, refer to events presumably dating in 1819 and 1820, respectively. However, the monuments in themselves do not constitute remains or evidence of those events.

Table 1.6 – Overview of case studies

Study type(*)	Subject	Research themes	Chapter
Case study (n=1)	Antarctic tourism	Cumulative impacts, application of the precautionary principle to tourism regulation and management	3
Case study (n=7)	Recent transformation of historic features	Site formation processes (Schiffer, 1983, 1985)	4
Cross-case study	Antarctic scientific bases	Legal designation as "heritage": application of Rubbish Theory (Thompson, 1979)	5
Case study (n=1)	Tourism at the airship mooring mast at Ny-Ålesund, Svalbard	Tourism as a behavioural system; use of the cultural heritage for tourism: tourism attraction systems (Leiper, 1990)	6
Case study (n=1)	Tourism at the abandoned quarry in Ny-London, Svalbard	Tourism as a behavioural system; use of cultural heritage as tourist attractions; research methods; management	7
Cross-case study	Cultural heritage tourism in Antarctica and Svalbard	Tourism as a behavioural system; use of cultural heritage as tourist attractions; research methods; management	8

(*) A case study comprises one or a few cases studied intensively; a cross-case study comprises many cases studied more generally (Guerring, 2007:20).

2 Research methods

- I know your methods, sir, and I applied them. Before I permitted anything to be moved I examined most carefully the ground outside, and also the floor of the room. There were no footmarks.
- Meaning that you saw none?
- I assure you, sir, that there were none.
- My good Hopkins, I have investigated many crimes, but I have never yet seen one which was committed by a flying creature. As long as the criminal remains upon two legs so must there be some indentation, some abrasion, some trifling displacement which can be detected by the scientific searcher. A. Conan Doyle (1989: 949-950).[1]

In the Border Patrol, tracking is called cutting sign. "Cutting" is looking; "sign" is evidence. No technology is involved. J. Tietz (29 November 2004: 90).[2]

2.1 INTRODUCTION: UNOBTRUSIVE RESEARCH METHODS

During the conduct of tourism at polar historic sites people and material culture overlap temporarily and spatially as a single behavioural phenomenon. "People" may be represented by individual tourists or groups of tourists, or by other kinds of visitors. "Artefacts" may refer to historic elements present at the sites; to the elements brought by tourists such as photographic cameras; and to site features altered by tourism or other cultural processes. Finally, there are external factors represented by the environmental features surrounding historic sites, and by the natural processes affecting them. Identifying methods of research suited for use in investigating the interactions of these various elements required some research in itself. This is reflected in one of the research questions listed in Chapter 1:

- What opportunistic, low technology research methods can be used to assess the status and change of polar historic sites and the effect of tourism at those sites?

However, prior to formulating that question there was a previous, even more basic question:

- How can the effects of tourism at polar historic sites be studied?

Much recent research into polar tourism favours as a primary research method the use of interviews with, and questionnaires completed by, tourists and other stakeholders, complemented in some instances with fieldwork (Bauer, 2001; Maher *et al.*,

---

1 Conan Doyle, A. (1989): *Sherlock Holmes, The complete stories*, Wordsworth Editions.

2 Tietz, J. (2004): Fine disturbances. *The New Yorker*, November 29, 2004, p. 90-101.

2003; Haase *et al.*, 2007; Powell, 2008; Lamers *et al.*, 2010; Basberg, forthcoming). In this study, an integrated approach was used to combine data about behaviour, material traces and perception elements of visitation at historic sites, with an emphasis on the former two. Integrated perspectives are based on the premise that "neither interview-surveys, nor observations of participants are sufficient, either by themselves or in combination, to adequately describe, analyse, and understand the way social systems both behave and evolve." (Rathje, 2001: 63-76). Rathje further suggested triangulating mental (perception), behavioural, and material records to gain a multi-dimensional perspective of the social systems under study (63-64). These methods, collectively, form part of unobtrusive research methods used in the social sciences (Webb *et al*,1966; Lee, 2002) and the bases of the archaeology of contemporary society (Rathje, 2001). In brief, these methods combine ethnographic observations with the "archaeology of the ethnographies" (Gorecki, 1985).[3]

Unobtrusive measures are data collected without the direct elicitation of information from a respondent or informant. Data collected in this way avoid problems of data quality that arise because of reactivity i.e. when people change their behaviour because they are aware of being studied (Lee, 2002: 143) or those that result when people report events as they *think* they are rather than as they actually are (Rathje, 2001; Bernard, 2002).

Webb *et al.* (1966) developed three broadly thematic approaches to non-reactive research: physical trace measurement, simple observation, and archival information. This typology of data *sources* may be recast in more active terms as a typology of data *acquisition* in data that can be found, data that can be captured, and data that can be retrieved (Lee, 2002: 13). Table 2.1 follows this typology of research methods to classify the various research methods used for this research.

2.2 *MODI OPERANDI*

Tourism activities in the Polar Regions mostly take place in fairly isolated and remote places. Tourism is largely ship-borne, albeit not exclusively; and tourists are regularly taken ashore to visit places of interest. Some places that are fairly accessible and have significant natural, scenic or cultural values have become tourism destinations where landings take place on a regular basis. Other sites are less frequently visited. Consequently, there are two basic platforms to conduct observations of tourist landings at historic sites: researchers can be part of a tourism contingent and follow its itinerary (the "mobile" approach), or alternatively they can be based at a landing site and observe successive landings as they take place (the *in situ* approach).

3 From a perspective of ethnoarchaeology, Gorecki (1985) suggested studying contemporary sites with different known periods of abandonment to assess how processes of site formation influenced the deterioration and visibility of sites over time, which would enable one / the researcher to make generalizations about these processes. A similar concept has been applied here, conceptually, to study historic sites in the Polar Regions affected by a range of recent and contemporary processes including tourism.

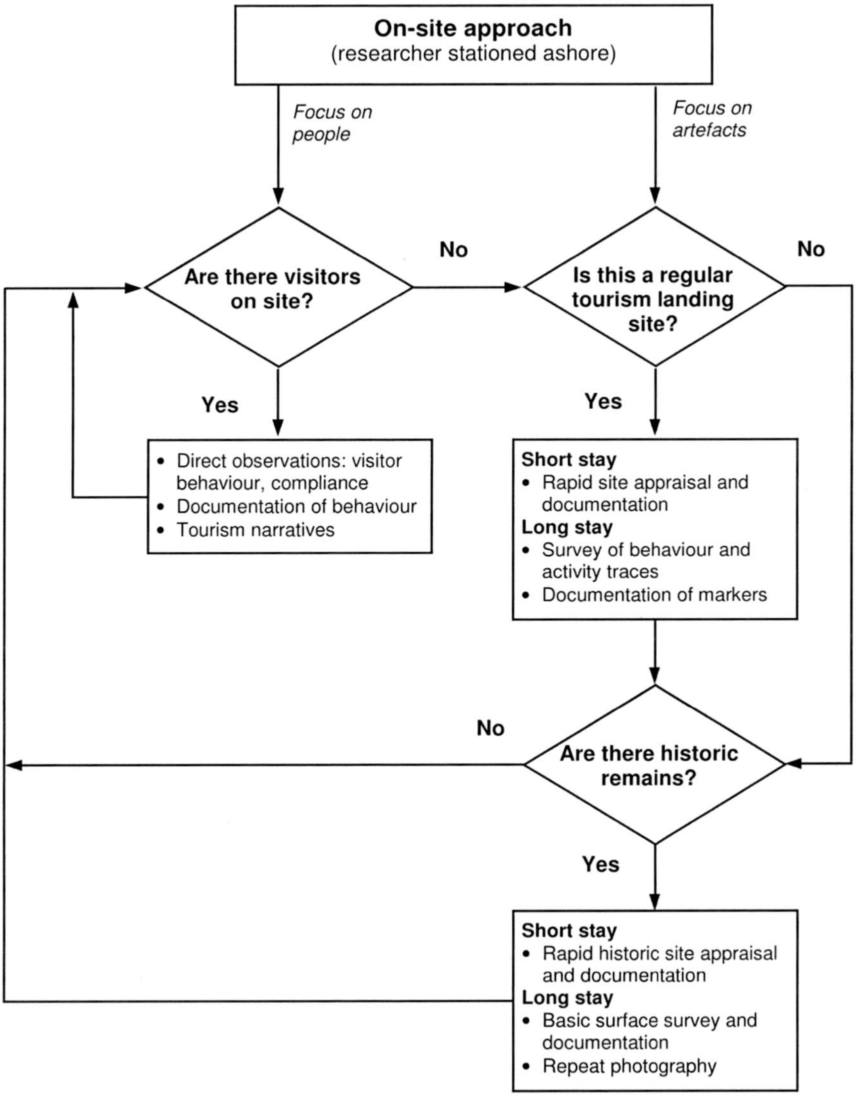

Fig. 2.1 – Flowchart of field research methods according to site type and fieldwork conditions: Researcher travels with tourist group.

In the "off-site" or "mobile" approach the researcher follows one particular group of visitors through different sites. Access to the group does not need to be negotiated since the observer is part of that group, for instance as a tourist or guide. Unless the researcher has a previous knowledge of the site, or the opportunity to return, the experience of each particular site is limited to the duration of the visit, which is usual-

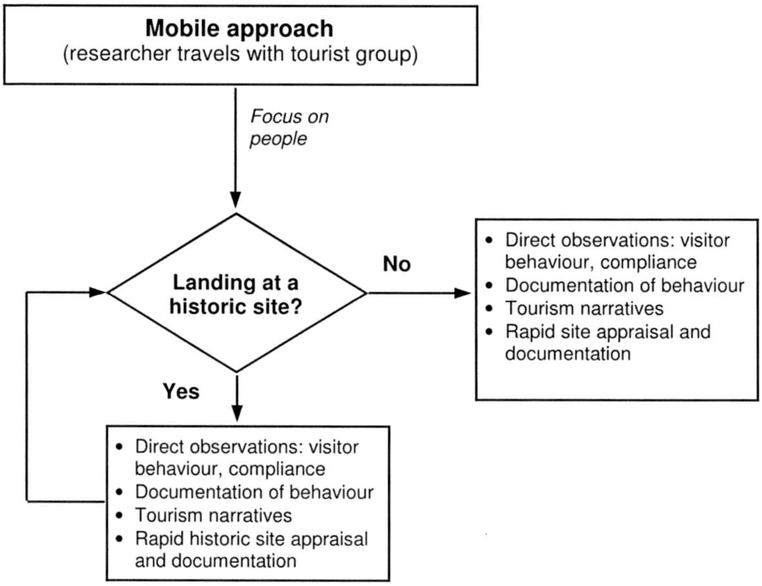

Fig. 2.2 – Flowchart of field research methods according to site type and fieldwork conditions: Researcher stationed ashore.

ly of less than two hours (e.g., Splettoettser *et al.*, 2004). In that time the researcher has to reconnoitre the site as well as conduct behavioural observations on the visitors. The selection of landing places is left to the expedition leader and ship's captain, who make decisions based on the weather, ice conditions, and other practical factors. In mainstream tourism landings have often an emphasis on wildlife and nature, so that in any given cruise few if any cultural heritage sites may be visited. Plainly this is not the case on cruises focused on the polar cultural heritage.

In the "on-site" or *in situ* approach the researcher is based at a location and observes successive landings by different tourist groups (e.g., Davis, 1998). The advantages and disadvantages are reversed in relation to the "mobile" approach: there is greater depth of site knowledge and a greater variety of visitors to observe. This may make possible the identification of site-specific patterns of behaviour, areas of activity, or consequences of behaviour that may not be apparent when observing a single landing. However, the number of tourist landings to any particular site during fieldwork may be limited, thus reducing the number of observations that one is able to make.

Both the "mobile" and "on-site" approaches were used for this study. Broadly speaking, fieldwork followed two research tracks – one focusing on people (visitor behaviour) and another focused on artefacts (both historic and contemporary) (Figs. 2.1 and 2.2). The selection of sites and research methods was made on the basis of the options available given a range of logistic and practical constraints. These included whether

Fig. 2.3 – Field equipment used in this research for general documentation (notebooks, pencils, digital and analogue photographic cameras, and digital voice recorder); surveys (GPS, Brunton compass, 3m and 30m measuring tapes, and pedometer); and direct observations (binoculars, manual tally counter). Photograph by R. Roura 2011.

access to the sites was dedicated or opportunistic access to sites; the duration of fieldwork; and the type and scale of each site.

The main research methods employed were basic opportunistic walk-over site surveys; direct observations; and repeat photography. Complementary research methods were semi-structured interviews with key informants; participant observation; the examination of tourist weblogs; and analysis of tourism statistics. In its conception, albeit not necessarily in specific methods, the approach used here is based on the tradition of rapid appraisal used in some disciplines of applied social sciences (e.g., Kumar, 1993; Beebe, 1995). Field equipment was basic and light-weight (<3kg) (Fig. 2.3).

2.3 SITE SURVEYS

2.3.1 Features surveyed

Surveys focused on historic features and on evidence of their transformation by the effects of natural and cultural processes (Schiffer, 1987, 1995; Buchli and Lucas, 2001; Goldberg and MacPhail, 2006; Rapp and Hill, 2006). The surveys did not aim to obtain information about historic activities and behaviour per se, but rather to describe the historic environment as the "scenario" in which contemporary tourism takes place. Within the sites, key historic features were documented systematically, and judgement was used to identify and locate traces of visitation and of other site formation processes.

The visible traces of human behaviour and activity were also examined (Bernard, 1998: 332). The focus here was not on the structure of behaviour itself – as identified in direct observations – but on the *consequences* of that behaviour (Martin and Bateson, 1983: 57). A trace is:

> *...any enduring mark left in or on a solid surface by a continuous movement. Most traces are of one or other of two kinds: additive and reductive. A line drawn with charcoal on paper, or with chalk on a blackboard, is additive, since the material of the charcoal or chalk forms an extra layer that is superimposed upon the substrate. Lines that are scratched, scored or etched into a surface are reductive, since in this case they are formed by removal of material from the surface itself. (Ingold, 2007:43).*

This description illustrates how many of the features investigated in the field were produced. Instead of being produced by a hand on paper with a writing implement, the traces studied here were made by site formation processes and agents and were traced on the landscape. Traces may also be referred to accretion measures, which arise from the deposit of material in some setting; or erosion measures, which are produced by selective wear in some material (Lee, 2002: 143). The type and location of visitation traces – to the extent that these can be identified above the overall "noise" of other site formation processes – show how cultural remains might be exposed to visitor behaviour (Fig. 2.4).

Throughout the research, material remains were not handled but examined as found. No excavations were carried out, and no materials were collected from the sites.

2.3.2 Survey checklist

Unravelling how tourism affects a historic site requires also the assessment of the status of and changes in sites that are affected by a range of site formation processes. A rapid appraisal checklist was produced to guide site surveys (Table 2.2). It took elements from several checklists used for official Antarctic inspections (Chapter 5); archaeological surveys (Kipfer, 2007); and visual monitoring techniques (Campbell *et al.*, 1993). From a management perspective, the checklist was structured on the basis of a condition-pressure-response framework,[4] which refers to:

- The *condition* of important historic remains of a site;
- The major *pressures* exerted on the site by the environment or people (or site formation processes in the terminology of behavioural archaeology);
- The *responses* to either the existing condition or to changes in the condition of a site resulting from pressures i.e. management responses.

Initially the checklist was conceived as a form that could be filled in directly in the field to produce consistent observations and records. For this it was designed to fit on a two-sided A5 page. However, it soon became apparent that a single checklist was inadequate to encompass all the circumstances of fieldwork – the different kinds of sites visited, of survey intensity, etc. Eventually the checklist was used primarily as an *aide mémoire* to guide observations, and as a reporting framework.

2.3.3 Survey intensity

Survey intensity, defined by the number of hours of fieldwork amount of effort and detail of documentation varied from site to site depending on the scale and complexity of each site and on logistic considerations. The time available on site ranged from only 1-2 hours to 5-10 working days. These factors resulted in a continuum of surveys with different degrees of survey intensity.

At one end of the spectrum, rapid appraisal (or reconnaissance survey) consisted of a rapid walk-over assessment and documentation of key features. Site knowledge acquired in previous visits to some of the sites facilitated this work. As noted above, in some instances behavioral observations were carried out simultaneously with these surveys.

At the other end of the spectrum, a fuller survey enabled the description and documentation of the site in more detail. The survey outcomes were either a sketch map or 3D sketch (without instruments) or a simple compass map (carried out with Brunton geological compass and tape) with a GPS positioning mark (Fig. 2.5) and site description. Repeat photography was used in some sites to identify changes to specific features that had taken place in the recent past.

Survey methods were adapted to the fact that field work was conducted single handled. This meant that some tasks that are normally conducted by two people – such as extending a measuring tape – were conducted by a single person. For this reason,

---

4 This is a simplified version of the DPSIR framework – Drivers, Pressure, State, Impact, and Response–often simplified to PSR, which is a used for reporting on environmental issues. The condition-pressure-response framework is currently used by the Australian Antarctic Division to assess and manage the impact of its activities, including impacts on cultural heritage. A similar framework has been used to in the Environmental Impact Assessment for Ny-Ålesund (Sander *et al.*, 2006) and is used for state of the environment reporting in Norway, including Svalbard.

Fig. 2.4 – Activity traces (footprints and informal rock cairns) around the reassembled cemetery at Whalers Bay, Deception Island. Photo by L. Hacquebord © 2005.

and for economy of time, counting steps (using a pedometer or a mechanic hand tally counter) and photographic documentation were preferred over other techniques such as measuring and drawing.

2.4 OBSERVATIONS

2.4.1 Direct observations

Direct observations of people's activities allow one to establish what people actually *do*. Observations were used to develop a behavioural repertoire of tourism activities at historic sites. Observation involves sampling, that is, selecting what, when and where to study a phenomenon, but also what should be noticed and what may be ignored (Lee, 2002: 44). Observation methods applied in this research were based on the experience of observing polar tourists, first applied by the author at Deception Island in 2002, complemented by a literature review (Webb *et al.*, 1966; Martin and Bateson, 1986; Veal, 1992, Sanger, 1996; Lee, 2002;), particularly ethnographic observations of tourists, both globally (Bruner, 2005) and in a polar setting (Ecuador, 2007). Tourism observation techniques were further refined elsewhere in the Arctic (Roura, 2009) and outside the Polar Regions. The primary function of direct observations was to observe behaviour, but in some instances they were used simultaneously to document tourism narratives, and to monitor compliance with regulations.

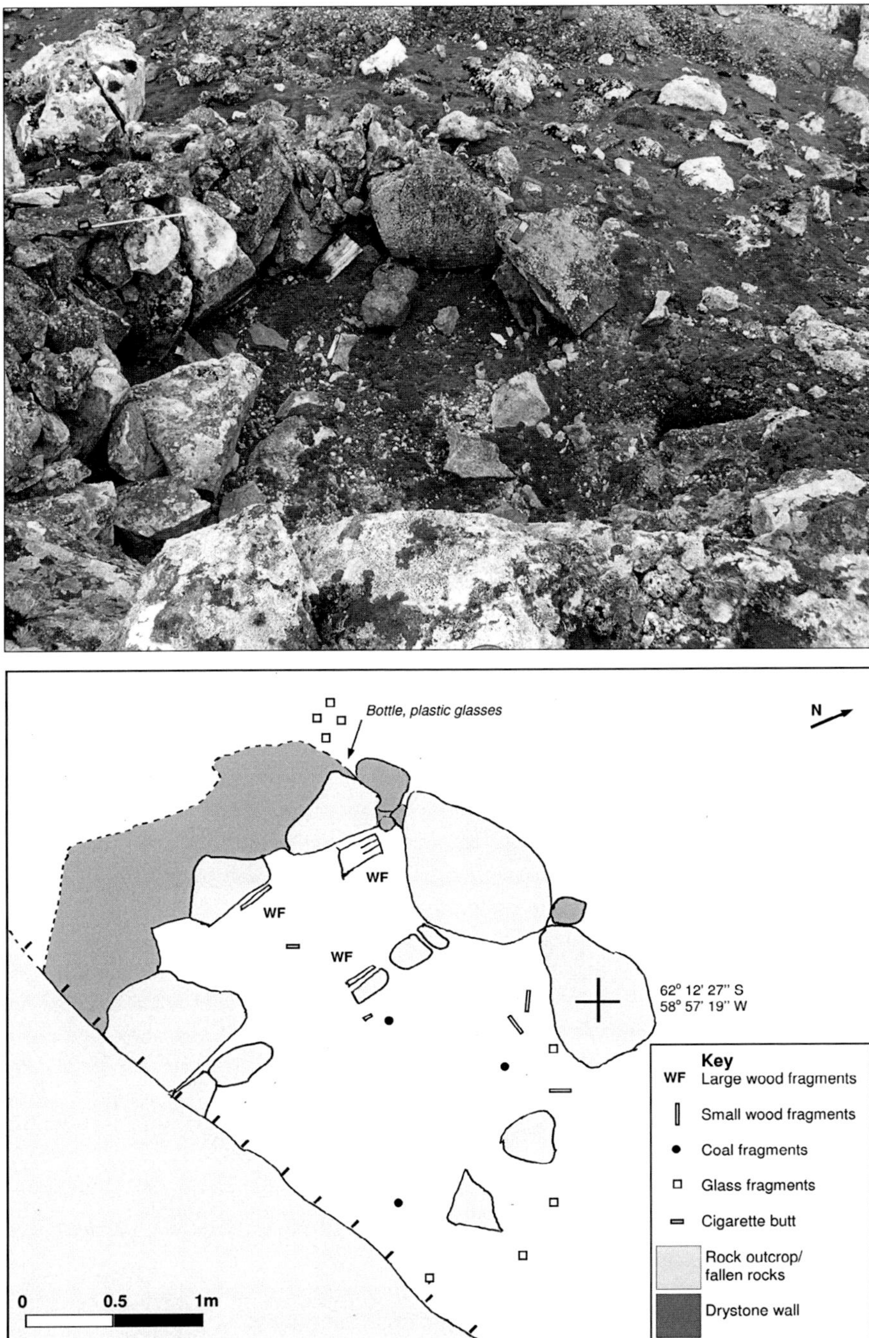

Fig. 2.5a-b – Dry-stone wall adjacent to a rock outcrop at King George Island, Antarctica; possibly a shelter used by sealers in the 1820s (a); sketch made with Brunton compass and tape, with GPS position mark (b). Site formation processes include vegetation growth and rock fall, as well as contemporary uses of the site and nearby area. Photo and sketch by R. Roura 2006.

Fig. 2.6 – Tourist behaviour. A visitor (top left) steps back to take a photograph – and stands on top of an artefact (an awning lying on the floor of the physical laboratory at the Discovery *Hut in Ross Island, Antarctica). Photo by author 2007.*

For the purposes of this research observations on tourist landings at historic sites were of greatest interest. However, opportunistic observations were also conducted of tourism activities at other kinds of sites (i.e. not historic sites) and of other kinds of visitors (i.e. not paying tourists). These observations served as controls and helped to build an understanding of tourism as a behavioural system.

A distinction should be made between sampling rules and recording rules (Martin and Bateson, 1983; see also discussion in Lee, 2002: 44). Sampling rules refer to the elements to be viewed during the observation period. Recording rules refer to how these are recorded. Observations followed an *ad libitum* [Latin for *as desired*] sampling pattern. This means that no systematic constraints were placed on what was recorded or when. Rather, whatever was visible and seemed relevant at the time was recorded (Martin and Bateson, 1983: 84). This included spatial, temporal and behavioural patterns of visitation by groups and individuals; and in particular interactions between visitors and material remains (and hence the exposure of material remains to visitation–see e.g. Fig. 2.6). Observations allowed the identification of, *inter alia*, the elements of a historic site that generated greater tourism interest; basic types of behaviour; and cause and effect relationships between visits and visitation traces. Site representations as narrated by tour guides and interpretive material were also documented, whenever possible.

The etic-emic distinction is relevant to the meaning of observations. The personal experience as described by the person engaging in the behaviour is the emic approach. On the contrary, the behaviour as described and categorised by an observer is the etic approach (Pearce, 2005:3-4). In this research direct observations were conducted from the perspective of an etic approach, that is, tourist behaviour was described regardless of the apparent motivations or perspectives of the individuals. For instance,

Fig. 2.7 – Tourist group observed from a distance. An enlarged photograph (below) shows that most tourists are standing around the tour guide, and that a few tourists are sitting on the monument to the left of the group. Photos by author 2007.

I may have recorded that an individual was looking away from the tour guide during an on-site lecture about the history of the site, but did not speculate as to why this was the case. Observations can be conducted from a certain distance away (which facilitates observations about group behaviour) or from within the group (which emphasises observations about individual behaviour). Both approaches were used in this research (Fig. 2.7 and 2.8), which provided complementary perspectives of individual and group behaviour at specific sites.

Observing a group from the distance provides the observer with anonymity, and therefore reduces issues of reactivity. This may change once the researcher enters a group. A group of 50 or more tourists appears anonymous to a newcomer, but those

Fig. 2.8 – A tourist group observed from within. A tour guide is reprimanding a tourist for leaning on a protected metal structure. This exchange is clearly apparent in the blow up photograph below. Photos by author 2007.

within the group may notice the presence of a new face in its midst, particularly after they have spent few days cruising together. When observations were carried out from within the group, access was negotiated on case-by-case bases with the expedition leaders, who in all instances allowed me to observe the visit. Plainly this was not an issue when I was acting as a tourist and therefore part of a tourist group.

My presence was not concealed from the tourists, although efforts were made to blend with the group rather than to stand out. For instance, observations were memorized and photographed rather than written down, thus outwardly my behaviour mimicked that of the tourists. I stood facing the tour guide rather than other tourists, while scanning the group periodically. Accessing the groups from outside did not seem to

Fig. 2.9 a-b – Behaviour and behaviour traces.

create significant reactivity issues. Tourist visits at particular sites generally follow a pre-determined pattern that did not seem to be altered by having a researcher within the group. Some tourists gave me an initial passing glance or even photographed me, but they soon turned their attention to other matters.

At present, historic sites in the study areas that are regularly used for tourism purposes receive anywhere from a few hundred visitors per year (e.g. historic huts in Ross Island, Antarctica) to over 10,000 visitors per year (e.g. Whalers Bay, Deception Island, South Shetland Islands). Visitor statistics provide a measure of how representative each visit is in the context of the total number of visitors to a site in that particular year. Typical figures for the groups observed in this research were estimated at 3-5% of all annual visits for most sites.

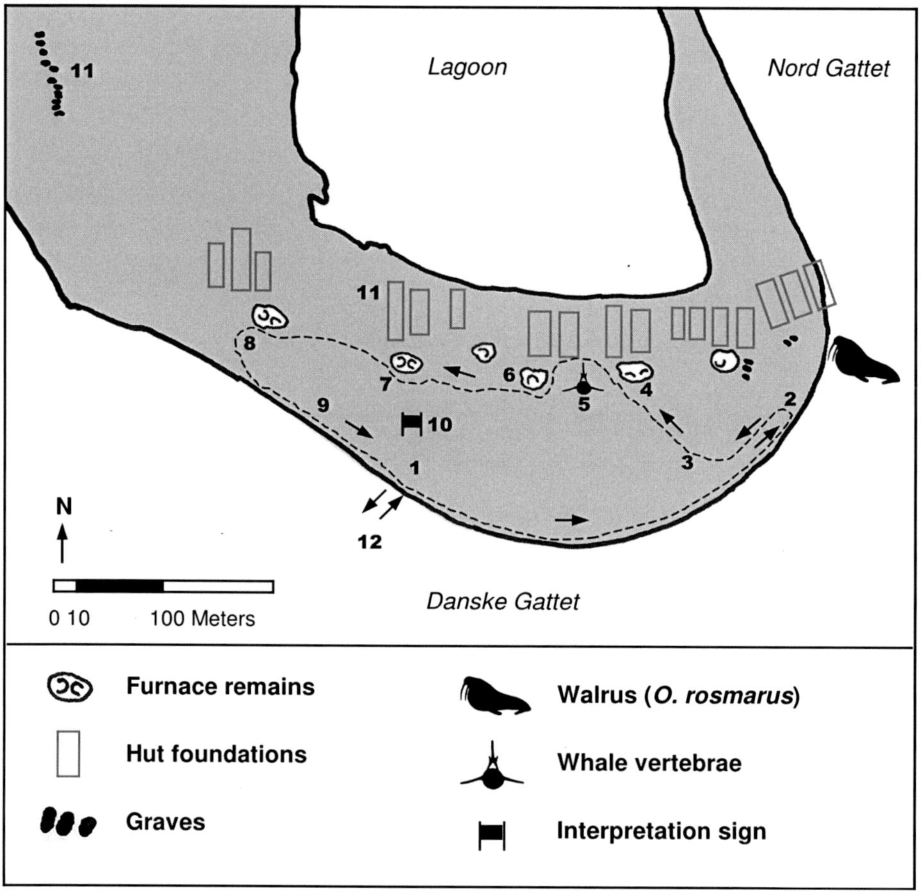

Fig. 2.10 – Map illustrating a landing at Smeerenburg, Svalbard, in which both historic and natural attractions were the focus of the visit (Chapter 8). Base map from Hacquebord et al. (2003).

A detailed report on each tourist landing was written down as soon as possible after the visit. Observations were first written as they were remembered, and then were re-ordered using a field observations checklist (Appendix 1) as a standard reporting framework. As noted above, in some instances reconnaissance site surveys were carried out simultaneously with a description of a tourist visit, which involved gathering a considerable amount of information in a short period of time. The information committed to memory was "downloaded" as soon as possible afterwards back in the camp or ship. Going through the rapid appraisal checklist and through the record of digital photographs assisted in remembering events and ensured consistent reporting.

Descriptive field notes were written in an ethnographic style. For this, guidance was taken from the work of Emerson *et al.* (1995). The outcome was a free-flowing text describing events as they were observed that could later on be turned into a more structured description and coded as required (Bernard, 2002: 365-389; Ryan and Ber-

Fig. 2.11 – Compliance with boot-cleaning regulations prior to entering an Antarctic historic hut. Photo by author 2007.

nard, 2003). For instance, the following notes describe an event observed during a landing at Ny-London, Spitsbergen (Fig. 2.9 a-b):

> *On the way back ...to Camp Mansfield, walking along a stream, we found a number of artefacts on the streambed – what seemed to be an enamel teapot lid, a terracotta bottle about 200ml in capacity, and a small [white] glass bottle some 8cm long. The guide picked up the former to photograph it noting that it was "a beauty". He returned the first bottle to the ground and picked up the second bottle, which was full of mud and possibly stuck to the sediments. The bottle broke in several pieces before he could lift it from the ground. "It was already busted" – he said, but I thought that this had not been the case.*

Sketches of the visits were made illustrating the sequence of the landing and the visitors' activities (Fig. 2.10). These allowed identifying the activity area used by particular groups of tourists.

2.4.2 Participant observation

Participant observation is a method used to collect data in a relatively unstructured manner in naturalistic settings by researchers who observe and/or take part in the

Fig. 2.12 – A photograph taken unobtrusively with a digital camera while it was hanging from my neck. It illustrates tourists walking on multiple paths as they approach a historic feature. Photo by author 2007.

activities of the people being studied. The method is characterized by the fact that the observer is continuously in touch with the person or group observed, observing, listening, and documenting his or her observations (Bogdewic, 1992; Dewalt *et al.*, *1998*; Bernard, 2002). The *method* includes the explicit recording and analysis of the information obtained from participating and observing: "If you didn't write it down in your field notes, it didn't happen" (Dewalt *et al,* 1998:270).

As applied in this research, and to the extent that it was applied at all as a distinct method, participant observation concerned my *role* in a certain context, which provided a platform to conduct observations in topics relevant to this thesis. In the course of this research I acted as a field tutor in a post-graduate course in Antarctica; as an environmental expert in several Antarctic settings, both in Antarctica and elsewhere; and as a tourist on cruises in Svalbard. The purpose was not to study particular social groups, but rather to observe *from within* discussions or activities relevant to this thesis. As long as these activities lasted, I played the dual role of researcher and of whatever particular role I had at the time. Field notes were produced as outlined previously.

2.4.3 Monitoring compliance through observations

Several of the historic sites investigated in this research have specific management plans and codes of conduct. These typically establish required forms of behaviour and

Fig. 2.13 – Photographic analysis. The enlargement (below) showed a man holding on to building remains, a behaviour that had not been noticed in the field. This action did not leave any obvious traces, but it could have affected a more fragile feature. The tourist had some difficulty walking. Photo by author 2007.

access restrictions, for instance, what visitors may or may not do at a site, or which parts of the site may be visited and which are off-limits. Some of these are legally binding instruments while others are non-mandatory guidelines. Through visitor observations at these sites it was possible to identify the extent to which the regulations were implemented and their apparent effectiveness. (Fig. 2.11).

Fig. 2.14a-b – Set of photographs illustrating changes to a building at Whalers Bay, Deception Island Island, Antarctica in 1993 (left) and 2005 (right). The photographs have not been specifically collected for monitoring purposes but are nevertheless similar enough as to allow comparisons Upper photo by R. Roura, photo below by L. Hacquebord)

2.5 PHOTOGRAPHY

Photography was primarily used as a documentation aid and as notes to record behavioural or site-specific observations. From that perspective it would be unnecessary to classify photography as a method in itself. However, in some instances photography was a primary rather than a complementary data-gathering tool. In brief opportunistic visits to some sites photography was essential to document the sites, and sometimes a substantial amount of visual data was generated in a relatively short time. In addition, photography was used to generate before/after data sets used to assess and interpret time serial changes to historic sites.

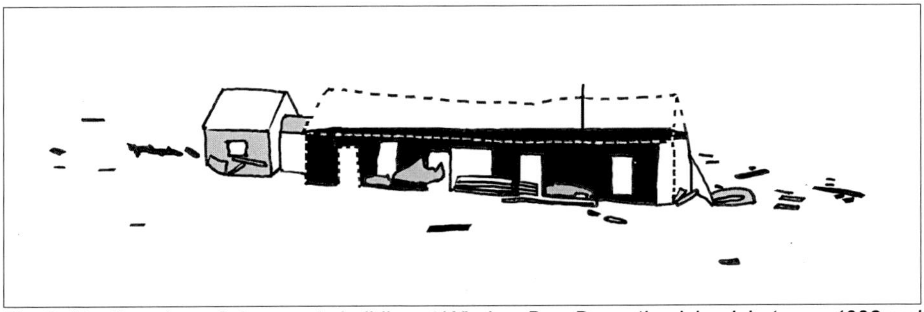

Fig. 2.15 – Overview of changes to building at Whalers Bay, Deception Island, between 1993 and 2005. Sketch based on a photograph from 1993. Key: White = no changes; grey = changes within the frame; dashed line = removed from frame; black = added to frame. Part of the roof has blown away and the outer wall in the foreground has collapsed, exposing part of the building foundations and an inside wall, as well as covering the area with debris. Sketch by R. Roura.

2.5.1 Photography as a documentation aid

The use of photography for research was broadly based on the work of Collier and Collier (1967). A "shooting list" was developed to organise the documentation of site surveys, which approximately followed the outline of the rapid appraisal checklist (Appendix 1). The sites were walked over systematically, and features of interest were photographed using a small but powerful digital camera (Nikon Coolpix 10 megapixels with a 3x zoom), with an analogue camera (Nikon F2 with 80-200mm zoom) used as a complement or back up.

While conducting tourism observations each photograph had as its purpose to illustrate a certain type of behaviour. When surrounded by tourists it is generally possible to take photographs discreetly and, simultaneously, quite openly, since many tourists are taking photographs too. A small digital camera (weighing less than 250 grams) made it possible to take photographs unobtrusively, sometimes "shooting from the hip" rather than taking the camera to the eye level (Fig. 2.12).

Most photographic analysis is a search for patterns and the definition of their significance (Collier and Collier, 1967:195). Analysing photographs was as important as collecting them and in some instances it provided additional information that had not been apparent when the photograph was taken (Fig. 2.13). Reviewing photographs is useful to minimise bias caused by "inattentional blindness" through which only those objects and details that receive focused attention are noticed and remembered (Simon and Chabris, 1999).

2.5.2 Repeat photography

A photographic method was developed to monitor the transformation of historic sites in the Polar regions (Roura, 2008; Chapter 4). It combines data collection based on repeat photography (Hall 2001, 2002) with the analysis of field data, based on the concepts of site formation processes developed by M.B. Schiffer (1987, 1995). This

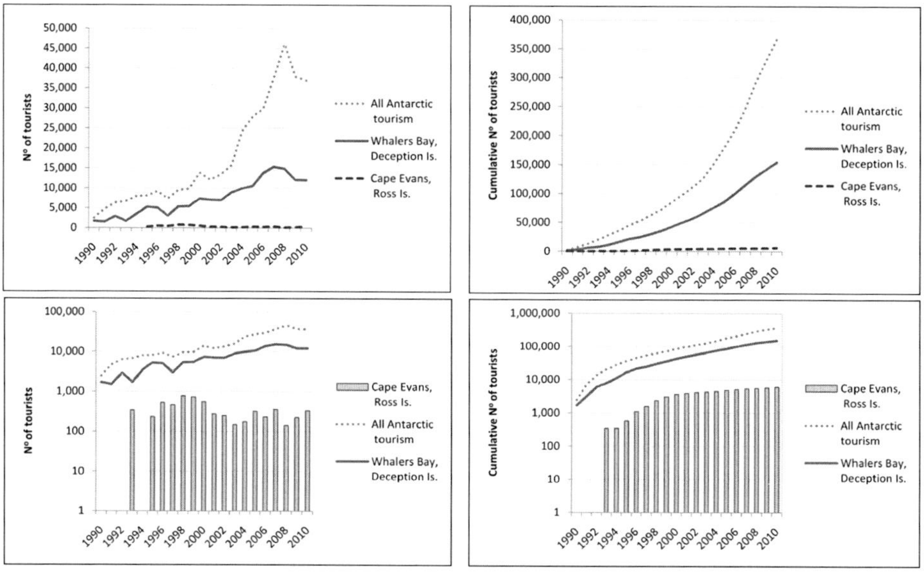

Fig. 2.16 – Four different representations of visitor numbers to two Antarctic historic sites compared with Antarctic ship-borne tourism as a whole, 1990-2010. Whalers Bay at Deception Island receives many more visitors than Cape Evans in Ross Island and more regularly. Visits at both sites, however, reflect the overall trajectory of tourism. Graphs by R. Roura on the bases of IAATO statistics.[6]

method is described in detail in Chapter 4. Figs. 2.14a-b and 2.15 show an example from Deception Island.

2.6 DOCUMENTARY RECORDS

For the most part, the corpus of information used for this research dated from the recent past, up to two or three decades ago. Original published accounts from the earlier occupants of the sites were used in some instances to complement field observations. However, archival research was generally not the focus of this research.

Two kinds of documentary records were used: regular records, and personal and episodic records.

2.6.1 Regular records

Regular records – that is, records that are collected on ongoing bases independently from this research – included tourism statistics and visitor books. Tourism statistics provide information on the level of tourism activity at particular sites and are an indirect

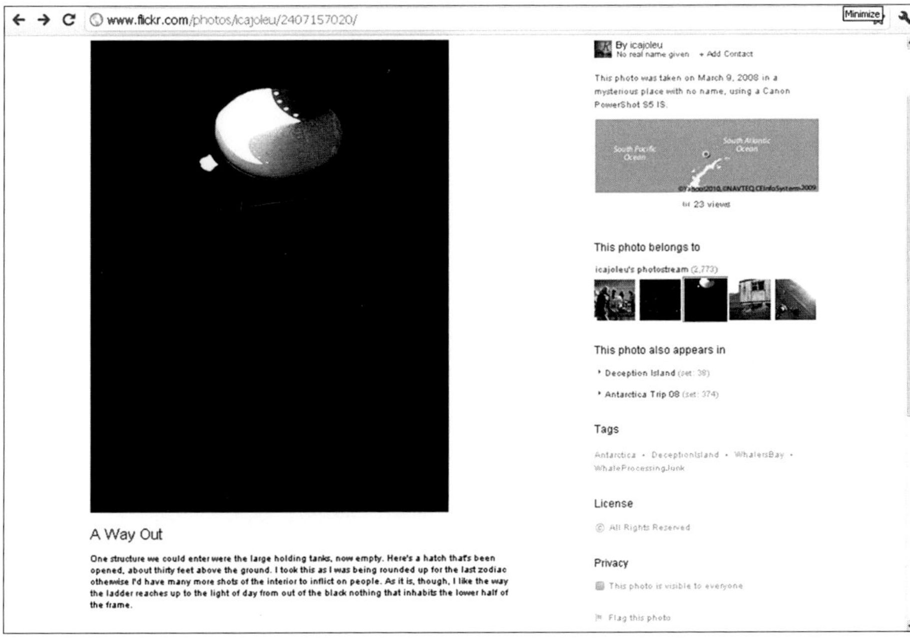

Fig. 2.17 – Photograph and accompanying blog from a personal website describing a landing at Whalers Bay, Deception Island, Antarctica, in 2008: One structure we could enter were the large holding tanks, now empty. Here's a hatch that's been opened, about thirty feet above the ground. ....I like the way the ladder reaches up to the light of day from out of the black nothing that inhabits the lower half of the frame.[6]

indicator of visitation pressure at those sites. Tourism statistics for Antarctica are available from the industry trade association, the International Association of Antarctic Tour Operators (IAATO).[5] In data made available to the public from 1989 onwards, details can be found of the number of tourist landings at specific sites. Data for the first decade or so list only passenger numbers, while more recent data also list the numbers of staff and crew. Different place names and coordinates have been used over time although these have now been standardized. Tourism statistics for Svalbard date back to the 1990s although the coverage is discontinuous and the data do not differentiate tourists from other kinds of visitors. Prior to 2001 data on the number of disembarkations in various locations are not certain, although subsequent data are reliable (Governor of Svalbard 2006a: 6). Some tourism statistics have been analysed providing information about visitation patterns to different locations (Geitz, 2004; Naveen *et al.*, 2006; Lynch *et al.*, 2009).

Tourism statistics were analysed to identify overall tourism trends (Chapter 3) and to assess landing patterns at historic sites (Chapter 8). Identifying landing patterns at historic sites required identifying which landing sites contain historic remains. This is

5 http://www.iaato.org

Fig. 2.18 – Most Antarctic HSMs are accompanied by a plaque with an official description of the site in the four official languages of the Antarctic Treaty, which reflects consensus by Antarctic Treaty Parties about the sites and their history. Photo by author 2007.

Fig. 2.19 – An Italian plaque at the airship mooring mast near Ny-Ålesund commemorating the "Amundsen-Ellsworth-Nobile" transpolar flight of 1926. A Norwegian plaque located nearby refers to Norwegian and US involvement in this expedition but makes no reference to the Italian contribution (Chapter 6). Photo by author 2007.

unambiguous in well-known sites, but less clear in smaller or lesser-known sites. Further, it is necessary to know the modus operandi of tourism landings at specific sites. It may be that tourists are taken to a certain site where historic remains exist but their activities do not focus on these remains. Finally, some sites – such as some active Antarctic research stations – have a mixture of contemporary and historic elements, and it cannot be assumed that historic remains constitute the main tourist attraction at those sites.

Different kinds of graphs were used to analyse tourism statistics (Fig. 2.16). Graphs enable the comparison of visitation levels between different sites, and this can be integrated with site-specific information such as behaviour and activity traces. Different graphs provide different perspectives on tourism patterns. Histograms or line graphs showing total tourist numbers illustrate the evolution of a site as a tourism destination. Cumulative graphs provide useful insights about the number of tourists that have visited specific sites, which assists the assessment of potential cumulative impacts. The use of logarithmic scales enables to compare visitation patterns at sites with substantially different numbers of visiting tourists.

Visitor books provide a daily record of visitation and an insight into the reaction of visitors to historic sites. For practical reasons visitor books were documented only if sufficient time was available, and documentation was restricted to the observation period i.e. the comments made by the same tourist groups whose behaviour had been observed.

2.6.2 Personal and episodic records

There is a wealth of personal and episodic records about historic sites in both Antarctica and Svalbard available from a range of official and unofficial information sources. These include conservation or management plans, brochures, and websites. This information may be used in a number of ways to assess the status and change of historic sites.

Many official Antarctic documents are available on the website of the Antarctic Treaty Secretariat.[8] This contains the official Antarctic Treaty System database that records instruments of different kinds approved by the ATCM; ATCM records; official inspection reports;[9] reports of Antarctic Treaty Meetings of Experts and other expert reports; management plans for protected areas; site specific guidelines, the official list of Historic Sites and Monuments, and other records.

Other Antarctic websites[10] and the Government of Svalbard website[11] provided additional background information relevant to this research. In particular, the Government

---

6 http://www.flickr.com/photos/icajoleu/2407157020/sizes/m/

7 http://www.iaato.org/tourism_stats.html, retrieved March 2011.

8 http://www.ats.aq.

9 Conducted under Art. VII of the Antarctic Treaty and Art. XIV of the Protocol

10 http://www.cep.aq; http://www.scar.aq.

11 http://www.sysselmannen.no

of Svalbard has published a range of documents specifically addressing tourism in Svalbard and the monitoring and management of historic sites (Government of Svalbard, 1999a, 1999b, 2006a, 2006b).

For both regions there is additional information available from a range of non-governmental organizations including the ICOMOS International Polar Heritage Committee, heritage trusts (which in the Antarctic undertake the management of some sites),[12] environmental groups (ASOC, Greenpeace International, and WWF), the tourism industry associations (IAATO and AECO),[13] and dedicated projects such as the Antarctic Site Inventory.[14]

2.6.3 *Internet blogs*

Direct observations enable the researcher to study and classify behaviour (etic approach) but they do not provide insights into the perceptions and motivations of tourists (emic approach). Personal websites posted by polar tourists on the internet (web logs or blogs) provide a valuable insight into visitors' perceptions. Blogs provide tourists' representations of "being there" in the Antarctic. The content of some blogs both illustrate and explain behaviour and as such they bridge the etic-emic divide (Roura, 2010; in prep.).

The tourism experience is one of a fundamentally visual nature (Urry, 1990: 145). Some blogs are entirely visual with minimal text, and others are the opposite. The information may include, for instance, what aspects of a site a visitor found interesting (it may be, for instance, wildlife rather than historic features); which parts of a site were visited during a landing; and the type of activities and behaviour in which tourists engaged. The most useful blogs contain details about when, where and why photographs were taken, who the photographer is, and contextual information about the cruise, the landing, etc. In addition, blogs may provide information valuable in documenting tourist behaviour and awareness of (and compliance with) existing regulation. For example, the caption of Figure 2.17 suggests that the photographer was not thinking about the technical or historic interest of the oil tank, but rather about its aesthetic and experiential values. Furthermore, when that photograph was taken visitors were not allowed to enter the tanks or anyother building at the abandoned whaling station.[15]

2.6.4 *Personal collections*

Prior to this research I had been involved in different kinds of Antarctic fieldwork starting in 1980-81. During that period I conducted seven expeditions to the Antarctica, visiting dozens of sites of past and contemporary activity in the Antarctic Peninsula and in the Ross Sea region. Some of these sites were at the time, or have been since, des

---

12 http://www.heritage-antarctica.org/

13 http://www.iaato.org; www.aeco.org.

14 http://www.oceanites.org.

15 (ATCM) Resolution 2, 2008 *Site guidelines for visitors.*

ignated as historic sites. From this experience a number of records were available from which data useful for this research could be obtained, including personal diaries, expedition reports in which I was the author or co-author (e.g. Greenpeace, 1991, 1992, 1994), some scientific publications (e.g. Roura, 2004), and photographic collections.

2.6.5 Markers

Following D. MacCannell (1976: 109-110) a *marker* is information about a tourist sight. This includes written narratives, such as inscriptions in memorial plaques or interpretation signs, and verbal narratives such as the stories told by tour guides. A distinction can be made between the content of site markers and the vehicle in which the text is contained. This difference may be regarded as academic (if not pedantic), but the characteristics of the vehicle and the text it contains provide somewhat different information. With regards to research methods, documenting the vehicle of a marker (such as stories told by tour guides, or by others with local expertise) were also recorded whenever possible.

2.7 INTERVIEWS

Interviews provided complementary information to unobtrusive research methods. Interviews were primarily conducted on the early stages of this thesis and guided later research. Interview topics depended on the interest and expertise of the informers in what concerned historic sites; polar tourism; and issues concerning specific sites or regions.

Two types of interviews were conducted: semi-structured interviews, and unstructured interviews. In semi-structured interviews the approach entails having a mental or written checklist for the interview, but allowing the interviews to be open ended as well as following up on the unexpected (Bernard, 2002). In this study, semi-structured interviews were usually pre-arranged; lasted about an hour; focused on a discrete list of key topics; and were recorded and/or written down. Unstructured interviews were opportunistic and usually brief; focused on one or two key topics; and were recorded mentally or jotting notes on any piece of paper that was available. In all cases, a proper written transcript was produced as soon as possible after the interview.

The interviews were mostly conducted in English; other languages were used occasionally, and a Norwegian interpreter was used once. Appendix 3 lists the most important formal interviews conducted as part of this research.

2.8 STUDY SITES

Four field seasons were carried out between 2005 and 2008 in Svalbard (two trips) and Antarctica (two trips). A fifth field season was a 2002 trip to Antarctica that was also included as part of this research. In Svalbard fieldwork was conducted in Longyearbyen (2006); on day cruises in the Ice Fjord (2006); on a weeklong cruise to NW Spitsbergen (2007); and during stays at Ny-Ålesund and Ny-London in Kings Bay (2007). In

Antarctica fieldwork was conducted on an opportunistic basis during stays at Deception Island (2002), King George Island (2006), and Ross Island (2008) (Table 2.2). These sites are discussed in further detail through various case studies. Complementary information was obtained by visiting polar museums and other places associated with the cultural heritage from Antarctica and Svalbard and that are located outside these regions (Table 2.3).

2.9 CLOSING REMARKS: A FOREST OF METHODS, A SEA OF DATA

The quotes opening this Chapter – from a Sherlock Holmes story; and from a form of traditional indigenous knowledge now used by law enforcement agencies in the USA-Mexico border – capture some of the issues concerning tourism impacts: polar tourists visiting historic sites are bound to leave some traces; it is all a matter of being able to see them. Let it be stressed that these analogies have some limitations – polar tourism is not a criminal activity, and only some forms of gross misbehaviour by tourists might be illegal. Furthermore, some deductions by the famous fictional detective are implausible (Berger, 2009: 17-20). The analogy, however, covers the link between behaviour, the material consequences of behaviour, and the potential of observations to detect evidence on the terrain.

In this study, investigating how to research the impact of tourism at polar historic sites was a research topic in itself. Unobtrusive methods provided a suitable set of research tools to choose from as mandated by the circumstances of each fieldwork season. Practical considerations required that in some instances fieldwork had to be opportunistic rather than dedicated. In addition, fieldwork was conducted single-handedly, which is unusual in the Polar Regions. These factors influenced determined the choice of research methods used in different study sites. In turn, the selection of methods influenced the type of spatial, visual and written data that were obtained.

Choosing social science research methods have been likened to walking in a forest (Sobo and de Munck, 1998: 13-14); using multiple research methods to collect information puts the researcher at risk of drowning in a sea of disparate data. Behavioural and material culture data constituted the bulk of data collected for this research. These data were evaluated from the perspective of behavioural archaeology for each case study. The outcome is an integrated perspective of tourism at historic polar sites in the context of other site formation processes.

2.10 TABLES CHAPTER 2

Table 2.1 – Unobtrusive research methods

Data acquisition approach	Research method
Found data	Site surveys – different kinds
Captured data	Behavioural observations
Retrieved data	Repeat photography
	Regular records e.g. tourism statistics
	Personal and episodic records e.g. internet blogs

Table 2.2 – Fieldwork checklist

FIELD CHECKLIST Rapid Appraisal of Cultural Heritage Sites in the Polar Regions © R. Roura, Arctic Centre, University of Groningen, The Netherlands		
ASSESSMENT DETAILS	**1. Site name and coordinates**	
	2. Date	**3. Time in/out**
	4. Weather conditions	
SITE TYPE	**5. Is there material evidence of human activity?** ☐ Yes, from past activities ☐ Yes, from contemporary activities ☐ Yes, from both past and contemporary activities ☐ Yes, but their age/origin is uncertain ☐ Not obvious/not visible ☐ Other	
	6. What are the main remains of past activities? ☐ Several buildings/structures ☐ A single building or structure ☐ Small features ☐ Maritime remains ☐ Ecofacts ☐ Grave/s ☐ Religious symbols ☐ Monuments, plaques ☐ Other	
	7. What is the site like? Observe ☐ Surface conditions, soil type, vegetation cover ☐ Estimate of site size ☐ Visibility of site ☐ Size and shape of site, feature, cluster ☐ Clusters or patterns of artefacts on a site	
SITE FUNCTION	**8. What is the original function of the site?** ☐ Expedition ☐ Industrial ☐ Research ☐ Other ☐ Unknown **9. What activities currently take place at the site?** ☐ Research ☐ Tourism ☐ Conservation/management ☐ Other ☐ Unknown	
NATURAL PRESSURES/ PROCESSES	**10. What are the natural pressures/threats to the site?** ☐ None apparent ☐ Biological activity ☐ Fluvial action ☐ Freeze-thaw effects ☐ Slope processes/rock fall ☐ Waves ☐ Weathering ☐ Wind action ☐ Other	

PEOPLE (CULTURAL) PRESSURES/ PROCESSES

1. What are the pressures/threats to the site by people (cultural processes?)

☐ None apparent	☐ Souvenir collection
☐ Graffiti	☐ Trampling damage
☐ Inappropriate repairs	☐ Destruction or vandalism
☐ Inappropriate uses	☐ Vehicle traces
☐ Littering	☐ Overall wear and tear
☐ People crowding in places	☐ Other

IMPACTS/ EFFECTS

2. What are the main impacts/effects as per 11, 12?

☐ Minor/none	☐ Exterior damage
☐ Erosion	☐ Interior damage
☐ Burial	☐ Other
☐ Structural damage	

3. Is damage:

☐ Localized and minor	☐ Occurred (past)
☐ Widespread and minor	☐ Ongoing (present)
☐ Localized and major	☐ Potential (future)
☐ Widespread and major	☐ Other

MANAGEMENT

4. Evidence of measures to manage the site?

☐ No go areas
☐ Established walking routes
☐ Interpretation signs
☐ Active conservation
☐ Other

5. Is there evidence of conflicts over the use of the site?

☐ Yes (specify)
☐ Perhaps (specify)
☐ No

6. Are there any national or territorial symbols on site?

☐ Flags
☐ Plaques
☐ Symbolic monuments
☐ Other

OVERALL SITE ASSESSMENT

7. What is the apparent overall condition of this site?

☐ Good	☐ Standing
☐ Fair	☐ A standing ruin
☐ Poor	☐ A ruin

8. What is apparent integrity of this site?

☐ Low
☐ Medium
☐ High

9. What is the <u>main</u> pressure/threat as per 10, 11?

10. How immediate is the main pressure/threat?

☐ Immediate < 5 years
☐ Short term 5-10 years
☐ Longer term – More than 10 years

CHANGE

11. What are the overall trends/changes for this site:

☐ Don't know	☐ Minor/no changes
☐ Improving	☐ Changes in natural values
☐ Stable	☐ Changes in heritage values
☐ Declining	☐ Changes in activities type, scale

SKETCH

12. Include:

☐ Shoreline/topographic setting
☐ Landing location
☐ Site proximity to water and distance to it
☐ Site relationship to drainage
☐ Exposure/slope direction (compass bearing) and degree/percent
☐ Path followed by visitors
☐ Places where visitors congregate
☐ Features that may explain the circulation of visitors
☐ North (magnetic and/or geographic) and estimated scale
☐ GPS position point/s

Table 2.3 – Fieldwork overview

Key locations	Dates	Key objective	Logistics	Chapters
Deception Island, South Shetland Islands, Antarctica • Ruins of whaling station, British base at Whalers Bay • Ruins of Chilean research station at Pendulum Cove • Tourism landing sites at Telefon Bay, Baily Head	3-22 Feb.2002.	Developing and testing tourism research methods	Dirección Nacional del Antártico; Argentine Navy (Base Decepción); ASOC	3, 4, 5, 8
Fildes Peninsula, King George Is., South Shetland Islands, Antarctica • Numerous remains of early 19th century sealing camps along ca. 25 km of coastline • Sites of unknown age, probably mid-20th century • Miscellaneous monuments and memorials	30 Jan. – 12 Feb. 2006	Assessing the effect of site formation processes on historic sites	Russian Antarctic Expedition (Bellingshausen Station); University of Jena, Germany; ASOC	4, 5, 8
Ijsfjorden, Svalbard • Day tourism at mining settlements of Pyramiden, Barentsburg • Miscellaneous sites in Longyearbyen	14-25 Sep. 2006	Familiarization with Svalbard; interviewing authorities and tour operators; behaviour observations	Arctic Centre, RUG	8
NW Spitsbergen • Ny-London (Peirson Harbour): Abandoned marble quarry • Ny-Ålesund: Airship mooring mast; old mining area • NW Spitsbergen: tourism landings at Dutch whaling station remains at Smeerenburg; trappers huts; sites with no or minor historic remains	14 June – 12 July 2007	Assessing: interactions of tourism and historic sites; effect of site formation processes on historic sites; recent changes to historic features; recording tourism narratives	Arctic Centre, RUG; Oceanwide Expeditions (MV *Akademik Multanovskiy*); Netherlands Arctic Station; Governor of Svalbard.	6, 7, 8
Ross Island, Antarctica • Remains from the "heroic era" of exploration (huts, memorial crosses), and various monuments at Hut Point Peninsula & Cape Evans • IGY building at Scott Base, Pram Point	20 Dec. 2007 – 4 Jan. 2008.	Behaviour observations; rapid site appraisal and documentation	Antarctica New Zealand; Gateway Antarctica, University of Canterbury, New Zealand.	4, 5, 8
Antarctic Treaty Consultative Meetings • ATCMs XXVIII to XXXIII; • Other meetings of the Antarctic Treaty System.	Two days – two weeks duration	Participant observation	ASOC	5, 8

Table 2.4 – Visits to polar museums and off-site cultural heritage

	Location	Polar heritage
2004	Polar Museum, Trømso, Norway	Extensive collection of polar historic remains.
2004	Prison Museum, Ushuaia, Argentina	Display of artefacts from the Swedish Antarctic Expedition 1902-03.
2006	Scott Polar Research Institute, Cambridge, United Kingdom	Display of RF Scott's documents and other "heroic era" artefacts.
2006	Monument to the S.S. *Yelcho*, Puerto Williams, Chile	Monument containing the original bow of the Chilean navy steam tug S.S. *Yelcho,* used in the rescue of the 1910-1913 Imperial Trans Antarctic Expedition from Elephant Island.
2006	Museo Salesiano, Punta Arenas, Chile	Display of artefacts from early 19th century sealing camps in the South Shetland Islands
2006	Polar Museum, Longyearbyen	Extensive collection of polar historic remains dating to the 17th century.
2007	Christchurch Museum, New Zealand	Permanent Antarctic collection, which includes original items from "heroic era" expeditions and original buildings from the IGY Cape Hallett Station.
2007	Monument to RF Scott, Christchurch, New Zealand	Statue of RF Scott sculpted by his widow, Kathleen Scott, dedicated in 1917.
2007	Port of Lyttelton, New Zealand	Port of departure of several "heroic era" Antarctic expeditions to the Ross Sea, an association still exploited in Antarctic tourism.
2007	Mining Museum, Ny-Ålesund, Svalbard	Display of items and photographs from Ny-Ålesund's coal mines.
2008	Reales Jardines Botánicos, Madrid, Spain	"Atrapados en el Hielo" (Trapped in the Ice), temporary exhibition of Frank Hurley's original photographs taken during the 1910-1913 Imperial Trans-Antarctic Expedition.
2008	Arctic and Antarctic Museum, St. Petersburg, Russia	Extensive collection of historic remains.
2008	Icebreaker *Krasin*, St. Petersburg, Russia	Original ship used in the rescue of the survivors of the airship *Italia* (1928). Permanent exhibition on the *Italia* rescue.
2009	Museo Corbeta *Uruguay*, Buenos Aires, Argentina	Original ship from the Argentine navy used in the rescue of the Swedish Antarctic Expedition 1902-03.
2010	*Fram* Museum, Oslo, Norway	Original ship used by Fritjof Nansen and Roald Amundsen.

Table 2.5 – Overview of research methods

KEY RESEARCH METHODS AND DATA SOURCES	INFORMATION PROVIDED	BEHAVIOURAL SCALE	SPATIAL SCALE	TEMPORAL SCALE	UNCERTAINTY/ DATA QUALITY	USE IN THIS RESEARCH
Direct observations	Visitor-artefact interactions (exposure); activity area used by visitors; compliance with regulations	Individual visitors, groups of visitors	Activity area within landing site	Duration of particular visits (usually <2 hours)	Reactivity by visitors; observation consistency and "inattentional blindness" by observer	21 sites (>30 groups of visitors observed)
Site surveys	Site status; evidence of site formation processes (including tourism)	Variable – original occupants to recent visitors (individuals and groups)	Historic sites (10 m^2 to 1 km^2)	Period since abandonment (usually > 10 years)	Issues of missing archaeological record and "equifinality"	All sites, varying intensity of effort (<2 hours to 40+ hours)
Repeat photography	Recent change to historic sites	Variable – original occupants to recent visitors (individuals and groups)	Individual features part of a larger historic site (1-100 m^2)	Period between before/after photograph (usually 5-20 years)	Assesses only obvious macroscopic changes in the foreground; comparability of before/after photographs	12 sites
Site statistics	Visitation trends over time	All visitors to specific sites	Landing sites defined by coordinates	Period since data collection began (usually 10-20 years)	Inaccurate or incomplete data; differentiation visitors/other visitors may be unclear	Most key sites

Table 2.5 continued

KEY RESEARCH METHODS AND DATA SOURCES	INFORMATION PROVIDED	BEHAVIOURAL SCALE	SPATIAL SCALE	TEMPORAL SCALE	UNCERTAINTY/ DATA QUALITY	USE IN THIS RESEARCH
Weblog analysis	Explicit, implicit information about visitor activities and behaviour; visitor-artefact interactions (exposure); compliance with regulations	Bloggers (single visitors, couples, etc)	Activity area within landing site	Period covered in blog entries (days to hours)	Available for frequently visited sites; reflect events as perceived/recalled by bloggers	Pilot use

3 Regulating Antarctic tourism and the precautionary principle†

ABSTRACT

Tourism in the Antarctic is increasing rapidly and it is expected that it will increase further in coming years. This Chapter discusses the issue of international regulation of Antarctic tourism; the assessment, prevention and regulation of cumulative impacts; and the status and possible practical relevance of the precautionary principle in the management of Antarctic tourism. Regulating Antarctic tourism is more than an environmental issue, with economic, practical and political implications. Adequate regulation of Antarctic tourism may be regarded as a challenge to the Antarctic Treaty System. The measures that have been adopted so far and most of the options under discussion should be characterized as "conditions" that do not address the more fundamental questions in relation to Antarctic tourism in the medium and long term. In our view, the developments in Antarctic tourism and related uncertainties regarding potential cumulative impacts may, now or in the near future, require the adoption of *restrictions* to Antarctic tourism activities to prevent irreversible damage in the future. This Chapter discusses the potential value of the precautionary principle for Antarctic management and refutes the assumption that application of the principle should always result in extreme solutions, such as a total ban on tourism in Antarctica.

3.1 INTRODUCTION

Antarctic tourism is a rapidly growing industry. From 1958 until 1987, an average of fewer than 1,000 tourists visited Antarctica each season. In the 1993-1994 season, the tourists visiting Antarctica outnumbered the scientists for the first time.[1] In recent years (1999-2003), between 13,000 and 15,000 tourists made landings in Antarctica (IAATO, 2003),[2] and during the last season (2003–2004) this number increased by 45

1 For an overview of the development of Antarctic tourism until 1993, see Enzenbacher (1992, 1994).

2 For Antarctic tourism statistics, see also the Web site of the International Association of Antarctic Tour Operators (IAATO), at <http://ww.iaato.org/tourism_stats.html>.

† Published as:

Bastmeijer, K. and Roura, R. (2004): Regulating Antarctic tourism and the precautionary principle. Current Development Note in *The American Journal of International Law 98*: 4, 763-781. Reproduced with permission.

percent to more than 19,500 (Fig. 3.1) (IAATO, 2004a). The estimate of total passengers for the 2003-2004 season, including those not landing, is over 27,000. This trend is likely to continue. In recent years, much larger ships have entered the market (see e.g. Hemmings, Fall 2000: 38-39),[3] and during the last season, "fly-sail" or "fly-cruise" operations were started: tourists are brought to the Antarctic by aircraft (avoiding the discomfort of the Southern Ocean) and then make excursions on yachts or ships (see e.g. Norway, 2001).[4] By global tourism standards, the numbers of tourists are very small. It should be remembered, however, that the Antarctic resident population in summer does not exceed 5,000.

Not only has tourism in the Antarctic intensified, but also the diversity of tourist activities has increased. Besides the "classic" ship-based tourism on small and medium-sized vessels – now also conducted with ships with a capacity of up to 500 passengers – other forms of tourism currently include yachts, the fly-sail operations mentioned above, large vessels (sometimes carrying over 1,000 passengers), land-based tourism (whether supported by air or by sea), and non-landing overflights. Activities conducted in the Antarctic today on a commercial basis include helicopter excursions, skiing expeditions, mountain climbing, snowboarding, kayaking, marathons, and scuba diving (Bastmeijer, 2003). Government-sponsored tourism has occurred intermittently for a number of years (IAATO, 2004a), although the details of these operations are sketchy. In addition to tourism, other types of commercial and non-commercial nongovernmental activities are being developed. Photography and film and art projects are well-known examples, but more recently various other activities have been conducted, such as the collection of meteorites for non-scientific purposes (see e.g. SCAR, 2000; New Zealand, 2001).[5]

To aid in understanding the development of tourism in Antarctica – and more particularly, to be able to make general predictions about the future development of Antarctic tourism – it is useful to have some knowledge of the development of nature-based tourism worldwide. The Québec Declaration on Ecotourism, adopted at the World Ecotourism Summit in 2002, notes the "growing interest of people in travelling to natural areas, both on land and sea."[6] This trend is richly documented and considerable attention has been devoted to determining which factors have caused it. For instance, at the 2003 World Park Congress it was explained that "[f]or the last 50 years advances in airline technology, energy efficiency, information technology, human social welfare and education led to increases for long distance travel" and that "[s]uch travel affects

3 See, e.g., Hemmings (Fall 2000: 31, 38-39) (noting that "there will be pressures to move to larger, less specialized ships").

4 See the Web site of Antarctica XXI ("The First Air Cruise to Antarctica"), at <http://www.antarcticaxxi.com/>. It has been noted that the establishment of air links to Antarctica introduces new tourism opportunities. See, e.g. Norway (2001).

5 See also CEP Resolution 3 (2001), *Collection of Meteorites in Antarctica*, available at <http://www.cep.aq/>.

6 World Ecotourism Summit, Québec Declaration on Ecotourism (May 22, 2002), available at <http://www.world-tourism.org/sustainable/IYE/quebec/anglais/quebec-eng.pdf>.

many parks" (Eagles, 2003). Figures provided by various organizations on nature-based tourism in regions of the world underline the trend described above. Thus, in Europe, according to the World Tourism Organization, "the number of tourists ... is expected to double in the next 25 years" and "most of the increase ... will come from alternative forms of travel not involving the classic 'sun and sand' tourism."[7] Research also indicates that the more remote places are becoming increasingly popular. For instance, the European Environmental Agency states with regard to the Arctic that "[i] n recent decades, tourism has become a major industry in some areas of the region."[8] About 40,000 tourists visit Svalbard each year,[9] and a comprehensive set of regulations was recently developed to protect the wilderness there from adverse impacts of tourism.[10] The increase in the diversity of tourist activities in "wilderness areas" may be illustrated by the "World Ice Golf Championship," which will be held from 24 to 28 March 2005, in Greenland, "the only golf tournament in the World where golfers play between huge icebergs."[11]

These general notions and the arbitrary selection of concrete examples indicate that the developments in the Antarctic are manifestations of a worldwide trend, and that they are very likely to continue. Predictions of future growth in the Antarctic tourist industry indicate that most categories will increase –such as large vessels and land-based tourism – whereas only the categories of smaller vessels (carrying fewer than fifty passengers) and yachts will remain consistent or increase just slightly, respectively (IAATO, 2004b). On the basis of this overview of the developments in Antarctic tourism, this Chapter examines the international regulation of Antarctic tourism. After discussing one of the main management issues with respect to Antarctic tourism – the assessment and prevention of cumulative impacts – we introduce the Antarctic Treaty System and analyze the existing instruments to address cumulative impacts. The Chapter next inquires into the status and possible practical relevance of the precautionary principle to the management of Antarctic tourism. Finally, the international debate on Antarctic tourism since 1991 and the measures that have been adopted by governments and industry to prevent adverse impacts by tourist activities in the

7 European Commission, Using Natural and Cultural Heritage to Develop Sustainable Tourism 1, 3, available at <http://europa.eu.int/comm/enterprise/services/tourism/studies/ecosystems/heritage.htm> (last modified 26 June 2003).

8 European Environment Agency, The Arctic Biogeographical Region – Warming up and changing? §2.3.4, available at <http://reports.eea.eu.int/report_2002_0524_154909/en/page001.html> (accessed 22 September 2004).

9 Travelling in Svalbard, at <http://www.sysselmannen.svalbard.no/travel_en.htm> (last modified 14 September 2004).

10 Svalbard Environmental Protection Act, Act of 15 June, 2001, No.79, Norway. Under the Act, various regulations have been adopted, including the Tourist Regulations of 1997. For the Act and the Tourist Regulations, see Laws and Regulations, available at <http://www.sysselmannen.svalbard.no/laws_enviro_en.htm>.

11 The World Ice Golf Championship, at <http://www.greenland-guide.gl/icegolf/> (accessed 22 July 2004).

Antarctic are described, particularly with a view to evaluating the extent to which the precautionary principle is being applied.

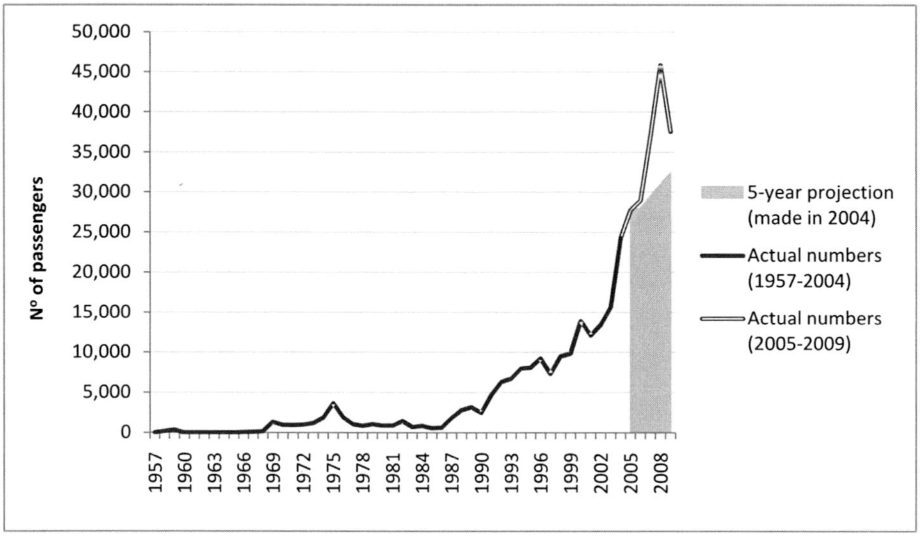

Fig. 3.1 – Antarctic seaborne tourism 1957-58 to 2003-04 and projections and actual data to 2008-09 (Based on Bastmeijer and Roura, 2004:764). *

* The lines are actual figures. The grey bar indicates IAATO's projected figures. The data set only applies to seaborne tourists participating in mostly commercial tourism operations. This includes passengers on large ships (> five hundred passengers), standard ships, and yachts (some of which may not carry paying passengers), whether landing or not. For the period 1998-99 to 2002-03 there are some discrepancies between different IAATO papers as regards the actual tourist numbers. In these cases, the lower figure was used. The total passengers figure will approximately be 5 percent or higher when adding the figures for tourists on fly-sail operations, non-landing overflights, air-supported landings, and several unreported types of tourism, including government-sponsored tourism, which are not included here (e.g., for the 2003-04 season the total number of passengers estimated (27,662) is 12 percent higher than the total of seaborne-only passengers (24,281)). In addition, the total number of people transported to Antarctica by tour operators is higher since these figures exclude officers, staff, and crew.

Sources:

- Actual data to 2003-2004 and projections: Bastmeijer and Roura (2004 and refs.);
- Actual data 2004-05 to 2008-209: IAATO statistics at <http://www.iaato.org/tourism_stats.html>, accessed February 2011.

3.2 ANTARCTIC TOURISM AND CUMULATIVE IMPACTS

Recent developments in Antarctic tourism raise concerns about the risk of cumulative impacts on the intrinsic values of the region. Tourism patterns since the early 1990s[12] indicate a trend toward the concentration of visits in a relatively limited number of Antarctic sites and a parallel trend toward visiting new sites. For reasons of accessibility, cost, and tourist attraction, most tourism is currently conducted in the region of the Antarctic Peninsula, although some other areas are in the process of development. All of those areas, which are often ice free, are usually biologically rich or otherwise have outstanding aesthetic, wilderness, historic, or scientific value, or a combination of those values, and are potentially susceptible to cumulative impacts. Furthermore, the tourism season, which usually extends from November to March, coincides with the peak of the breeding season for many Antarctic species. The detection, minimization, and management of cumulative impacts is therefore an issue closely associated with tourism.

Cumulative impacts are "the results of additive and aggregative actions producing impacts that accumulate incrementally or synergistically over time and space."[13] In the Antarctic context, cumulative impact has been defined as the impact of combined past, present, and reasonably foreseeable future activities. These activities may occur over time and space (De Poorter and Dalziell, 1996).[14] Conceptually, all activities in a certain area contribute to the cumulative impacts in that area, although as regards Antarctica interest has centred on the possible cumulative environmental impacts of commercial ship-based tourism, which currently constitutes the bulk of the industry (see e.g. Hofman and Jatko, 2001). In many locations, the activities of tour operators and national Antarctic programs or other operators take place simultaneously, but in other locations tourism is dominant and potentially the main contributor to cumulative impacts (IAATO, 2002).[15]

Possible cumulative impacts associated with ship-based tourism have been identified. These include impacts on landscape, fauna, flora, historical artefacts, and science programs and support activities in the areas visited, or on nearby marine areas. Potential impacts may include, among many others, footpath development, littering, damage to floral assemblages, declines in the number and sizes of breeding colonies,

12 Tourism Statistics, at <http://www.iaato.org> (accessed 23 September 2004).

13 This is the definition used by the International Association for Impact Assessment, taken from Vanclay and Bronstein (eds.) (1995).

14 The Protocol on Environmental Protection to the Antarctic Treaty, infra note 31, refers only to "cumulative impacts in the light of existing and known planned activities," see Annex I, Arts. 2(1)(b), 3(2)(f), but the ATCM Guidelines for EIA in Antarctica, available at <http://www.cep.aq/>, state: "A cumulative impact is the combined impact of past, present, and reasonably foreseeable activities. These activities may occur over time and space and can be additive or interactive/synergistic ..." (CEP, undated; accessed 23 September 2004).

15 "There was emphasis on the importance of addressing the possible cumulative impacts of ship-borne tourism, particularly at sites where there are regular landings."

appropriation of historic artefacts, interference with research activities, and coastal or marine pollution (IAATO, 2002: 5-7). These impacts may result from a variety of processes: for example, the incremental effects of the crowding of activities in time or space; effects occurring after a time lag or away from the causal activity; effects of the fragmentation of habitats; and "nibbling" effects of numerous small impacts. Discussion of the diverse typology of cumulative impacts falls beyond the scope of this Chapter. For present purposes, it suffices to note that cumulative impacts may affect a broad group of Antarctic values, develop through complex processes, and have effects that are not immediately apparent or easily linked to their cause. Site characteristics influence the nature and severity of potential cumulative effects. For instance, a sensitive site that is logistically convenient to tour operators is potentially more vulnerable to cumulative impacts than an equally sensitive site that is less accessible to visitors. In addition, some sites have characteristics that may influence cumulative impacts for reasons that are unique to the tourism activity, such as distinctive or novel features, and thus are more likely to attract visitors than areas that lack such novelties (IAATO, 2002). In addition, tour operators may give preference to sites where the "wilderness experience" can be maximized (IAATO, 2002).

Overall, a network of sites where tourism activity occurs is being established throughout the Antarctic region. Cumulative impacts may occur at some of these locations, particularly where both environmental sensitivity and tourism interest are high. It should be noted that even for Antarctic shipborne tourism the debate on cumulative impacts is largely focused on the terrestrial environment and on the sites where tourists disembark. The impacts of reaching the landing site, including those of long distance shipping and ship-to-shore transport, are generally ignored. This inattention contrasts with concerns raised by some nongovernmental organizations about cruise tourism in the Arctic and elsewhere (see e.g. Hertz and Davis, 2002; IAATO, 2002; WWF, 2004).[16]

3.3 THE ANTARCTIC TREATY SYSTEM AND EXISTING LEGAL INSTRUMENTS TO ADDRESS CUMULATIVE IMPACTS

The described developments in Antarctic tourism and the related concerns regarding cumulative impacts prompt the question of responsibility for addressing these concerns. Many wilderness areas in the world are protected against the adverse impacts of human activities by the state that enjoys sovereign power over the territory concerned. Antarctica, however, is not subject to undisputed territorial sovereignty.

16 "WWF is concerned about the overall impacts on the marine and coastal environments of the Arctic caused by the increased traffic of all types of vessels, and the variability in tour operator and tourist conduct on the voyages." (WWF, 2004). For a detailed report, see, for example, Herz and Davis (2002). Awareness about the effects of cruising in the Antarctic marine environment may be increasing. See IAATO (2002: 5): "It was pointed out that cumulative impacts upon the marine environment are an issue that may also warrant attention."

3.3.1 *The Antarctic Treaty and Environmental Protocol*

During the first half of the 20th century, seven states laid territorial claims to parts of the continent, but these claims became the subject of international disputes. In 1959 the claimant states and five other states involved in Antarctic research signed the Antarctic Treaty.[17] A central element of the Treaty is the "agreement to disagree" of Article IV: the position of each contracting party with regard to the legal status of Antarctica is respected and the contracting parties agree to manage Antarctica collectively. Safeguarding international peace and ensuring the freedom of scientific research are the two pillars of the Antarctic Treaty. It was also agreed that other states wishing to become contracting parties to the Treaty and able to demonstrate their interest in Antarctica by conducting "substantial scientific research activity" could receive "consultative status."[18] Today, there are twenty-nine consultative parties (ATCPs, 2004).[19] The consultative parties discuss Antarctic management issues at the Antarctic Treaty Consultative Meetings (ATCMs) where decisions are made by consensus.[20] Since the Treaty was promulgated, several other conventions and more than two hundred recommendations have been adopted, a set of instruments collectively known as the Antarctic Treaty System (see e.g. Auburn, 1982; Scott, 2003; Molenaar, 2003).[21]

With the adoption of the Protocol on Environmental Protection to the Antarctic Treaty (Protocol) in 1991, protection of the Antarctic environment became the third pillar of the Antarctic Treaty System.[22] By signing and ratifying the Protocol, its contracting parties "commit themselves to the comprehensive protection of the Antarctic environment and ... designate Antarctica as a natural reserve, devoted to peace and science."[23] The Protocol entered into force in January 1998 and applies to most human activities South

---

17 Antarctic Treaty, Dec. 1, 1959, 12 UST 794, 402 UNTS 71.

18 Antarctic Treaty, Art. IX(2).

19 At the twenty-seventh ATCM, May 23-June 4, 2004, Ukraine became the twenty-ninth Antarctic Treaty Consultative Party (ATCPs, 2004: paragraph 25).

20 ATCMs are hosted by the consultative parties in alphabetical order in English. At present, the ATCMs are held annually and last two weeks. At ATCMs, business is conducted in four official languages and by means of working groups, of which there are currently four: Liability, Legal and Institutional Matters, Tourism, and Operational Matters.

21 For a comprehensive discussion of the Antarctic Treaty System, see, among others, Auburn (1982). For a discussion of recent institutional developments in the Antarctic Treaty System, Scott (2003). For a discussion of the Convention on the Conservation of Antarctic Marine Living Resources, see Molenaar (2001).

22 Protocol on Environmental Protection to the Antarctic Treaty, Oct. 4, 1991, 30 ILM 1455 (1991) [hereinafter Protocol].

23 Protocol, Art. 2. To date, thirty-two states have become contracting parties to the Protocol: all twenty-eight consultative parties, plus three contracting parties to the Antarctic Treaty that do not have consultative status (Canada, Greece, and Romania). (Note October 2010: Three non-consultative parties – Belarus, the Czech Republic, and Monaco – have subsequently adhered to the Protocol).

of 60° S of latitude. Fundamental environmental principles are laid down in Article 3 of the Protocol, making clear that the values it protects include "the Antarctic environment and dependent and associated ecosystems and the intrinsic value of Antarctica, including its wilderness and aesthetic values and its value as an area for the conduct of scientific research."[24] Engaging in mineral resource activities for other than scientific purposes is forbidden,[25] and all other activities must be subjected to a prior environmental impact assessment (EIA).[26] Permit requirements must be in place for taking or harmfully interfering with Antarctic flora and fauna, introducing non-native species into Antarctica, and entering Antarctic Specially Protected Areas.[27] Furthermore, provisions on waste management on land and at sea must be respected.[28] The Protocol prohibits damaging any historic site or monument.[29] It also calls for the conclusion of an annex on liability for environmental damage.[30] A Committee for Environmental Protection (CEP) was established to advise the ATCM on implementing the Protocol and furthering the protection of the Antarctic environment.[31] On the basis of advice from the CEP or on its own initiative, the ATCM may adopt additional measures in accordance with Article IX of the Antarctic Treaty.[32]

3.3.2 Relevant instruments under the Protocol to address cumulative impacts

With the exception of some designated protected areas representing a very small percentage of the Antarctic continent (see e.g. Hansom and Gordon, 2000),[33] no part of Antarctica is in principle off limits to the tourism industry. At present, access of tour

24 Protocol, Art. 3(1).

25 Protocol, Art. 7.

26 Protocol, Art. 8 and Annex I.

27 Protocol, Annexes II, V.

28 Protocol, Annexes III, IV.

29 Protocol, Annex V, Art. 8.

30 Obligations as regards a liability annex are laid down in Article 16 of the Protocol. Negotiations on a liability annex started in 1992 and agreement is expected to be reached within the next two years. The chairman of the working group on liability stated that it was his "aim to conclude negotiations on the Annex by the ATCM in Stockholm" (to be held June 6-17, 2005). (ATCPs, 2004: paragraph 72). (Update October 2010: Annex VI of the Protocol on Liability Arising from Environmental Emergencies was approved by means of Measure 1 (2005) It has not yet become effective.).

31 The CEP was established by Article 11 of the Protocol. Its functions are stated in Article 12.

32 Protocol, Art. 10 (1)(b).

33 In 1997 the total land area that was protected was 790 square kilometers, equivalent to approximately 0.007% of the continental area of Antarctica (Hansom and Gordon 1998: 270). This area will have increased somewhat since the entry into force of the Protocol in 1998.

operators to Antarctica is primarily controlled by EIA requirements.[34] The EIA – in parallel with environmental monitoring[35] – constitutes the key instrument of the Protocol to detect, minimize, and manage the environmental impacts of tourist activities. Article 8, paragraph 2 of the Protocol makes explicit that the EIA obligations also apply to tourist activities.[36] Although concerns were expressed about the development of tourism during the negotiations on the Protocol, "the prevailing view was that the Protocol and its Annexes should be designed to address generically all activities in Antarctica" (Richardson, 2000).

The Protocol stipulates that cumulative impacts be considered as a component of the EIA process,[37] although the level of analysis this "consideration" requires is not defined. Broadly speaking, impact assessments can be defined as "the process of identifying the future consequences of a current or proposed action."[38] To assess cumulative impact, an EIA should consider the effects of a proposed activity and those of other activities in the area (Cooper and Sheate, 2004: 5-6). Both past and present activities, as well as those that can reasonably be expected in the foreseeable future, are to be taken into account. Thus, the role of cumulative impact assessment in composing an EIA is "to extend the temporal and spatial boundaries of the assessment and set the assessment within the context of the wider area or region." (Cooper and Sheate, 2004: 6).

The criteria of intensity and duration of the impacts were designed to determine the level of EIA required under the Protocol. The three-tiered EIA system of the Protocol refers to whether impacts are less than, no more than, or more than "minor and transitory." The greater the intensity and duration of impacts, the greater the detail and scrutiny required in the EIA. Cumulative impacts, if and when they occur, may be "minor or transitory" in the terminology of the Protocol, but they may also be more serious. For example, tourism could plainly be a factor in introducing or translocating alien species or diseases, but this process is not well understood or monitored. Although the risk may be low, the potential consequences may be severe, and the impacts are difficult

34 At the Antarctic Tourism Workshop held by Antarctica New Zealand on 23 June 2000, two teams debated the statement that "Market Forces and Environmental Impact Assessment are enough to manage tourism." The chair declared the negative team the winner by a narrow margin. Antarctica New Zealand (2000) (In file with author).

35 See, e.g., Protocol, Art. 3(2)(e).

36 Article 8(2) of the Protocol provides:
Each Party shall ensure that the assessment procedures set out in Annex I are applied in the planning processes leading to decisions about any activities undertaken in the Antarctic Treaty area pursuant to scientific research programmes, tourism and all other governmental and non-governmental activities in the Antarctic Treaty area for which advance notice is required under Article VII(5) of the Antarctic Treaty, including associated logistic support activities.

37 Protocol, Art. 3(2)(c) and Annex I, Arts. 2(1), 3(2)(f). For an analysis of its application to tourism, see Hemmings and Roura (2003: 13), and Kriwoken and Rootes (2000: 138).

38 International Association for Impact Assessment, Home Page, at <http://www.iaia.org>.

to assess in an Antarctic EIA. In our view, the assessment of cumulative impacts is one of the gaps in the existing EIA process as applied to Antarctic activities. As stated by W. Bush, "[T]he cumulative impact of less significant activities is likely to be the main loophole by which activities can proceed without adequate caution" (Bush, 2000a: 21, 25). Conceptual constraints also hamper the application of EIA to tourism, as envisaged by the Protocol. These derive from applying an EIA system that was developed for scientific activities and associated logistics at a few discrete sites, where environmental reference states can be established and impacts monitored by the operators themselves, to the transient, fast-moving, multisite activities that characterize contemporary Antarctic tourism. Current impact assessment processes are not well suited to evaluating cumulative impacts (Hemmings and Roura, 2003).[39] Additional constraints are practical, and concern the way EIA requirements are implemented by tour operators. The latter drew up less than 20 percent of the Initial Environmental Evaluations (IEEs) produced over the period 1991–2002 (Hemmings and Roura, 2003: 17). Recent analysis of tourism EIAs suggests that "[k]ey aspects of EIA, including scoping, critical assessment, monitoring and auditing are either poorly developed or absent" (Gee, 2004). No Comprehensive Environmental Evaluations, which are the highest level of EIA under the Protocol, have been produced for Antarctic tourism, even though some IEEs describe such tourism in certain instances as "involving multiple cruises within a season, and across several years, involving transport of many hundreds (possibly thousands) of persons to tens of different sites spread across a huge area, where diverse activities will take place. This is a substantially more complex scenario than anticipated for IEE application." (Hemmings and Roura, 2003: 19). In addition, tourism EIAs tend to focus on the assessment of impacts resulting from a single season or from a few seasons at most. They do not usually, to our knowledge, address the impact on sites becoming tourism "destinations," that is, sites that will continue to be visited repeatedly by tourists on a seasonal basis for the foreseeable future and that will therefore be exposed, in a sense, to permanent human activity.

Several proposals that additional tools be used to evaluate Antarctic tourism have been put forward. These include Strategic Environmental Assessments, regional assessments, and cumulative impacts assessments (see e.g. ASOC, 2000; Kriwoken and Rootes, 2000). The International Association of Antarctica Tour Operators (IAATO) conducted a "programmatic" EIA for its 1997-1998 season, which complied with U.S. domestic legislation (Kriwoken and Rootes, 2000: 145-146), and some industry IEEs have involved the operations of five or more companies. However, questions remain as to whether these "programmatic" EIAs better evaluate cumulative impacts or are simply "large" EIAs meant to cut down on the proponent's paperwork.

39 A similar concept was expressed by Argentina at the seventh CEP meeting and twenty-seventh ATCM (CEP, 2004: paragraphs 74, 75).

Environmental monitoring after the activity complements the assessment of impacts prior to the activity.[40] Over the years, several attempts have been made to monitor the impacts of tourism visits on specific sites (e.g. Giese, 1998; Patterson *et al.*, 2003); a non-profit organization, Oceanites Inc., conducted a major program in tourism landing sites on the Antarctic Peninsula (USA, 2004).[41] This work is financially or logistically supported by some Treaty parties and the tourism industry (UK, 2004a). The scope of this Chapter does not permit a detailed analysis of environmental monitoring in Antarctica, which is an ongoing subject of debate and intersessional work.[42] However, the monitoring programs currently in place, although worthwhile in their own right for inventorying the flora, fauna, and other characteristics of various sites visited by tourists, will not easily establish clear cause-and-effect linkages between tourism activity and environmental changes. There are simply too many potential causes of environmental change, such as the natural variability of indicator species and climate change, in addition to the potential contribution of human activities, including tourism.[43] Thus, it remains to be seen whether current monitoring approaches will provide the answers needed in a timely manner to inform decisions on the management of tourism.

We conclude that EIA and monitoring obligations under the Protocol are necessary, but not sufficient, to address the issue of cumulative impacts adequately (see e.g. Antarctica New Zealand, 2000:13).[44] Gaps in knowledge can be seen in relation to potential adverse cumulative impacts of tourist activities, and current monitoring initiatives cannot yet fill these gaps. For instance, what is the status of wildlife populations at sites frequently visited by tourists? Has this status changed in any way as a result of tourism? These basic questions can perhaps be answered with any certainty about only a few sites frequently visited by tourists. Furthermore, uncertainties abound with regard to the actual and potential impacts of tourism and the future development of the Antarctic tourist industry.

---

40 Conceptually, both impact assessment before the activity and monitoring of impacts after the activity are indeed part of the EIA process. In the Antarctic context, somewhat confusingly, there are monitoring initiatives that take place independently of EIA processes, and some EIA processes that do not include a monitoring component.

41 "In ten seasons of fieldwork, beginning in November 1994, the Antarctic Site Inventory has demonstrated an ability to reach Antarctic Peninsula visitor sites frequently and cost-effectively. Through February 2004, Inventory researchers have made 570 visits to 89 Peninsula locations." (USA, 2004).

42 CEP (2004), Annex 6.

43 Presentation by Oceanites Inc. at the ATME, Norway (22-25 March 2004).

44 According to one of the opponents, "[EIA] simply evaluates impacts from single projects or expeditions. It is rarely good at assessing cumulative impacts and does not ask relevant public policy questions about alternatives and the desirability or otherwise of a proposal." (Antarctica New Zealand, 2000: 13).

These uncertainties and gaps in knowledge raise questions about the status of the precautionary principle in Antarctic management and the practical relevance of the principle to the Antarctic tourism debate.[45]

3.4 THE PRECAUTIONARY PRINCIPLE

3.4.1 Reasons for applying the precautionary principle to Antarctica

The Protocol does not clearly oblige the contracting parties to take the precautionary principle into account in decision making.[46] It can be argued that such an obligation derives more implicitly from particular provisions of the Protocol (e.g., Art. 3(2)(c)). However, it can also be argued that since the relevant parts of Article 4 of the Convention on the Regulation of Antarctic Mineral Resource Activities (CRAMRA) – reflecting the precautionary principle – were not transferred to Article 3 of the Protocol, the negotiating states did not accept an obligation to take the precautionary principle into account in implementing the Protocol. Nonetheless, various strong arguments can be made for applying the precautionary principle to the management of human activities in Antarctica:

- Application of the principle would harmonize with the designation of Antarctica as "a natural reserve, devoted to peace and science," in Article 2 of the Protocol (Scott, 2001: 966).
- The precautionary principle has been codified in other international or regional agreements on the protection of natural areas (Verschuuren, 2003; Cameron, 2001: 121, 133).
- Application of the precautionary principle would also be consistent with the proactive approach of the Antarctic Treaty System. In the past, the consultative parties of the Antarctic Treaty System adopted legal instruments concerning human activities without knowing whether these activities would be initiated or whether they would result in significant impacts on the Antarctic environment (e.g., the Convention on the Conservation of Antarctic Seals and CRAMRA) (A.D. Hemmings, personal communication to authors, May 2001).[47]
- The Protocol itself also contains specific prohibitions that are (at least partly) based on the precautionary principle (e.g., the prohibitions on mineral resource activities (Scott, 2001) and keeping dogs in Antarctica).

---

45 For discussions on the precautionary principle, see O'Riordan, Cameron, and Jordan (eds.) (2001); Scott (2001) and Verschuuren (2003).

46 On the status of the precautionary principle under the Protocol, see Bastmeijer (2003: 293-98); and Scott (2001).

47 For the conventions, see Convention on the Conservation of Antarctic Seals, 1 June 1972, 29 UST 441; Convention on the Regulation of Antarctic Mineral Resource Activities, 2 June 1988, 27 ILM 859 (1988) (CRAMRA has not entered into force, see Bastmeijer, 2003: 44-46).

J. Cameron notes the difference between prevention, which applies to known threats, and precaution, which applies to uncertain threats (Cameron, 1993). In the Antarctic context, not all activities carry the same degree of uncertainty. While there are inherent risks to operating in Antarctica, and not all Antarctic operators work on the basis of the same level of operational experience or standards, most "traditional" Antarctic operations are characterized by a relatively low degree of uncertainty. They are usually conducted by tried and tested methods. The implementation of Protocol requirements, such as education and training, contingency plans, and the application of mitigating measures, can contribute to minimizing the risks and impacts of an activity. However, as the uncertainties and risks increase, for instance as a result of "novel" activities or the rapid growth of a particular activity, the considerations change. In the Antarctic context, this formula applies as much to tourism as to subglacial lake research and the establishment of air links. The uncertainty that attaches to the direct, indirect, and cumulative impacts of these activities is not always of a low level. Under some circumstances, applying the precautionary principle becomes necessary so as to act in accordance with the objectives and spirit of the Protocol.

3.4.2 Practical Implications of Applying the Precautionary Principle

Both domestic and international discussions on the subject clearly indicate the importance of showing the practical implications of applying the precautionary principle. For example, should any uncertainty about the environmental impacts of tourist activities in Antarctica automatically lead to the prohibition of Antarctic tourism? The answer is no, as this interpretation would transform the precautionary principle into an absolute norm, which is not its purpose as developed in international and domestic environmental law (Verschuuren, 2003: 37-41; Scott, 2001: 965). Its aim is to ensure that "uncertainties" about the impacts of an activity (or set of activities) are weighed in the decision-making process. The decision is not only based on the socioeconomic advantages of a project or activity and any adverse environmental impact that is 100 percent certain to take place; gaps in knowledge and risks are also taken into account.[48] Generally speaking, a reasonable chance that serious adverse impacts will take place combined with questionable socioeconomic importance or available alternatives will push the "pointer" to requirements for additional precautionary measures or even to a "no go" or "not yet" decision. Applying the precautionary principle in the Antarctic tourism debate may theoretically result in a range of measures that may be appropriate for different spatial and temporal scales. It may result, for example, in imposing certain general conditions on Antarctic tourism as a whole, and specific restrictions regarding

48 Cameron defines risk as an "amalgam of the probability of an event occurring and the seriousness of the consequences should it occur. Thus, a high-risk strategy is one that either combines a relatively high probability with relatively innocuous consequences or one that combines a relatively low probability with relatively serious consequences." Cameron (1993: 7).

particular regions or sites.[49] Thus, the precautionary principle may be applied as a practical instrument to the management of tourism, which may result in imposing certain restrictions on tourist activities, without having to limit its application to extreme measures such as the banning of Antarctic tourism altogether.[50]

In our view, some of the practical results implied by application of the precautionary principle to Antarctic tourism might include:

- Improving the applicability of EIA to tourism;
- Generally improving the process of prior assessment of cumulative impacts rather than expecting that ex post monitoring will provide all the answers;[51]
- Prohibiting tourist activities in potentially sensitive sites where environmental monitoring is lacking or insufficient (e.g., Baily Head on Deception Island in the South Shetland Islands, one of the twenty most visited tourist sites on the entire continent because it contains one of the largest Antarctic penguin colonies, which is situated at the center of a narrow valley, forcing visitors to circulate through dense concentrations of seabirds);
- Establishing temporal or spatial limitations for certain sites as required by their specific values and characteristics, including limiting the number of visitors where appropriate (e.g., Hannah Point on Livingston Island in the South Shetlands, a frequently visited site with a high degree of biodiversity including sensitive species such as [breeding] giant petrels, where visitors tend to disperse over the entire area rather than concentrate on a single access route);
- Concentrating tourist activities in designated areas to be managed in such a way that these activities may be conducted in a sustainable manner;
- Preventing access to any site that has not been visited before, so as not to expand the number of sites exposed to a human presence; and adopting restrictions on the permissible types of tourist activities.

49 It has been suggested that some penguin species may habituate to human presence and that "[f]or this reason, managers may suggest that visits be conducted in areas that have consistently received some level of tourism, rather than concentrating on "expedition touring", in which the focus is to visit new colonies and retreat into areas rarely visited." Patterson *et al.* (2003: 306).

50 See Scott (2001: 965) (stating that "it is now clearly recognized that the precautionary principle does not have to be applied in an absolutist fashion"); see also Laws (1991) (holding that "[i]n view of the facts, one unrealistic proposal-that all activities be considered high risk until proven not to be-is quite unacceptable").

51 This may include the use of Strategic Environmental Assessment approaches. See, e.g., ASOC (2000); see also Norway (2004).

3.5 THE ANTARCTIC TOURISM DEBATE 1991-2004: IS THE PRECAUTIONARY PRINCIPLE BEING APPLIED?

3.5.1 *The Antarctic tourism debate 1991-2003*

The discussions on tourism in Antarctica at the ATCMs since 1991 show that, generally speaking, the consultative parties consider tourism in Antarctica to be legitimate "use."[52] Although tourism does not belong to the same category as the activities at the core of the Antarctic Treaty and the Protocol – the preservation of peace, international cooperation, scientific research, and comprehensive environmental protection – the total prohibition of tourism in the Antarctic has never been discussed. However, as early as 1991, various consultative parties expressed the view that the Protocol could not regulate Antarctic tourism adequately. At the sixteenth and seventeenth ATCMs in 1991 and 1992, these states proposed that a separate annex on tourism be added to the Protocol (Chile *et al.*, 1992; Richardson, 2000: 77), but consensus on the need for such an annex could not be reached. At the eighteenth ATCM in 1994, consensus was reached on a non-legally binding approach." (ATCPs, 1994: paragraph 59) by the adoption of Recommendation XVIII–1, which includes both guidelines for visitors to the Antarctic and guidelines for the organizers of nongovernmental expeditions. Between 1994 and 2001, "tourism and nongovernmental expeditions" was a separate agenda item at the ATCMs, but the general character of the discussions indicates that the issue was not regarded as a priority and no substantial additional legal measures to regulate Antarctic tourism were adopted (Richardson, 2000: 79-82).

Since 2001, the question of additional measures has received closer attention. That year, at the twenty-fourth ATCM, it was "noted that there is an increase in the diversity of tourism activities, which may present new management challenges," and the consultative parties agreed on "the importance of appropriate management of Antarctic tourism." (ATCPs, 2001: paragraph 106). At the ATCMs in 2002 and 2003, several contracting parties and observers submitted papers on the management of Antarctic tourism, but the discussions were brief and no legally binding measures were considered" (ATCPs, 2003: paragraphs 129-152).

3.5.2 *Self-regulation in Antarctica: The International Association of Antarctica Tour Operators (IAATO)*

For some of the consultative parties, the existence of a tourism industry association capable of taking self-regulatory initiatives may have constituted one of the reasons for postponing discussions on the management of tourism at the ATCMs. IAATO, which was founded in 1991 by seven tour operators, represents the interests of the Antarctic tourism sector (Splettstoesser, 2000). It is "dedicated to appropriate, safe and environ-

52 For a summary of the tourism debate since 1991, see ASOC (2002); Bastmeijer (2003); Richardson (2000).

mentally sound private-sector travel to the Antarctic."[53] Currently, sixty-nine organizations are members (in different categories) of the association (IAATO, 2004c). It takes many initiatives to increase the environmental awareness of its members. The IAATO bylaws include the obligation to respect the relevant provisions of the Protocol – which is particularly meaningful for members in states outside the Antarctic Treaty System – as well as various additional requirements on such matters as the qualifications and experience of expedition staff, and the numbers of visitors allowed to make landings in the Antarctic.[54] In addition to the bylaws, specific guidelines, for example, the Marine Wildlife Watching Guidelines, have been adopted.[55] An organization that applies for membership must take an IAATO observer aboard during one of its expeditions. After assessing whether the tour operator meets IAATO standards, the observer produces a report that forms the basis for the IAATO annual meeting to decide on the application for membership.

IAATO has expert status at the ATCM and is represented on some national delegations. While the work of IAATO is generally well regarded, the discussions at the twenty-seventh ATCM appear to indicate that an increasing number of consultative parties question whether the Antarctic environment can be adequately protected through self-regulation.

3.5.3 Outcome of the XXVIIth ATCM, Cape Town, 2004

The discussions on the adoption of additional measures for regulating Antarctic tourism continued at the twenty-seventh ATCM, which took place in Cape Town from 23 May to 4 June 2004.[56] The issues considered in Cape Town – following discussions in March 2004 at an Antarctic Treaty Meeting of Experts (ATME) on Antarctic tourism organised by Norway (Norway, 2004)[57] – included the need for an accreditation system for Antarctic tour operators, the establishment of a tourism database, the prohibition of structural tourism facilities in Antarctica, safety and self-sufficiency, and the development of site-specific guidelines.

Unlike previous ATCMs, the twenty-seventh meeting dealt with Antarctic tourism as an integral part of the agenda and made it the subject of a separate working group.[58]

53 About IAATO: Objectives, at <http://www.iaato.org/objectives.html> (accessed 24 September 2004).

54 IAATO bylaws, 2 May 2004, available at <http://www.iaato.org/bylaws.html>.

55 IAATO, marine Wildlife Watching Guidelines, available at http://www.iaato.org/wildlife.html>(last modified January 2003).(IAATO

56 This section includes references to the Draft XXVII ATCM Report (ATCPs, 2004). The final report is available at http://www.ats.aq.

57 Norway (2004). This report and the papers submitted to the twenty-seventh ATCM by consultative parties and experts constituted the basis for the discussions on Antarctic tourism in Cape Town and are available at <http://www.ats.aq/>.

58 The Tourism Working Group established at the twenty-seventh ATCM held a four-day meeting during the second week of the ATCM (31 May - 3 June 2004).

A key element of the debate was the growing acceptance that in the face of current developments in tourism, some regulatory action might be necessary.[59] Various parties seem to consider accreditation as a possible next step in the regulation of tourism. However, the meeting in Cape Town did not take a decision on this issue. In the near future, parties will probably consider options as to how accreditation would work, specifically the role of the tourism industry (through IAATO), as well as how the chosen accreditation scheme would relate to authorizations processed under domestic legislation implementing the Protocol.

The establishment of structural tourism facilities in Antarctica, and its implications regarding legal jurisdiction and the acquisition (or assertion) of property rights, was debated briefly (see 3.8 below). Australia proposed the establishment of an Antarctic tourism database, to be based on existing databases. Although most consultative parties supported the idea, a decision was postponed because of questions about financial aspects and the legal protection of confidential information under domestic legislation.

"Adventure tourism" a term that loosely encompasses small, independent expeditions, some commercial, some private, and is usually applied to entities that are not members of IAATO was also debated at Cape Town. Adventure tourism constitutes a small part of the overall tourism industry, but it concerns many parties because of its implications for search and rescue and associated costs. However, not all parties accepted that adventure tourism was "the problem," as some portrayed it. Rather, the focus of the debate remained on tourism in all its forms, including mainstream tourism. Perhaps the most significant effect of this discussion was the deletion of the term "adventure tourism" from Resolution 3 (2004) *Tourism and non-governmental activities: Enhanced co-operation amongst Parties*, and its application instead to "Antarctic tourism" as a whole.[60]

The discussions in Cape Town resulted in the adoption of some measures under Article IX of the Antarctic Treaty, including Insurance and Contingency Planning for Tourism and Nongovernmental Activities in the Antarctic Treaty Area. In conjunction with this measure, the ATCM adopted Resolution 4 (2007), *Guidelines on contingency planning, insurance and other matters for tourist and other non-governmental activities in the Antarctic Treaty Area*, with an annex containing guidelines for those organizing or conducting the activities. The measure and guidelines are designed to promote safety and self-sufficiency of tourist and other nongovernmental expeditions. These guidelines supplement those attached to the 1994 Recommendation XVIII-1. Furthermore, the resolution on enhanced cooperation among parties was adopted to prevent operators from exploiting jurisdictional loopholes in their domestic laws. In addition to these concrete decisions, the ATCM established several intersessional contact groups

---

59 According to nongovernmental organization experts at the ATCM, "[t]he most obvious aspect to the Cape Town debate was that nobody now disputes that tourism (and non-governmental activity-in the ATCM these are always conjoined twins) is increasing, may pose problems and needs regulation." ASOC (2004: paragraph 59).

60 All measures, decisions, and resolutions that were adopted by the twenty-seventh ATCM are available at <http://www.ats.aq>

to continue the work on accreditation schemes, cumulative impacts, and environmental monitoring.

Overall, the tourism discussion in Cape Town – as previously in Norway – focused on various technical responses, rather than on the development of a comprehensive policy.

3.6 CHARACTERIZING THE CURRENT MEASURES: CONDITIONS OR RESTRICTIONS?

The discussions at the twenty-seventh ATCM show that the parties take the issue of tourism management in Antarctica seriously. Thus, the Antarctic tourism debate has entered a new stage. While at the previous ATCM in Madrid some consultative parties "stated that tourism activities were legitimate and that their regulation and management should be achieved through effective implementation of existing legal instruments" ATCPs (2003: paragraph 130) at the Cape Town meeting there was general agreement that certain issues, at least, needed additional regulation. However, the decision-making process is slow; the legal and regulatory framework for tourism has not developed as fast as the industry itself. The tourism debate and the measures adopted are characterized by an approach that "sets conditions"; substantial restrictions have not been imposed or seriously discussed. Whether this approach will enable the consultative parties to take the necessary steps in time to prevent adverse impacts on the Antarctic environment and system of governance remains to be seen. This question is difficult to answer because many related questions are still open. Are Antarctic tourism activities causing any adverse impacts on the Antarctic today? How will Antarctic tourism develop over the next ten years? Will potential problems be solved in a timely fashion by industry self-regulation?

With regard to particular issues, the tourism sector itself has adopted guidelines and procedures that go beyond the Protocol. Examples include the current industry requirement of no more than a hundred passengers ashore at one time, and a staff-passenger ratio of 1:20; guidelines on the cleaning of boots after each visit to an Antarctic site to prevent the introduction and translocation of alien species and diseases; and guidelines on watching marine wildlife.[61] IAATO also plays a role in promoting compliance with the Protocol, particularly in view of the limited human and financial resources made available by some contracting parties to enforce their domestic Antarctic implementing legislation. However, whatever the industry's demonstrable concerns as regards minimizing environmental damage, plainly tour operators have to operate profitably in Antarctica. According to our observations, most measures that have been adopted by the tourism industry should also be characterized as conditions to engaging in Antarctic tourist activities (Hofman and Jatko, 2001: 10),[62] conditions that do not substantially restrict Antarctic tourism initiatives. For instance, the limitation to no more than a hundred passengers ashore at one time does not pose ser-

61 See IAATO bylaws, May 2, 2004, available at http://www.iaato.org/bylaws.html.

62 The workshop determined nine possible impact avoidance and mitigation measures, only a few of which could be considered precautionary.

ious restrictions in practice, as it is also justified for various practical reasons. IAATO may find it difficult to adopt such restrictions, given the commercial interests of its members, customer demands, and the dynamic nature of the industry. It may even be forced to weaken its bylaws to maintain its influence as the industry continues to grow and change (see e.g. ASOC, 2002b).[63] For example, until 2001 IAATO required that its members carry no more than four hundred passengers on their vessels, but the association recently removed this upper limit to enable companies with larger-capacity vessels to join (IAATO, 2001: paragraph 2.2.4).[64] The Antarctic Treaty Parties should be aware of the possible limitations of self-regulation, particularly in the long term.

3.7 IS THE PRECAUTIONARY PRINCIPLE BEING APPLIED?

In our view, the above discussion justifies the conclusion that the consultative parties and the industry are applying the precautionary principle only to a limited extent. Most of the possible practical implications of the precautionary principle – identified above – are not being seriously discussed. Although improving the applicability of EIA to tourism receives attention and initiatives are taken to improve monitoring, we believe that the speed and character of the tourism developments in the Antarctic require a more fundamental discussion. At an international symposium on the future of the Antarctic Treaty System held in 1995, one of the participants stated that:

> *the system has taken pride in the fact that we were almost always looking forward to the future and taking decisions before events occurred. I think that now we are uncertain and hesitate to address the problems in advance (Jackson, 1995).*

In our view, the manner in which Antarctic tourism has been dealt with in the past decade appears to support this statement. Certainly, the consultative parties and other stakeholders should continue their work on solving concrete and pressing concerns, but – at the same time – they should adopt a more strategic approach to addressing Antarctic tourism, taking into account the precautionary principle. An important step

63 ASOC (2002) states: "The present [tourism] industry-and its association-has particular historic roots in the small, owner-operated, ship-based, polar-specific companies of the late 80s ... As the industry diversifies, however, the viability of this sort of association comes into question. It faces difficulties whichever way it turns. Hold onto existing small-ship terms of membership, and newer and larger operators cannot join – and the association loses influence ... Jettison problematical entry requirements to encourage new operators to join, and you maintain the influence, but invariably lower standards."

64 IAATO (2001: paragraph 2.2.4) (stating that "IAATO has embraced the changing nature of Antarctic tourism and would encourage support amongst Antarctic Treaty Parties as a possible way forward. Under the above categories of Membership, IAATO has been inclusive yet taken the precautionary approach until further scientific or aesthetic evidence proves otherwise.")

would be to develop scenarios concerning the development of Antarctic tourism in the next ten years and seriously discuss the measures that would fit these scenarios.

The first step in a more serious application of the precautionary principle in regulating Antarctic tourism may be made in the near future. Some "signals" point to a growing awareness of the need to use a precautionary approach to tourism management. The report of the ATME reflects the expert participants' acknowledgment that EIA and monitoring do not provide all the answers in time: "[T]he Meeting recognized the need to improve the assessment and monitoring of cumulative impacts, and considered that monitoring also had to provide for management responses if and when concerns arise, taking into account the precautionary approach." (Norway, 2004: paragraph 12). Furthermore, at the seventh meeting of the CEP in Cape Town, the use of the precautionary approach was discussed in the context of managing tourism landing sites. While recognizing the value of monitoring in the long term, the committee debated whether site-specific guidelines might be required for certain frequently visited places.[65] These guidelines would have some precautionary measures built in. Specifically, it was suggested that daily visitation time limits be established at some highly sensitive sites. According to the draft CEP report, "It was noted that no scientific advice is currently available to inform these restrictions, but that the approach taken in the guidelines is necessarily of a precautionary nature." (CEP, 2004: paragraph 175). Since only a hundred passengers can land at any site at the same time, such limits would implicitly impose a cap on the number of tourists.[66] One could question the need for ongoing tourist visits at sites defined as highly sensitive and instead recommend stricter precautionary measures, such as keeping such sites off-limits to tourism until sufficient information to inform management decisions becomes available. However, regardless of the specific merits of this proposal, the committee's discussion appears to indicate some willingness to take precautionary action before there is conclusive evidence of damage resulting from tourism. The proposal will be further debated at the twenty-eighth ATCM, to be held in Stockholm in 2005.[67]

---

65 The United Kingdom stated in a working paper in 2004:
The collection of data at many of these sites continues. It is anticipated that such long-term information may assist the Parties ultimately in determining whether management plans under Annex V to the Protocol may be required for a number of these sites. In the interim, *we believe that the behavioural code set out in Recommendation XVIII-1 could be significantly enhanced by incorporating into it a set of Site-specific Guidelines.* United Kingdom (2004: 3a) (emphasis added).

66 The proposed time limits were of six to eight hours per day, depending on the environmental sensitivity of the sites. In practice, these time limits would have an effect on tourist landing sites only if the number of tourists exceeded several times current numbers.

67 For a summary update of the Antarctic tourism discussions since the Cape Town ATCM, see Epilogue.

3.8 REGULATING ANTARCTIC TOURISM: A CHALLENGE TO THE ANTARCTIC TREATY SYSTEM

Before concluding this Chapter, we wish to stress that the regulation of Antarctic tourism is more than an environmental issue. It is multifaceted, with dimensions that implicate financial liability, search and rescue, and complex political matters concerning, among others, the use of Antarctic territory subject to unresolved sovereignty claims and the intrusion of influential economic interests into the Antarctic Treaty System. Unbalances in any of these dimensions may trigger a response from the Antarctic Treaty System; they may be exploited by individual states to promote their own agendas or they may be ignored at the expense of the stability of the entire system.

Adequate regulation of Antarctic tourism represents a challenge to the Antarctic Treaty System: "tourism is the first large-scale multinational activity on land in Antarctica that is not under direct government control or largely government funded, and thus in many senses the tourism issue poses the first real test of international governance of private activities in Antarctica" (B. Oxman, personal communication to authors, 19 July 2004). For instance, the increase in Antarctic tourism and the multinational character of most tourist activities will raise more and more jurisdictional questions (see e.g. Bastmeijer, 2003: 434-438; 455-456). The Treaty and the Protocol do not clarify the issue of jurisdiction with regard to most persons in Antarctica (Bush, 2000b; Bastmeijer, 2003).[68] The lack of clarity in this respect seems to be accepted by the consultative parties as a consequence of Article IV of the Treaty; however, the developments in Antarctic tourism will probably force the consultative parties to discuss this sensitive issue. Research has revealed substantial differences in the jurisdictional scope of application of the domestic Antarctic legislation of the various contracting parties to the Protocol, and individuals may try to use these differences to limit governmental intervention in their proposed activities (Bastmeijer, 2003: 132-153; 455-456). The resolution on enhanced cooperation among parties, adopted at the twenty-seventh ATCM, may foreshadow a broader discussion of these issues. Moreover, the increase in activities by states that are not contracting parties to the Treaty and the Protocol is a concern. For example, the ATME in Norway "considered the challenges introduced... when tour companies and/or tourist ships operate from third party States," adding that "[i]n this context the current lack of a mechanism whereby the ATCM could engage with such third Parties was noted." (Norway, 2004: paragraph 30). While this quotation refers to the power of the ATCM, it should be noted that the jurisdiction of

68 Provision is made only with regard to "Antarctic officials" under Article VIII(1) of the Treaty. The possibility under Article IX(1)(e) of the Treaty of adopting further measures regarding "questions relating to the exercise of jurisdiction in Antarctica" has never been acted upon. See Bush (2000: 309, 312) (observing that "[t]he Protocol thus leaves unstated those for whom each party to the Protocol is responsible with regard to ensuring compliance; national legal systems cannot afford to allow such vagueness"); see also Bastmeijer (2003: 104-22) (stating that there are strong arguments for the interpretation that each contracting party is to ensure that its measures to implement the relevant provisions of the Protocol address all activities identified in Article VII(5)(a).

the contracting parties to regulate Antarctic activities is not limited to the activities that are organized within their territories. Without prejudice to the positions of states with respect to territorial claims, various bases of jurisdiction under international law provide each contracting party with the power to apply its domestic legislation to Antarctic activities conducted by its nationals, activities involving ships and aircraft flying its flag, and activities proceeding from its territory. International law also leaves room for more creative approaches, such as the application of domestic legislation to persons who conduct substantial activities within the contracting parties' territory with respect to the solicitation of customers for Antarctic expeditions organized elsewhere. Various arguments support the view that each contracting party is in fact obliged to use a comprehensive approach in determining the jurisdictional scope of application of its implementing legislation. For instance, Article VII(5)(a) of the Treaty already obliges each contracting party to give the other contracting parties advance notice of "all expeditions to and within Antarctica, on the part of its ships or nationals, and all expeditions to Antarctica organized in or proceeding from its territory." The explicit reference in several key provisions of the Protocol to this reporting obligation (e.g., those on EIA, environmental principles, waste management, and emergencies) appears to indicate that contracting parties should apply their implementing legislation to these activities.[69] This interpretation is also in harmony with the agreement, contained in Article X of the Treaty and Article 13, paragraph 2 of the Protocol, that "[e]ach of the Contracting Parties undertakes to exert appropriate efforts, consistent with the Charter of the United Nations, to the end that no one engages in any activity in Antarctica contrary to," respectively, the principles or purposes of the Treaty and the Protocol. In addition, questions may arise about the assertion of property rights by private individuals or commercial entities, such as through the long-term use of a site, which may include the construction of structural tourism facilities.[70] The establishment of such facilities in Antarctica and how it may affect legal jurisdiction and the acquisition (or assertion) of property rights are of concern to some parties, including some claimant states. At the ATME, New Zealand proposed prohibiting such facilities, and was supported

69 Protocol, Arts. 3(4)(a), 8(2), 15(1)(a) and Annex III, Art. 1. For a comprehensive discussion of this issue, see Bastmeijer (2003: 115-18).

70 The case of Pendulum Cove on Deception Island is an example of the assertion of the right to use a site. This is one of the sites most visited by tourists in the whole Antarctic Peninsula, where they swim in volcanically heated waters. Tourist activity may in some instances interfere with the functioning of highly sensitive instruments used to measure volcanic activity. Indeed, it is the different manifestations of the same volcanic activity that draw both scientists and tourists to the site. During the discussions concerning the establishment of an Antarctic Specially Managed Area at Deception Island, tour operators stated that they have been visiting this site for over thirty years and that they should not be excluded from doing so in the future (R. Roura, personal observation, February 2002).

by several (claimant and non-claimant) consultative parties.[71] While various claimant states have expressed concern about certain tourism and jurisdictional issues, positions may not necessarily be clear-cut or adhere to a claimant-non-claimant divide. Different states may promote their interests in different ways. For instance, Chile – a claimant state that established the first Antarctic hotel adjacent to one of its stations in the 1980s (Hansom and Gordon, 1998: 253) – said at the ATME that "in its opinion the Antarctic Treaty System did not prohibit building a facility and indicated that according to Chilean legislation a public facility could be leased or made over as a concession to a tour operator." (Norway, 2004: paragraph 37). In Cape Town, there appeared to be little interest in discussing this issue in detail, but New Zealand is likely to submit a more developed paper at the next ATCM in Stockholm.

3.9 CONCLUSION

Tourism in the Antarctic is increasing rapidly. Within less than fifteen years, it has increased from a total of approximately 4800 tourists in 1990-1991, to more than 19,500 landing passengers in 2003-2004 (IAATO, 2004b: 17; Hemmings and Roura, 2003).[72] In view of the worldwide growth of nature-based tourism and specific developments in the Antarctic tourism sector (e.g., the start of fly-sail operations), Antarctic tourism is expected to increase even more. The members of the Antarctic Treaty System are actively discussing the need for measures in addition to the Protocol to manage Antarctic tourism. However, the measures that were adopted at the twenty-seventh ATCM and most of the options under discussion should be characterized as "conditions" that do not address the more fundamental questions facing Antarctic tourism in the medium and long term. The measures adopted by the consultative parties so far principally relate to the concrete problems that are experienced today (e.g., the costs of search-and-rescue operations for tourists in need of help) and for which sufficient proof is available. The additional measures adopted by the industry are intended to prevent or minimize the adverse impacts of tourism. Although IAATO's contribution deserves to be recognized, generally speaking, the measures that have been adopted do not substantially affect the possibilities for the current tourist industry to undertake the initiatives it wishes and to continue to grow.

---

71 See, for example, Norway (2004: paragraph 35): "Germany ... suggested that permanent land-based tourism was not consistent with either the Antarctic Treaty or the Environmental Protocol and was also not in line with German domestic law because all permits must be restricted to a specific period."; and Norway (2004: paragraph 36): "Norway expressed support for the New Zealand view that the issue of land-based tourism activities has not only environmental ramifications that need to be taken into account, but the matter was also a policy related [one] and raised issues such as sovereignty and jurisdiction."

72 Note that the estimate of total passengers for the 2003-2004 season, including those not landing, amounts to over 27,000 (*2003-2004 Summary of seaborne, airborne, and land-based Antarctic tourism*, available from <http://www.iaato.org/tourism_stats.html>, accessed 23 September 2004).

In our view, the conclusion is justified that the consultative parties and the industry – through IAATO –are applying the precautionary principle only to a limited extent. The above discussion on Antarctic tourism shows the potential value of the precautionary principle for Antarctic management and refutes the assumption that application of the principle always results in extreme solutions, such as completely banning particular categories of human activities. We do not advocate a total ban on tourism in Antarctica. However, in our view more strategic policy options and management measures should be considered to prevent irreversible damage in the future.

As mentioned above, these options and measures could be devised in response to scenarios projecting the development of Antarctic tourism for the next ten years. These developments and related uncertainties about potential cumulative impacts may require, now or in the near future, the adoption of restrictions on Antarctic tourist activities. Examples of such restrictions include limiting such activities at particular sensitive sites; prohibiting visits to "new," previously unvisited sites; prohibiting the construction of tourism infrastructure ashore; and prohibiting or limiting certain new types of activities, such as fly-cruise operations. In the longer term, it may even be necessary to consider concentrating tourist activities in designated, well-managed areas. The involvement of the industry in debates about the management of tourism is necessary and its initiatives on self-regulation are welcome, but the consultative parties should also be aware of the possible limitations of self-regulation in view of the commercial interests involved.

Ultimately, the objectives, purposes, and principles of the Antarctic Treaty and its Protocol should guide tourism management strategies and policy. The precautionary principle may prove to be useful and steer us through uncertainty.

3.10 ACKNOWLEDGMENTS

The authors wish to thank Dr. Alan Hemmings (University of Canterbury, New Zealand) for many inspiring discussions on Antarctic tourism issues, and Erik Jaap Molenaar for useful comments on an earlier manuscript. Ricardo Roura acknowledges the support of ASOC to attend the meetings mentioned in this Chapter.

4 Monitoring the transformation of historic features in Antarctica and Svalbard: Local processes and regional contexts†

ABSTRACT

Historical sites in Antarctica and Svalbard contain the material remains of past activities of exploration and exploitation of these regions. These sites have been subject to transformation by cultural and non-cultural (natural) processes since their abandonment to the present. For research and management purposes it is important to monitor and explain these changes. This article focuses on the transformation of historic features in Antarctica and Svalbard as assessed through repeat photography. Seven historical features were selected representing a range of site types and past and present site functions. Data collection was based on the opportunistic reproduction of photographs of historic features taken up to 20+ years previously. Data analysis was performed using the concepts of site formation processes developed by M.B. Schiffer (1983, 1987). Time-serial changes were observed in the seven photo-couples examined in the present instance. No feature degraded significantly during the monitoring period; rather, several features were restored in different ways. Changes were interpreted to result from a range of cultural processes (including conservation, research, and tourism) and natural processes (mainly wind action). Local changes take place in the context of broader regional developments in Antarctica and Svalbard. Despite the "time capsule" narratives about some sites, historical sites in the Polar Regions are dynamic entities that not only reflect the past as it once was but are also a window onto the present.

4.1 INTRODUCTION

The exploration and exploitation of the Polar Regions has left a legacy of material remains spread over the landscape. These include exploration remains such as base camps, wrecks of ships, airships, docks and mooring masts; exploitation remains including huts and depots of hunters and trappers, and ruins of structures associated with whaling, sealing, and mining; and graves, crosses and other memorials. To the casual visitor some of these sites may appear to be timeless and unchanging, a capsule of times past that has apparently been frozen in the polar environment. This

† Published as:

Roura RM (2010): "Monitoring the transformation of historic features in Antarctica and Svalbard: Local processes and regional contexts." *Polar Record 46*: 04, 289-311. Reproduced with permission.

perception is reflected, for instance, in narratives about huts from the early period of Antarctic exploration, or the "heroic era." "Visitors to the huts are fascinated by finding the legendary explorers" belongings preserved *as though they had just walked out*" (NHM undated; original emphasis). However, historical sites are subject to transformation by cultural and non- cultural (natural) processes since the abandonment by their original occupants right up to the present (Schiffer, 1983, 1987). While the polar environment protects many sites from the type of decay encountered in temperate regions, many have experienced gradual but significant deterioration (Blanchette *et al.*, 2002, 2004), which has sometimes been linked to regional warming (Barr, 2008). Human presence in the Polar Regions has increased over the recent past, with consequential pressures not only on the wilderness and natural environment (Tin *et al.*, 2008) but also on the tangible cultural heritage.

For historical and scientific research it is important to assess the status and change of historic sites, defined here as locations that demonstrate past human activity as evidenced by the presence of artefacts, ecofacts, features, structures, or other material remains (Kipfer, 2007: 50), and to unravel the different processes of site transformation. This knowledge may assist in informing decisions about site conservation and management.

This article focuses on the recent transformation of several historical sites in Antarctica and on the archipelago of Svalbard, as assessed through ground based repeat photography. Antarctica and Svalbard are inherently different but have a number of similarities that set them apart from of the rest of the Polar Regions. Both regions had no indigenous people and following discovery in 1596 in the case of Svalbard and no later than 1820 in the case of Antarctica they experienced successive waves of activity including exploration, natural resources exploitation, science, and tourism (Sugden, 1982). The sequence and tempo of events were different in each case and this is reflected in the material cultural record. The earliest known historical sites in Svalbard date from the 17th century, while those in the area covered by the Antarctic Treaty and related instruments date from the early 19th century (see e.g. Marstrander, 1999; Jørgensen 2005; Harrowfield, 2005; Stehberg *et al.*, 2008). Some historic events in Svalbard (such as mining, and active warfare during World War II) have a limited or no parallel in the Antarctic.

The territorial tensions that resulted from the *terra nullius* status of these regions were addressed by the 1920 Spitsbergen Treaty and the 1959 Antarctic Treaty. These treaties entered into force in 1925 and 1961, respectively. The former treaty provides for Norwegian sovereignty over the archipelago whilst maintaining equal rights for the nationals of all contracting parties, subject to some Norwegian laws (Mathisen, 1954; Ulfstein, 1996; Pedersen, 2009). Article IV of the Antarctic Treaty "freezes" territorial claims in Antarctica while the treaty is in force (Beck, 1986; Triggs, 1987).

Comprehensive environmental protection regimes have been established for both regions, and in each region the same instrument that protects the natural environment also protects historical sites. These are protected as "cultural monuments" in Svalbard and as "Historic sites and Monuments" (HSMs) in Antarctica. The 2001 Svalbard Environmental Protection Act automatically protects structures, sites, and movable historical objects predating 1946, as well as cultural features such as human graves and certain animal slaughtering sites, irrespective of their age. Cultural remains postdating 1946 that are of particular historical or cultural value may also be protected by a decision of Norway's directorate of cultural heritage. Under the 1991 Protocol of Environmental Protection to the Antarctic Treaty (the Protocol), Antarctic sites or monuments of recognised historical value may be designated as HSMs and also as part of protected areas (for details see Bizzarri, 2006; Chile, 2007a). Listed HSMs may not be damaged, removed or destroyed. In addition, remains predating 1958 whose existence or present location is not known have a degree of protection, but there is no automatic long-term protection of Antarctic cultural remains on the basis of their age alone. Past activity sites that are not protected as HSMs are subject to removal under the Protocol's waste management requirements.

This article aims to answer the following questions. What are the changes experienced by historical sites in the Polar Regions in the contemporary past? What are the causes of these changes? And what do these changes tell us about broader contemporary developments in the Polar Regions? The article first discusses repeat photography as a method of monitoring the transformation of historic sites, based on Roura (2008b). This is followed by seven case studies documenting changes of historical features in the recent past. The article then discusses recent developments in Antarctica and Svalbard that may be linked to site-specific changes. While the entire history of each site is briefly reviewed, the monitoring period focuses on the recent past (up to 20+ years). This has been a period of substantive change in Antarctica and Svalbard including the establishment of environmental protection regimes, the expansion of polar tourism, advances in heritage conservation, and emerging evidence of climatic changes. These and other contemporary processes may directly or indirectly contribute to the transformation of historical sites in the sense of greater protection or ultimate destruction. Knowledge about recent changes is particularly relevant to the contemporary management of historic sites, even though the interest in historic site transformation transcends management applications.

4.2 METHODS

4.2.1 Repeat photography as a research method

Historical sites in Antarctica and Svalbard may range from large, complex sites comprising many buildings, structures, and ruins, to sites with virtually no traces of past human activity. Change at a site can be rapid and obvious. A wood frame building may be destroyed in a storm, for instance, or the remains of an expedition camp may be salvaged or reused in another context. Change can also be gradual and subtle, as reflected in the "frozen in time" narratives about huts from the "heroic era" of Antarctic exploration, which suggest virtually untouched remains. Over time, however, cumulative changes may well result in significant site transformation. The opportunistic use of ground based repeat photography provides a relatively simple way of monitoring the transformation of these sites.

Repeat photography, or photo monitoring, means using photographs taken at a specific site to monitor conditions or change over time (Hall, 2001). This technique is often used to assess changes to natural and cultural landscapes (e.g., Rogers *et al.*,

1984; Ahlstrom, 1992; Nyssen *et al.*, 2009). The method may use photographs specifically collected for monitoring purposes, or the replication of historical collections. In addition to scientific assessments repeat photography has been used to highlight and publicise decadal or rapid environmental changes (Greenpeace, 2002; ICIMOD, 2008; EIS, n.d.). In the polar environment repeat photography, whether aerial or ground based, has been used for a broad range of purposes including assessing the effects of regional warming on glaciers (Fox and Cziferszky, 2008), lichen growth (Smith, 1995), the environmental footprint of contemporary activities (New Zealand, 2006), and the status and change of historical sites (Governor of Svalbard, 1999a). Analysis methods for ground based repeat photography include using grids or spatial analysis software to measure surface changes of the features of interest (Hall, 2001, 2002; Tape *et al.*, 2006; Roush *et al.*, 2007), expert rating of key indicators (Nyssen *et al.*, 2009), and broad whole photo comparisons (Tape *et al.*, 2006; Masiokas *et al.*, 2008). Overall, while the basic technique is the same, data collection and analysis methods may differ significantly depending on the purpose of monitoring and the quality of the photo couples.

Obviously repeatability is a key element in repeat photography. This requires a monitoring protocol and sup- port documentation that enables locating historic vantage points and replicating earlier photographs (Hall, 2001, 2002). Precise replication, that is, when photographs of a certain feature are taken from exactly the same camera position and direction using identical camera formats, may be achieved through putting in place two permanent markers, one indicating the camera position and the other the direction of the photograph from that position (see Hall, 2001 for details; see also Smith, 2007 for an alternative method). In contrast, "casual repeat photography" refers to the matching of photographs not specifically collected according to a monitoring protocol, but that nevertheless overlap sufficiently so that comparisons are possible (Ahlstrom, 1992; Cameron, 1999). In the Polar Regions photographs of important sites may not have been collected with monitoring in mind so that there is no record of vantage points or other technical information; site visits may be infrequent, opportunistic, and brief; and environmental regulations may discourage the establishment of permanent markers. Under those circumstances casual repeat photography may be a practical alternative. Photo couples alone, however, are insufficient to assess the transformation of sites; a conceptual framework for analysis is also required.

4.2.2 Data collection

Data collection for this research was based on the reproduction of existing photographs not specifically collected for repeat photography. Research focused on specific features representing a fraction of larger sites. Seven historic features were selected representing a selection of site types, conditions, and past and present site functions for which a set of comparable photo couples had been assembled (Table 1). Matching photographs were taken either in the context of dedicated fieldwork or during opportunistic visits to the sites. Additional matching photographs were found in personal collections, publications, and on the World Wide Web. The earlier or "before" photographs, dating back to up to 20+ years, were reproduced in order to obtain recent or "after" photographs (Table 2). The outcome was several sets of paired photographs or photo-couples, which were broadly comparable but not identical. The relatively short time period covered by repeat photography was meant to illustrate recent changes; indeed, some historic features had long been abandoned and were already ruins when monitoring began. The information contained in the photographs was complemented with on site observation of contemporary human behaviour and activities, site surveys of different kinds, interviews, and information from secondary sources. These methods have been summarised in Chapter 2 and are discussed elsewhere in this thesis. Site visitation data were obtained from IAATO (n.d.), New Zealand *et al.* (2006), and Governor of Svalbard (2006a).

4.2.3 Data analysis

The interpretation of changes was made using the concepts developed by Schiffer (1983, 1987) to identify the factors that create the archaeological record, or site formation processes. In cultural formation processes the agency of transformation is human behaviour. In non-cultural, or natural, formation processes the agency of transformation is provided by the processes of the natural environment (Schiffer, 1987: 7). Thus, any cultural or natural process acting on a historical site contributes to the transformation of that site and adds its own patterns over previously existing patterns. The effect of these processes would be to add to, remove from, or change the condition/integrity, location or orientation of artefacts within a historic site, and to introduce changes to the way individual artefacts relate to each other in space.

Data analysis evaluated noticeable macroscopic changes to individual components of historical features. An inventory was made of the separate objects that could be identified with certainty in both the "before" and "after" photographs. For this, 20×30 cm print outs of the photographs were used. This step identified a sub sample of the total population of cultural objects in the photographs. The contour of each of the objects included in the inventory was drawn in a sheet of acetate using a permanent marker. The lowest definition photograph of the set was used as a base for the drawing. Each of these objects was individually numbered. Buildings and other large features were counted as a single object.

Individual objects were classified on the basis of their observable attributes: material (and hence density), size, and shape. Each individual object was compared in the "before" and "after" photograph to assess whether or not it had changed over time. Changes were classified as additive, reductive or *in situ*. By this it is meant that objects had been added or removed to the photograph's frame, or that had remained within the photograph's frame but had changed in size, shape, form, position, or spatial location (Tables 4.3 and 4.4). In case of uncertainty (due to parallax or definition differences in the photo couple) the object was dropped from the analysis. The changes were drawn in the acetate sheet outline using a three-tone pattern, which eliminated unnecessary detail.

A judgement was made of the degree of disorder introduced by the changes through a whole photograph assessment (Table 4.4). Disorder, in this context, concerns the formal, spatial and relational arrangement of material cultural elements (as defined by Schiffer, 1983, 1987). For instance, greater disorder results from the breakage of individual objects (formal), or from disaggregating groups of objects that were previously together (spatial and relational). Greater order results from repairs or maintenance (formal), or the aggregation, alignment or stacking of objects (spatial and relational).

4.3 CASE STUDIES

The discussion now turns to actual examples of recent changes to historical features in Antarctica and Svalbard. The features covered approximately 1–100 m^2 in surface area and were in most cases part of larger, complex sites up to about 1 km^2 in surface area that were active at various times during 1820s–1950s. All but one of the sites discussed here were legally protected under applicable legislation at the time of writing.

4.3.1 Building foundations with cooking stove, Ny-London, Svalbard

Ny-London (Port Peirson) on Kings Bay, Svalbard (78° 57' 49" N, 12° 02' 56" E), is the site of an abandoned marble quarry of the Northern Exploration Company Ltd. of England. The quarry operated intermittently between 1911 and 1920. The venture was unsuccessful as the quarried marble blocks crumbled during storage due to the effect

Fig. *4.1a-b – House foundations with a cooking stove, Ny- London, Svalbard, in the late 1990s and in 2007. 1990s photograph by Governor of Svalbard. 2007 photograph by author.*

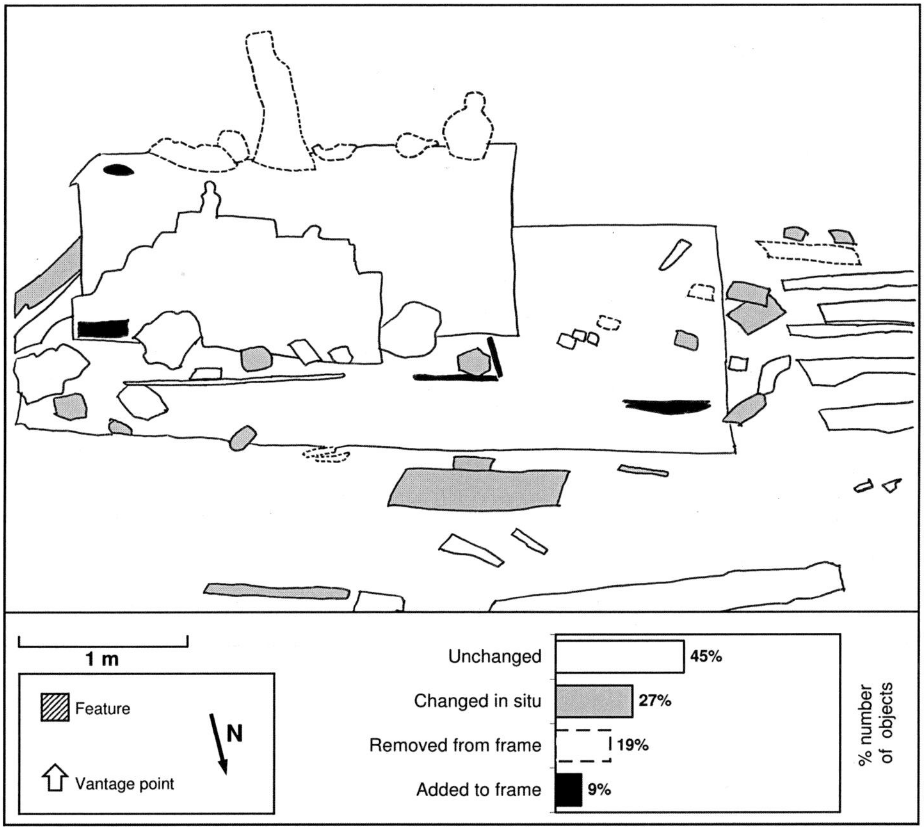

Fig. 4.2 – Overview of changes to house foundations with a cooking stove, Ny-London, Svalbard, between the late 1990s and 2007. Sketch based on photograph from the late 1990s. Key: white = no changes; grey = changes within the frame; dashed line = removed from frame; black = added to frame. Sketch by author.

of frost. Following a 1933 visit Ny-London was described as " ... a few tumble-down ... huts and machine houses, the relics of yet another unsuccessful mining venture" (Polunin, 1945: 90). The settlement's materials and equipment were scavenged for reuse elsewhere, and in the 1950s most buildings were taken to Ny-Ålesund where some of them are still in use. Currently most structures at Ny-London, except two wood frame buildings, are standing ruins or ruins. There are also two quarries, a gravel rail bed, refuse dumps, and artefact scatters. The two standing buildings are maintained and still used periodically for shelter and recreation by government field officers and residents of nearby Ny-Ålesund. Organised tourist landings also take place regularly through the summer. Visitor numbers have increased from less than 100 visitors in 1996 to more than 1,500 in 2006. The site is automatically protected under the 2001 Svalbard legislation (Governor of Svalbard, n.d.; Roura, 2008b).

A brochure produced in the late 1990s (Governor of Svalbard n.d.) contained nine photographs illustrating eight different cultural features of Ny-London. These

Fig. 4.3 – *Tourists visit Ny-London, July 2008. Note items on the top of the stove that were absent in 2007. Photograph by M. Loonen.*

photographs were reproduced opportunistically during fieldwork in July 2007 (Roura, 2008b). For the purposes of this article the analysis will focus on a single feature, a house foundation in which a cooking stove remains standing (78° 57' 49" N, 12° 03' 06" E) (Fig. 4.1). This feature stands out in the landscape and is close to the beach at which most visitors land and to the buildings used as field accommodation. It is therefore one of several cultural features of Ny-London where visitors tend to congregate. The definition of the "before" photograph was poor; consequently, the analysis focused only on the most obvious changes (Fig. 4.2). Among some minor changes to small objects, some objects placed on top of the stove visible in the late 1990s, which at some point had been used for shooting practice, had been removed from that position some ten years later. However, an on-site examination of the stove showed that in 2007 these objects were still on the site albeit not visible from where the photo-couple had been taken. Some objects were lying on the building floor and others had been stored inside the cooking stove. This suggested that the rearrangement of objects had resulted from a combination of wind action and human agency.

A photograph of the same site taken in 2008, that is, a year after the "after" photograph described above showed further changes to the site (Fig. 4.3). Some objects had been replaced on top of the stove, and other objects had been removed from sight or had changed position within the frame. Overall, the condition and integrity of the site did not change significantly between 1997 and 2007, or between 2007 and 2008. However, it is apparent that the site is not static; that it is regularly visited; and that cultural remains that had once been used as shooting targets are now put on display for visitors.

4.3.2 Entrance to "Josefine" coal mine, Ny-Ålesund, Svalbard

Ny-Ålesund (78° 55' 38" N, 11° 56' 00" E) is a former Norwegian mining settlement established in 1916. Between 1926 and 1928 it was the base for several attempts to reach the North Pole by air. Since 1964 Ny-Ålesund has operated primarily to support science, and it has now become an international research centre. Ny-Ålesund is run by Kings Bay AS, a Norwegian crown company. The former mining area is located approximately one kilometre south of Ny-Ålesund between the lower slopes of Mount Zeppelin and the sea. The area covers approximately 1.1 km^2 at the surface level and contains the remains of nine mines. The older mines, which are located uphill some 50–100 metres above the sea level, were developed between 1916–1923, while lower altitude mines were developed between 1941–1946. A mining disaster on 5 November 1962 resulted in the death of 21 people, and caused the end of coal mining at Ny-Ålesund (Hanoa, 1993; Arlov, 1996; Governor of Svalbard, 1999b).

Surface features in the mining area include some surviving structures of the mines, such as entrance points and ventilation shafts; transport infrastructure such as train tracks, skips and bridges; waste materials such as wood piles, and 200 litre fuel drums; artefacts such as shovels, borers and coal cars; landscape modifications such as gravel roads and rock dumps; and memorials to the dead. The mining area is also a gravesite for those miners who died there and whose bodies were not recovered. The area is protected under the provisions of the 2001 Svalbard legislation and by Kings Bay AS local land planning regulations. Nothing can be removed from the area and nothing can be built within its limits (B. Paulsen, personal communication to author, 9 July 2007). There is limited contemporary activity at the mining area itself, which is currently off limits for organised tourism. Recent management of the mining area has been based on the main principle of leaving things as they are. Between 1997 and 2004 the historic significance of cultural remains was assessed, some old dumpsites and contaminated sites that posed a hazard to the environment or to wildlife were cleaned up, and some historic features were restored (Sander *et al.*, 2006: 15–17).

In June-July 2007 the author carried out a pedestrian survey of the mining area, documenting the various surface features of the area, and interviewed mining veterans. In 2008 a colleague, M. Loonen, replicated nearly thirty photographs dating from 1998 (reproduced in Governor of Svalbard 1999b). This case study concerns the entrance of "Josefine" (78° 54' 56" N, 11° 55' 11" E), a coalmine established in 1921 and abandoned in 1924 (Fig. 4.4). The mine entrance itself is filled up with ice, but the wooden structure above is still standing. It is one of the best preserved mining structures in the area. In the "after" photograph it is apparent that, among other minor changes to surface remains, part of the entrance roof has been restored (Fig. 4.5). This is counterintuitive in the context of the old mining area, which is a complex, seemingly untidy site that suggests ongoing degradation rather than renewal.

To the casual visitor the chaotic appearance of the mining area raises questions as to why such place on the edge of the Arctic wilderness is protected at all. Pearson and McGowan (2000) discuss criteria for the protection of historic coalmines and while they focus on Australian mines the basic concepts are broadly applicable elsewhere. Historical coal mines may be significant because of the attributes or combination of attributes of the coalmine above ground features that make it a good example of a

Fig. *4.4a-b – Entrance to "Josefine" coal mine near NyÅlesund, in 1998 and 2008. The arrow in the right photograph shows a wooden peg planted by mining veterans in 2006 to identify the site. 1998 photograph by Governor of Svalbard. 2008 photograph by M. Loonen.*

mine of a certain location or period. Coalmine features that are rare, uncommon or of particular interest because of their good condition, age, or combination of features demonstrating the mining operations may also be factors for conservation (Pearson and McGowan, 2000). However, given the overall poor condition of the mining remains in Ny-Ålesund the conservation criteria does not seem to apply to the more recent mines in the area (dating from 1946), although they may apply to the older mines (dating 1916–1923), whose surface features are somewhat better preserved and automatically protected under Svalbard's 2001 legislation.

Pearson and McGowan (2000: 156–157) further note that "[t]he significance of mining features which satisfy or do not satisfy the above attributes might be enhanced by association with specific historical events (such as major mine disasters), technological innovations developed at the site, or strong or special meaning for a particular community or group." In this regard, the significance of the old mining area at Ny-Ålesund

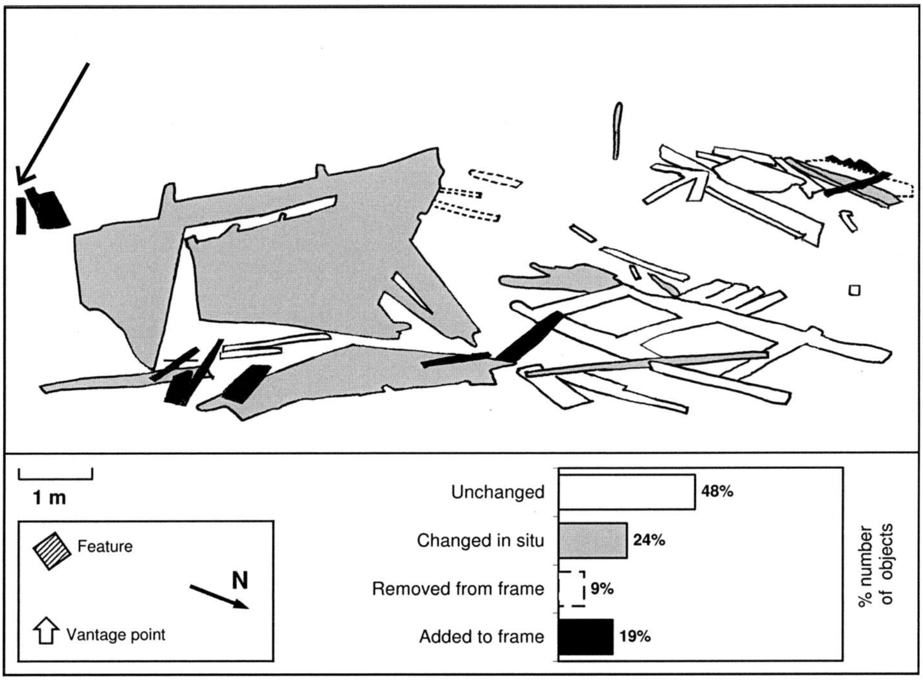

Fig. *4.5 – Overview of changes to the entrance to the "Josefine" coalmine near NyÅlesund between 1998 and 2008. The arrow in the 2008 photograph shows a wooden peg planted by mining veterans in 2006 to identify the site. Sketch based on 1998 photograph. Key: white = no changes; grey = changes within the frame; dashed line = removed from frame; black = added to frame.*

is manifold. First, 82 people died in mining accidents at Ny-Ålesund over the years, and as noted above the mining area still contains human remains. Secondly, the mining area itself is a reminder of significant events in the political history of Norway since the 1962 disaster resulted in the fall of the government. Finally, mining veterans (some of whom are survivors of the 1962 accident) constitute an active interest group that is concerned about the disappearance of both the physical remains of mining and the memory of the mining times. According to a mining veteran, nowadays residents and tourists alike know little about Ny-Ålesund's mining past. For them, Ny-Ålesund is primarily associated with exploration and science (L. Kristiansen, personal communication to author, 29 June 2007). Yet, for the miners the mining times were significant and represent a key part of their identity, even though some workers spent a relatively short part of their working lives in the coalmines at Ny-Ålesund (O. Bye, personal communication to author, 29 June 2007). As part of several initiatives to maintain the memory of mining times, in 2006 a group of veterans and their supporters marked with pegs and documented several sites of interest around the mining area so that their location would not be lost (Paulsen, 2006). One of those pegs, marking the entrance to "Josefine", is visible in Fig. 4.4b.

4.3.3 Cross at Observation Hill, Ross Island, Antarctica

Fig.4.6a-b – Cross at Observation Hill, Ross Island, Antarctica, in 1990 and late 2007. The cross is facing in opposite directions in each of the photographs, and it is shorter in 1990 than in 2007. Photographs by author.

The cross at Observation Hill (77° 51' S, 166° 41" E) was erected by members of the British Antarctic Expedition of 1910–1913, in memory of Captain Robert Falcon Scott's party, which perished on the return journey from the South Pole in March 1912. The cross is composed of two layers of jarrah (*Erythrophleum chlorostachis*), an Australian native wood, which are bolted together to double the thickness. Carved in the cross are the inscription "In Memoriam", the names and titles of the five members of the South Pole party, and the closing lines of Alfred, Lord Tennyson's *Ulysses*: "To strive, to seek, to find, and not to yield". The cross faces south onto the Ross Ice Shelf, the direction in which Scott's party left, never to return, and the original position in which the cross was placed on 1913:

> *Tuesday, January 22. Rousing out at 6 A.M. we got the large piece of the cross up Observation Hill by 11 A.M. It was a heavy job, and the ice was looking very bad all round, and I for one was glad when we had got it up by 5 o'clock or so. It is really magnificent, and will be a permanent memorial which would be seen from the ship nine miles off with a naked eye. It stands nine feet out of the rocks, and many feet into the ground, and I do not believe it will ever move. When it was up, facing out over the Barrier, we gave three cheers and one more. (Cherry-Garrard, 1937:567).*

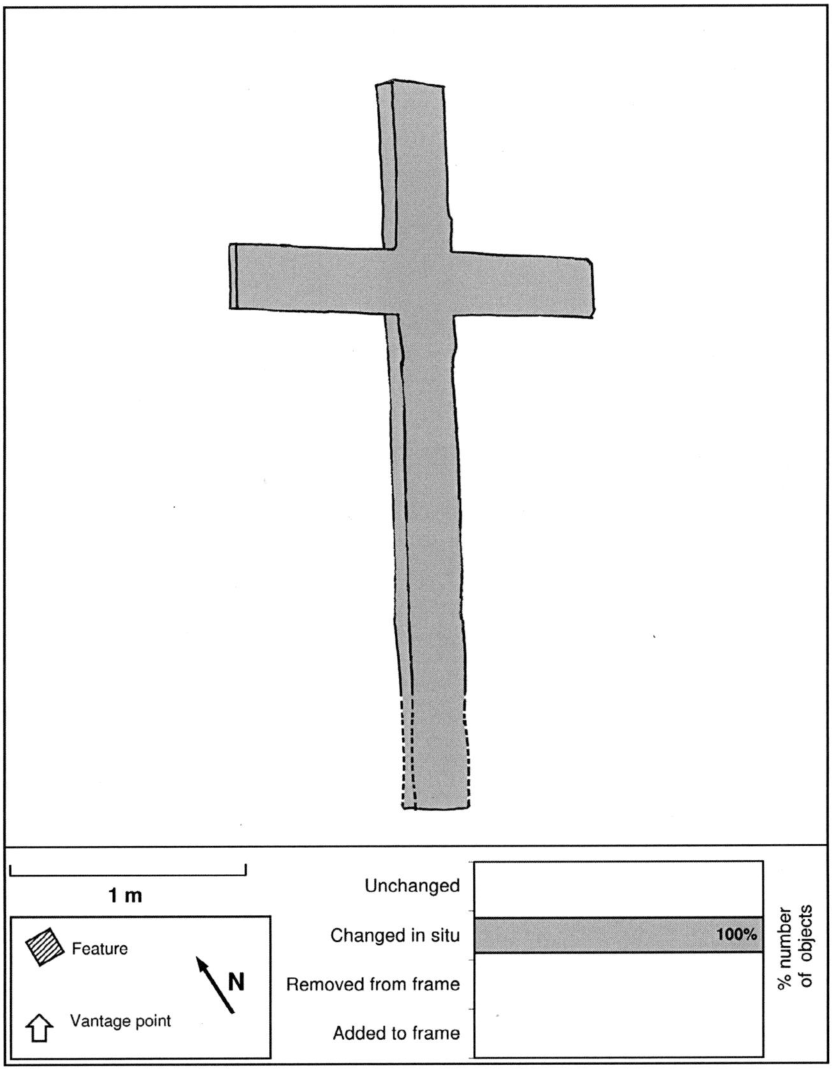

Fig. *4.7 – Overview of changes to the cross at Observation Hill, Ross Island, Antarctica, between 1990 and 2007. Sketch based on 1990 photograph. Key: white = no changes; grey = changes within the frame; dashed line = removed from frame; black = added to frame. Sketch by author.*

The cross has been protected as HSM 20 since 1972. The cross is adjacent to McMurdo station, the largest station in Antarctica, with some 250 winter residents and 1,000 summer residents. Observation Hill, as its name suggests, overlooks the surrounding area, and therefore is frequently visited by local residents. It is one of the few places in the area that residents are allowed to visit on their own. A few hundred tourists visit McMurdo every year, and presumably some of them are able to visit Observation Hill. In the 2007-2008 summer the author examined the cross and broadly replicated photographs taken in the course of several visits to the area through the 1990s. A visual examination suggested that the cross was in good condition, and it was apparent that it had been recently restored. The inscriptions are fading as a result of abrasion by wind transported ice particles and sediments. Some of the letters, which had originally been engraved in the wood and at some point painted white, are now protruding from the surrounding surface.

At some point in time the orientation of the cross was changed 180° from its 1913 position facing south, as described by Cherry-Garrard, to a north-facing position. The circumstances of this change are unknown to the author. The photo couple (Fig. 4. 6) shows that in 1990 the cross still faced to the north, while it now faces to the south, that is, it is back in its 1913 original position. In addition, the cross is now shorter than in 1990: the person standing under the cross in 1990 (Lillian Hansen) is approximately 1.7 metres tall, while the person standing in a similar position in 2007 (Chiu-Pih Tan) is some ten centimetres shorter. Estimating the distances between the highest point of the cross, and the top of the heads of the people below suggests that the cross was some 40cm longer in 1990 than in 2007.

The changes to the cross' orientation and integrity since 1990 (Fig. 4.7) may be explained by the fact that it was downed by hurricane force winds in June 1993. The cross was restored by members of the New Zealand Antarctic programme and transported back by foot to the top of the 230 m hill by a team from the nearby stations from the United States and New Zealand (NZAHT, 2008). There are some minor dents and scratches on the cross, some of which may be related to its downhill fall (or falls). At the back of the cross there are some faint traces of engraved graffiti. The cross stands on a concrete base, upon which some people have written down their names as the concrete set. These inscriptions are recent and presumably made by some of those who transported the cross back to the top of Observation Hill since they were made before the cement set. The fact that the cross is shorter now than in 1990 suggest that was broken off at the ground level when it fell.

Other traces suggested recent contemporary activity. The 2007 photograph shows a stain at the base of the cross, and in the field it was quite apparent that somebody had recently urinated there. This act could be an intentional desecration or just thoughtlessness. A few metres to the south of the cross there are two strings of Tibetan flags, which were not there in the 1990s, and which seem to reflect the travels, beliefs or fashions of McMurdo's inhabitants. The restoration and relocation of the cross suggest that its historical values are maintained and reinforced. However, the top of Observation Hill, because of its location and characteristics, seems to have acquired multiple functions that transcend the events of the "heroic era". These include being a place of recreation and refuge for McMurdo's residents; a place where contrasting religious symbols are placed; and even a suitable place for the exercise of bodily functions.

4.3.4 Scott's chart table at the Terra Nova Hut, Cape Evans, Ross Island, Antarctica

The *Terra Nova* Hut at Cape Evans (77° 38' S, 166° 24' E) was built in January 1911 by the British Antarctic Expedition in *Terra Nova*, of 1910-1913, led by Captain Robert Falcon Scott, RN. The Ross Sea party of Sir Ernest Shackleton's Imperial Trans-Antarctic

Fig. 4.8a-b – The chart table of R.F. Scott in the Terra Nova hut at Cape Evans, Ross Island, in 1990 and late 2007. Photographs by author.

Expedition of 1914-1917 subsequently used it as a base. Afterwards the hut was abandoned and, according to Harrowfield (2005), not revisited until 1947 (see also Roura, 1996). For the following decades the activities at the hut interspersed visitation by United States and New Zealand expedition members and sporadic attempts to clean up and restore the site (Harrowfield, 2005).

The hut is weathered but still structurally sound. However, over the years it has deteriorated markedly due to snow accumulation, modern interventions, and the decay or disappearance of artefacts, sometimes through souvenir collection (NZAHT, 2004). The hut and surrounding area contains approximately 8,000 artefacts, some of which have been described as "iconic", that is, " ... those elements of the building fabric or artefacts that are of such extremely high cultural heritage value that their loss would result in a serious diminution of the heritage value of the site" (NZAHT, 2004: 87). The

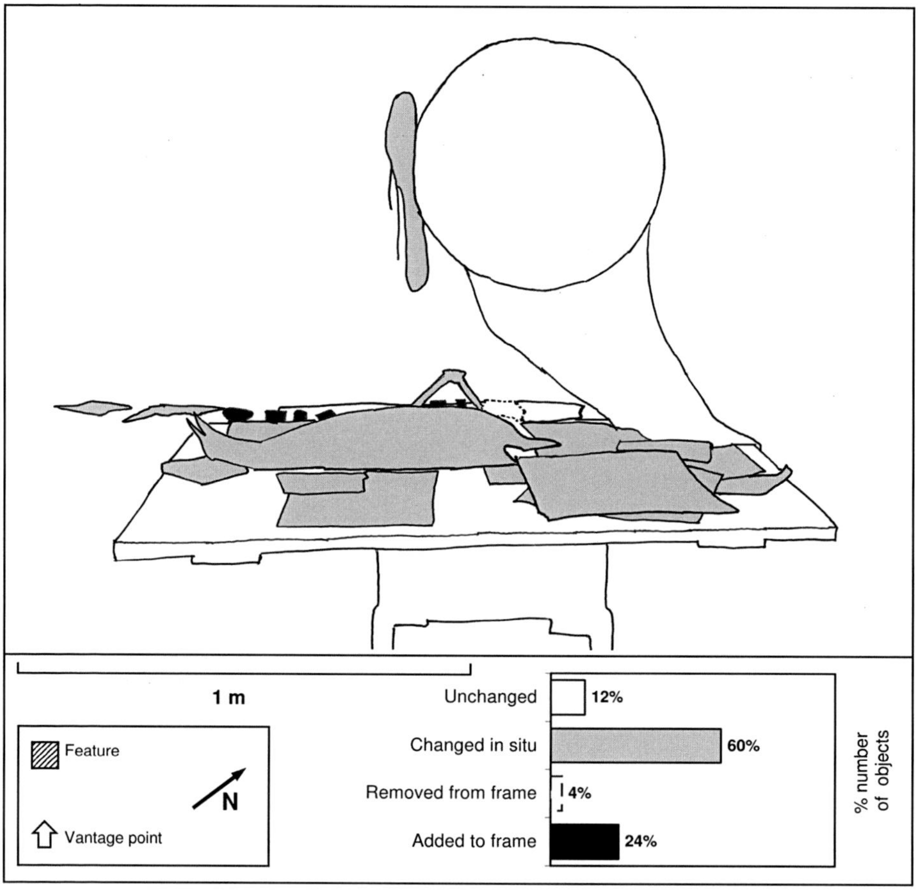

Fig. 4.9 – *Overview of changes at the chart table of R.F. Scott in the Terra Nova hut at Cape Evans, Ross Island, Antarctica, between 1990 and late 2007. Sketch based on 1990 photograph. Sketch by author. Key: white = no changes; grey = changes within the frame; dashed line = removed from frame; black = added to frame.*

hut has been protected as HSM 16 since 1972. Additional protection to the hut and surrounding area is provided by designation of the site as Antarctic Specially Protected Area (ASPA) 155 since 1997. Visitors require a permit issued by competent authorities to enter the ASPA. Visitor numbers are capped at 2,000 annually.

Conservation, research and visitation are the main contemporary activities at the hut. The hut is subject to a conservation project conceived and carried out by the New Zealand Antarctic Heritage Trust, a charity formed in 1987. Conservation work, which is its current phase began in 1999, is planned to last until 2014 (NZAHT, 2004; NZAHT, undated). Approximately 1200 people visited the *Terra Nova* Hut every year through the early 2000s, two thirds of whom were residents from McMurdo Station and Scott Base, located some 30 kilometres away. The number of tourists in organised cruises, however, is relatively low at about 400–500 per annum.

During a brief visit in December 2007 photographs taken in the course of several visits to the hut through the 1990s were broadly replicated (Fig. 4.8). One of those sets of photographs is that of Scott's chart table. Entering the private space of RF Scott, the inner sanctum of the *Terra Nova* hut, is arguably the climactic moment of a visit to the hut. The cubicle contains Scott's bed, bookshelf, his Mackintosh raincoat, and other personal objects. The chart table with the stuffed Emperor penguin lying on it and a collection of documents and other objects, under the faint light of a small double glazed window, occupies a dominating place in the cubicle. A well-known 1911 photograph by Herbert Ponting shows Scott writing his journal at the chart table while smoking a pipe. In that picture the table is clear of objects, barring Scott's journal and a tin of tobacco. The table Scott used in 1911 does not appear to be the same as the one occupying the same place decades later. (Wheeler (1999: 55) suggested that a replacement had been brought over from Shackleton's hut at Cape Royds).

Comparing the 1990 and 2007 photographs it is apparent that the somewhat disorderly arrangement of objects on the chart table has been replaced by a tidy "tablescape" (Fig. 4. 9). Cultural processes are clearly the main processes of change in this case, although they may have in turn been triggered by the effect of the environmental conditions within the hut on the objects lying on the table (soft rot, fungi, corrosion and other problems affect the objects inside the hut, see e.g. Blanchette *et al.*, 2002, 2004). Currently the hut is subject to an ambitious conservation programme that aims to stabilise, maintain and repair the hut and its artefacts. This may involve "... to undo inappropriate interventions during the modern era and reflect more authentically the heroic era use of the hut" (NZAHT, 2004: 101). Other features of the hut and many of its artefacts are now being restored and rearranged in comparable ways.

The changes illustrated in this set of photographs reflect carefully planned conservation actions that contrast with some of the *ad hoc* measures taken in the past (see e.g. Harrowfield, 2005). However, the ongoing conservation interventions have been criticised as too intrusive (Save the Huts undated). From this critical perspective the *Terra Nova* hut and other huts from the "heroic era" are now mediated environments rather than the time capsules they once were and still purport to be. The disassembling and reordering of objects may give the impression that the hut is now in the same state that it has been since the "heroic era", when this is no longer the case: "[t]he hut will be changed to the [New Zealand Antarctic Heritage] Trust's version of its history – not to any "authentic" 1909 state." (Save the Huts, undated). A further concern is the loss of value and information, in a scientific sense, which results from changes such as removing the patina of time from historic objects during the conservation process, the rearrangement of objects, and the disposal of objects in poor condition.

Regardless of the position adopted with regards to the conservation of the hut, what is perhaps most remarkable is that the measure of disorder in the arrangement of objects in a remote hut in the Antarctic, last inhabited in 1917, is not increasing but decreasing. This challenges the notion of what constitutes a "time capsule". Rather, the hut as it is now reveals much about contemporary notions of what the heroic past "should" look like and not solely about the actual "heroic era" of Antarctic exploration.

4.3.5 Former cemetery at Whalers Bay, Deception Island, Antarctica

Deception Island in the South Shetland Islands is a ring shaped island about 12km in diametre, with a narrow entrance into Port Foster, a central landlocked harbour (a flooded volcanic caldera). Activities in the island have included sealing (1820s), industrial whale processing (1912–1931), scientific base operations (from 1944), and organised tourism (from 1958). A 1969 volcanic eruption forced the evacuation of the island and caused a significant destruction of some active and abandoned buildings and other features.

Whalers Bay (62° 59' S, 60° 34' W) on the eastern part of Port Foster, contains the remains of a whaling factory, which was abandoned in 1931, as well as the remains of the British Base B, which was abandoned following the 1969 eruption. One of the whaling factory's features was a cemetery (approximately 62° 58' 38" S, 60° 33' 39" W) established in 1908. The cemetery contained the graves of 34 Norwegian, Swedish, Chilean, and Russian whalers buried there in the first part of the 20th century, and a British burial from the 1950s (Headland, 2002: 93) although some believe that the correct number of burials is 36 rather than 35 (J. Berguño, personal communication to author, 7 July 2009). A memorial was erected for ten men lost at sea, nine of which bodies were not recovered (Headland, 2002: 93). The cemetery was described thus following a visit in 1947:

> *A metal fence, with its door open for the rare visitors, surrounds the grounds ... in the centre [of the cemetery] a crumbling stele commands over five wavy rows of wind-weathered graves. There are humble graves with a coarse wooden cross, some with faint names and dates that nobody could read any longer ... A mound keeps the ashes [sic] of the victims of the Graham, sunk in 1924 ... The most moving grave among all is that of ... a [16 years old], just a child, who lies [there] in his earthy nest. (Orrego Vicuña 1948: 117–118; author's translation).*

During the 1969 eruption the cemetery was partly buried and partly removed by a lahar (a mud and ash flow caused by a volcanic eruption). Headland (2002) speculates that some memorial objects fell down and that some low lying grave markers may still be in their place beneath the surface of the volcanic deposits.

The remains of a coffin resurfaced from the ground years after the eruption, possibly not in its original location but where the lahar took it. A cross remembering Hans Gulliksen, a Norwegian carpenter, was sent by his relatives around 1972 to replace the original cross, which had disappeared during the eruption (Fig. 4.10a) (Barr and Downie, 2002). In February 2002 a wooden cross situated on a hillside next to Argentina's Base Decepción in Fumarole Bay, about ten kilometres northwest from Whalers Bay, was identified to be that of the Norwegian whaler Peder Knapstad (Fig. 4.11a). The cross was relocated to Whalers Bay and placed close to the coffin (Fig. 4.11b) (Barr *et al.*, 2004).

The site of the cemetery was designated as HSM 31 in 1972. All historic sites at Whalers Bay were subsumed into HSM 71 in 2005. Following its designation as an Antarctic Specially Managed Area (ASMA) in 2005 under the provisions of Annex V of the Protocol, Deception Island is managed internationally. Organised tourism landings

Fig. 4.10a-b – Gulliksen's cross at its original location in Whalers Bay, Deception Island, Antarctica in 2002 and after relocation within Whalers Bay in 2007. Photographs by A. Bendala.

frequently take place at Whalers Bay, with over ten thousand visitors landing there every summer. Other visitors to Whalers Bay include national programme personnel conducting scientific research or on recreational visits, and yacht crews. The island's management plan contains provisions for the conducting of tourism landings, complemented with non binding site specific guidelines.

Several comparable sets of photographs taken by the author and other visitors between 2002 and 2006 were collected for this case study (Figs. 4.10 and 4.11). Since 2001– 2002 Gulliksen's cross has been removed from its original location some 400m away and placed next to Knapstad's cross and the coffin remains (Fig. 4.12). Small

Fig. 4.11a-b – Knapstad's cross at its hillside location at Fumarole Bay, Deception Island, Antarctica, in February 2002, and just after being re-erected at Whalers Bay on 18 February 2002. Photographs by author.

rock cairns have been built under each cross in the past few years, adding to the rocks initially put in place to support them.

The changes observed between 2001–2002 and 2004– 2005 (Fig. 4.13) result from active management and tourism. Barr and Downie (2002) recommended that all mortuary elements should be placed together and apparently this has been acted upon (Figs. 4.10b and 4.12). The original cemetery has not been reconstructed but has been somehow recreated approximately where it used to be. The cairns most likely developed as a cumulative feature built by accretion. Visitors at polar historical sites follow a basic behavioural repertoire that includes making a record of the visit either by taking something from the site (e.g., photographs) or by leaving something on the site (e.g., a signature in a visitors book) (Chapter 8). It is interpreted that many of the rocks in these cairns would have been placed spontaneously by successive visitors to

Fig. 4.12 – The complex of mortuary features at Whalers Bay, Deception Island, Antarctica, early in 2006. Visitor footprints are apparent on the ground around the features. Photograph by L. Hacquebord.

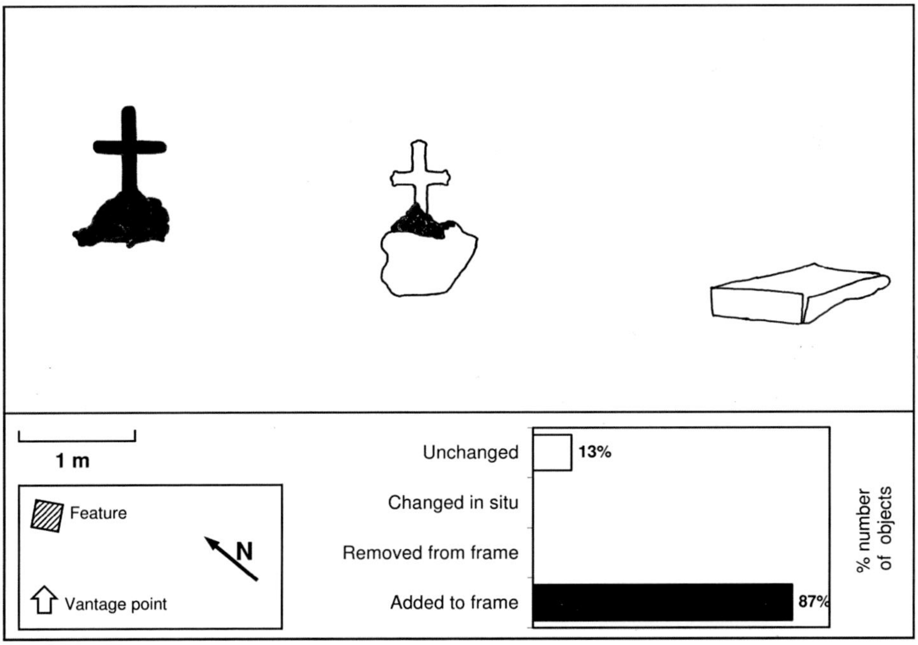

Fig. 4.13 – Overview of changes to the complex of mortuary features at Whalers Bay, Deception Island, Antarctica, between 2002 and 2006. Sketch based on 2006 photograph. Assumes that the rocks in the cairns were added to the frame. Sketch by author. Key: white = no changes; grey = changes within the frame; dashed line = removed from frame; black = added to frame.

Whalers Bay as a record of their visit and as a mark of respect and permanence (what Schiffer (1987: 80) calls "offertory shrine").

4.3.6 Sealer's shelter, Fildes Peninsula, King George Island, Antarctica

The remains of early 19th century sealing shelters in the South Shetland Islands are the oldest known sites in Antarctica (Stehberg *et al.*, 2008; see also Lewis Smith and Simpson, 1987). The South Shetland Islands were discovered in 1819. Almost immediately the populations of fur seals *Arctocephalus gazella* were intensively exploited. The first sealing period lasted only from 1819 to 1827, peaking in the 1820-1821 season, and ended with the commercial extinction of the fur seals, although later in the century the exploitation was restarted on several occasions (Pearson and Stehberg, 2006). The sealers" *modus operandi* has been described as follows:

> *Gangs of sealers were put ashore from a ship or sent off in smaller boats to seek beaches where seals congregated or to scrape and dry skins, while the ship went off to safer anchorage or to deploy more sealing gangs. The sealers had to improvise their own shelters while ashore, and it is the stone bases and timber roof beams of these shelters that are now found around the beaches (Pearson and Stehberg, 2006: 338).*

Many of the sealing sites at Fildes Peninsula in King George Island have been described by Stehberg (2004), Pearson and Stehberg (2006), and Stehberg *et al.* (2008) (see also Chile, 2007b). One of the best preserved sites in the area is known as Cuatro Pircas (62° 10' 00" S, 58° 58' 12" W). This is a complex of four drywall stone enclosures and associated artefacts located in one of beaches of Fildes Peninsula facing north to the Drake Passage. A photograph of the largest and best preserved structure taken in 1984 by R. Stehberg (reproduced in Stehberg, 2004: 69) was repeated by the author in 2006 (Fig. 4.14).

Unlike the other sites discussed in this article, to date the sealing sites are not protected by legally binding regulations. Cuatro Pircas is not subject to regular visitation but is within reach for visits for residents of nearby stations and tourists. Since 1985 and until its removal in 2004 a Brazilian field hut was located some 120 metres away from the site, so incidental visitation of the ruins was likely. In addition, Fildes Peninsula is one of the few sites in the Antarctic where there is tourism accommodation infrastructure ashore (Pfeiffer *et al.*, 2007; Bastmeijer and Roura, 2008). Tourist groups have sometimes camped at different sites of the peninsula (Pfeiffer *et al.*, 2007: 156). It possible, but not certain, that small tourist groups could have visited Cuatro Pircas. No obvious evidence of visitation was apparent in 2006.

Repeat photography shows that a number of subtle changes in the structure took place between the 1984 and 2006, even though the structure's basic frame remains mostly intact (Fig. 4.15). In the "after" photograph rocks in and around the wall have changed their location and spatial orientation; ceramic pot fragments have been removed; and a lower ground surface within the structure is apparent. An on-site surface examination discovered a range of exposed artefacts around this feature, including pottery fragments, two half buried leather soles, coal fragments, a barrel hoop and

Fig. 4.14. a-b – Sealer's shelter at Cuatro Pircas, Fildes Peninsula, King George Island, Antarctica in 1984 and 2006. The arrow on the earlier photograph shows fragments of pottery. 1984 photograph by R. Stehberg. 2006 photograph by author.

other metal fragments. Also found at the site, adjacent to a whale vertebra, were two wooden beams and two plastic fragments with lettering. These were interpreted to be what was left of an interpretation or protection sign.

As other sealing sites in the region, this site was likely to have been occupied early in the 19th century, possibly for a relatively short time only, and then has been abandoned ever since, albeit subject to a range of continuous and episodic processes of transformation. Cuatro Pircas was subject to an archaeological excavation in 1984, immediately prior to which the "before" photograph was taken (R. Stehberg, personal communication to author, 4 October 2007). This research activity would account for the

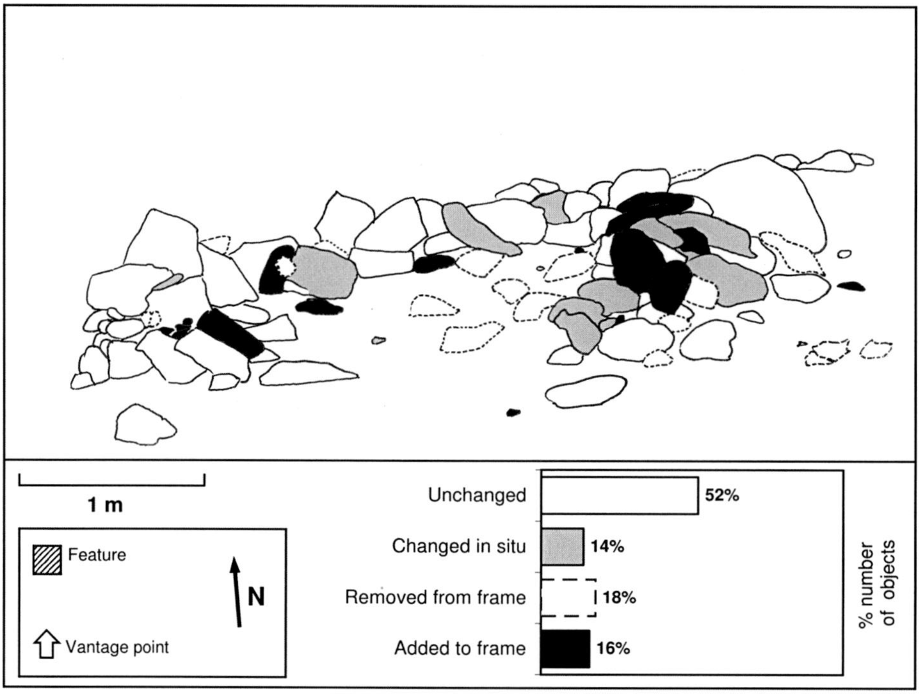

Fig. 4.15 – Overview of changes at the sealers shelter at Cuatro Pircas, Fildes Peninsula, King George Island, Antarctica between 1984 and 2006. Sketch based on 1984 photograph. Sketch by author. Key: white = no changes; grey = changes within the frame; dashed line = removed from frame; black = added to frame.

removal or changes in the location of individual rocks, the apparent reconstruction of parts of the structure, and the removal of ceramic artefacts. In addition, the lower level of the ground surface may be a result of a partial backfilling of the sandy soil after the excavation. Stehberg *et al.* (2008: 87) note that the site is " ... exposed to wind, snow loads, and the activity of seals and human visitors." Fur seals roam about and within Cuatro Pircas" structures and their activities could flatten the dry wall enclosures. Artefacts on the ground surface may have become exposed with the development of a lag deposit (or desert pavement) by effect of wind action, which removes finer particles and leaves behind coarser particles (Schiffer, 1987; Campbell and Claridge, 1989).

Fildes Peninsula is now one of the most developed ice free areas in the Antarctic, with four year-round stations, a hard rock airstrip, increasing levels of tourism and non-governmental activity (Pfeiffer *et al.*, 2007), and "cross-pollination" between different types of operators (Bastmeijer and Roura, 2008). The sealing sites are little known and have low visibility (Stehberg, 2004: 71), which in this dynamic context makes them particularly vulnerable. Discussions have been underway since 2004 to establish a management regime for the Fildes Peninsula area (see e.g. Brazil *et al.*, 2006). However, Cuatro Pircas and other sealing sites were not included in any debate on the management of Fildes Peninsula up until 2007. The sites were apparently unknown even by people with an extensive knowledge of Fildes Peninsula (e.g., Pfeiffer *et al.*, 2007 do not mention the sites). A management plan for the sites is currently being prepared (Stehberg *et al.*, 2008: 93). In practice, however, effective protection may still be some years away, as there is substantive disagreement between different stakeholders as to whether the best way of managing Fildes Peninsula is an ASMA, a code of conduct, or other means (see e.g. Germany and Chile, 2007). The dispersed nature of sealing remains poses further practical difficulties for their protection (see CEP, 2007: paragraph 154).

4.3.7 Boatshed at Base A, Port Lockroy, Goudier Island, Antarctica

Base A at Port Lockroy, Goudier Island (64° 49' S, 63° 29' W), is a Historic site and Monument described in the designation instrument (ATCM Measure 4, 1995) as "[o]f historic importance as an Operation Tabarin base from 1944 and for scientific research, including the first measurements of the ionosphere, and the first recording of an atmospheric whistler, from Antarctica. Port Lockroy was a key monitoring site during the International Geophysical Year of 1957/58." The base was closed between 1962 and 1994, during which time it fell into disrepair. In March 1991 the international environmental group Greenpeace visited the site and described it as follows:

> *The building seems to have been left much as it was when people lived there. The cupboards are stocked with food supplies. The living room is strewn with papers, including weather reports and postage correspondence files, dating back to the late 1950s ... Outside ... there was a pile of one hundred completely rusted-out barrels ... Behind the emergency shed (sic) there was an open rusted drum, which looked like it had been used as a brazier for burning general rubbish (Greenpeace, 1991: 21).*

In October 1991 the Protocol on Environmental Protection to the Antarctic Treaty was signed. At the time parties agreed that it was desirable for them to implement Annexes I-IV of the Protocol pending its entry into force (ATCPs, 1991). Early in 1993 Greenpeace revisited the site and reported that it had been partly cleaned up, in compliance with Annex III of the Protocol, although there were "...substantial quantities of twisted and tangled cooper and steel wire lying around the site, with three penguin carcases entangled around it" (Greenpeace,1994: 102). (The activists summarily packed the wire in boxes and stored them inside one of the buildings (personal observation, 31 January 1993)). Between 1995 and 1996, following a conservation survey, Base A was restored. In addition, in 1995 Base A was designated as HSM 61. Since November 1996 the station has been open to tourists and visitors during the summer season (BAS, 2001). Base A operates as manned "living museum". It has a gift shop and runs a busy post office " ... on behalf of the government of the British Antarctic Territory (*sic*) which donates a proportion of the post office revenue to the Trust ... Around 70,000 cards are posted each year for over 100 countries" (UKAHT, undated).

Unlike the other case studies, this one was based on "mining" various sources for information rather than recent fieldwork. The "before" photograph was taken by the author in 1993, and the "after" photograph, dating from 2005, was found on the

Fig. 4.16 a-b – Boat house of Base A at Port Lockroy in 1993 and 2005. 1993 photograph by author. 2005 photograph by © Eco-Photo Explorers (www.ecophotoexplorers.com/AntarcticGallery.asp, accessed April 2009. Reproduced with permission).

World Wide Web (Fig. 4.16). In 1993 the boathouse was apparently sound although weathered. By 2005 all fuel drums had been removed; the building has been painted and a wood and concrete pad added in front of its entrance; and a plaque by the door identifies the building as part of a historic site (Fig. 4.17). There are still penguins about the site, despite the increase in tourism, tourism visits have not caused a decline in penguin population at this site (Cobley *et al.*, 1999).[1] Both the "before" and "after" photographs show the masts of a sailing vessel standing behind the boathouse, suggesting ongoing visitation by yachts. In addition, a cruise ship is visible in the background of the "after" photograph. These details illustrate the well known expansion of tourism in the Antarctic since the early 1990s. Port Lockroy has become one of the most visited

1 However, longer-term studies indicate a decrease in breeding productivity at this site resulting from frequent visitation (Trathan *et al.*, 2008).

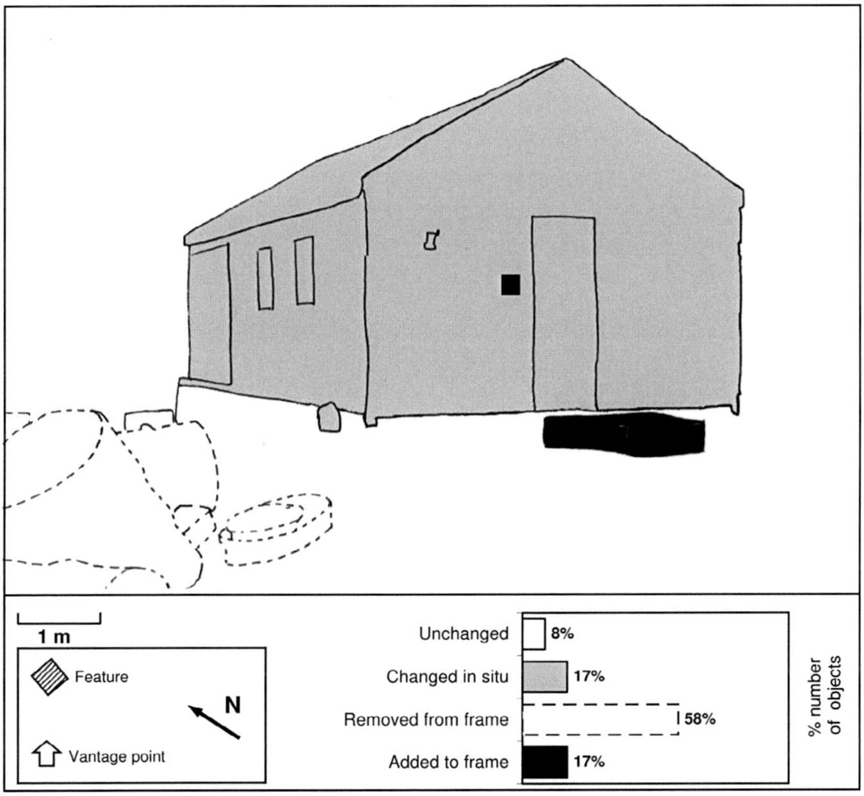

Fig. 4.17 – Overview of changes at the boat house Base A at Port Lockroy, Goudier Island, Antarctica, between 1993 and 2005. Sketch based on 1993 photograph. Sketch by author. Key: white = no changes; grey = changes within the frame; dashed line = removed from frame; black = added to frame.

sites in Antarctica, receiving over 15,000 visitors in 2004-2005, up from about 2,000 visitors in 1992-1993. More recent data reveal a continued growth (over 18,000 visitors in 2008-2009). Through its transformation from an abandoned base to a historical site frequented by tourists Port Lockroy illustrates the linkages between environmental protection, heritage conservation, tourism management, and the assertion of territorial claims.

4.4 DISCUSSION

4.4.1 Overview: Site-specific changes

Historical sites in Antarctica and Svalbard may appropriately be described using the term *palimpsest*. In its original usage "palimpsest" refers to pieces of a scroll or a book in which the original text has been scraped off so that the writing media could be reused. In archaeology and related disciplines this term is used to describe " ... a superimposition of successive activities, the material traces of which are partially destroyed or reworked because of the process of superimposition" (Bailey, 2007: 203). In this research time-serial changes were observed in the seven photo couples that were examined, and which corresponded to the most recent changes of an evolving palimpsest (Table 4.4). Plainly the analysis assessed obvious macroscopic changes, most objects in a historic site would have changed somehow over time by effect of, for instance, chemical or microbiological processes, but those changes may not be apparent in photographs such as those used here. Only the most conspicuous changes were detected, and mostly in objects located in the foreground of the photographs. Small objects moved around within the frame; others disappeared from or were added to the frame. Apparently random arrangements of objects were modified so that, for instance, objects were aligned, stacked, or put in display. Some large features were restored or reconstructed and other features were created.

Barring a feature that had fallen due to the wind and had been subsequently restored, no feature degraded significantly during the monitoring period. Rather, the condition of several features improved outwardly as a result of conservation actions. This assessment only describes the changes as observed; it does not evaluate whether these changes had positive or negative effects on, for instance, the site's historical values or authenticity. However, assessing and explaining changes to historic features is a first step towards identifying and managing undesirable effects.

The interpretation of changes was made on the bases of photographic material and additional site information. However, it may be difficult to link causes and effects of site transformation. The term "equifinality" describes the process of reaching the same final state from different initial conditions and in different ways (Lyman, 2004). For instance, an artefact may have broken as a result of any of a number of processes such as wind, freezing water, a footfall, but these may not always be identified when the only evidence is a broken artefact, or its photograph. In the features examined here cultural processes had more obvious effect than natural processes. Some changes were interpreted to result almost certainly from activities such as trampling, cairn building, stacking or storing of objects, building, and waste removal. These activities are associated with conservation and other forms of active management and, in some instances, with organised tourism. The natural process that had most likely caused changes to some of the historic features discussed here was wind action, and in one case animal activity. Whatever the immediate and proximal causes of change to individual sites, features, and artefacts, changes as assessed here reflect the broader context in which the sites have been used and managed in the recent past. The transformation of historical sites in Antarctica and Svalbard occurs in the context of recent developments in these regions. For instance, a stone in a cairn reflects a specific event, for example a visitor spontaneously placing a stone in a certain place. The cairn itself reflects a succession of such events. These changes, however minor or ephemeral, take place in the context of expanding polar tourism. Key regional developments of the recent past will be briefly reviewed below.

4.4.2 The regional context of local changes

Many historical sites in the Polar Regions look like abandoned places, yet they have become an arena in which contemporary activities take place in juxtaposition to historic features, which serve as background. Historical sites in Antarctica and Svalbard were formed by behaviour in past cultural systems and were at some point abandoned. Many sites remained abandoned for years or decades until they were subsequently relocated, while others remained within reach of visitors or were re- occupied. Some sites have been (and in some cases still are) the objects of archaeological or historical research. However, many sites are now part of a contemporary cultural system that determines how the sites are used and maintained. Present day site use (for instance conservation or tourism) usually differs with the original use (for instance research or whaling). Nevertheless, the past may still be an essential component of present day activities through, for instance, the use of material remains as tourism attractions, or narratives about historic events (Chapter 6). Conceptually, this situation can be usefully described using the notions of systemic context and archaeological context developed by Schiffer in 1972:

> *Systemic context labels the condition of an element which is participating in a behavioral system. Archaeological context describes materials which have passed through a cultural system, and which are now the objects of investigation of archaeologists (1972: 157).*

As noted previously, historical sites in Antarctica and Svalbard are protected by environmental legislation. In Svalbard the main criteria for protection is age, while in the Antarctic it is recognised historical significance. Additional criteria specific to each region apply to certain remains that do not meet these basic criteria but that nonetheless merit protection. The legislation balances sometimes uneasily the protection of both the natural environment and of the tangible cultural heritage. On the basis of this protection framework, active conservation efforts of historical sites have increased in both regions in the past decades, in parallel with environmental management, and these are apparent in several of the case studies presented here. Conservation efforts are mostly government supported but also point towards an emerging heritage sector. Conservation criteria differ both between and within regions. For instance, historic huts in the Ross Sea are conserved with a philosophy that determines that intervention may be substantive in some instances "...with major repairs where necessary, and the possible future replication of some elements to preserve the overall ambiance of the hut" (NZAHT, 2004: 10). This contrasts with the approach taken in Svalbard where many more sites are protected than in Antarctica, many of which are in principle left to "die in beauty". This concept means "...passive preservation, letting nature take its course without interfering in order to prevent decay" (Prestvold, 2003: 11). Nevertheless, some historic sites in Antarctica are essentially unmanaged; and conversely, some, but certainly not all, high priority sites in Svalbard are actively preserved (Governor of Svalbard, 2000). Overall, conservation has not been free of controversy and in some instances may be regarded as a significant factor of site transformation, but underscores the importance assigned to historical sites in both regions.

Archaeological research in Antarctica and Svalbard predates the period discussed here, but it has been a dynamic endeavour in the recent past (see e.g. Harrowfield, 2005 for the Antarctic and Jørgensen, 2005 for Svalbard – and has more recently been boosted by the International Polar Year 2007-2009. Excavation and sampling standards are higher now than they were in the past, on account of the legal and institutional arrangements that have emerged, but the challenge remains of how to promote research while ensuring the sustainability of the archaeological resources. Jørgensen (2005: 60) has made this point for Svalbard, but it is equally applicable to the Antarctic. A conservative or, perhaps, precautionary approach is needed for the management of archaeological resources in these regions. Apart from archaeology and related disciplines, scientific research and logistics at or near historic sites may also contribute directly or indirectly to the transformation of those sites. This may occur through inadvertent damage, infrastructure development, and (historically at least) souvenir collection (for examples see Harrowfield, 2005; NZAHT, 2004; Stehberg *et al.*, 2008). Overall, the theoretical argument over these sites will only increase in intensity as various processes, such as tourism and climate change, continue to degrade these sites and diminish their value to historical scientists.

The first tourism cruise to Spitsbergen was organised in 1871 (Conway, 1906: 302) while organised tourism in the Antarctic Treaty area began in the 1950s (Headland, 1994: 275). However, the 1990s marked an inflection point in which tourism began to increase significantly and became eventually a major presence (Viken and Jørgensen, 1998; Bastmeijer and Roura, 2004; Kolltveit, 2006). Svalbard's tourism expansion resulted in part from government policy (Viken and Jørgensen, 1998), and consequently it has been somehow regulated from the start through a combination of state regulation and industry guidelines. Further stringent restrictions on tourism at key cultural heritage sites have recently been discussed (Governor of Svalbard, 2009). Antarctic tourism has developed more spontaneously and has been only subject to certain conditions rather than to actual restrictions (Bastmeijer and Roura, 2004). This is only recently starting to change following several shipping accidents involving tourism vessels, including the sinking of the MV *Explorer* in November 2007. Tourism management at Antarctic historic sites combine generic and site-specific regulation, both binding and non-binding, which emerges from Antarctic Treaty states and from the industry itself. By and large historic sites in Antarctica and Svalbard are protected and managed for the values they represent, and tourism is a result of conservation than rather than its primary objective. However, a small number of historical sites, by design or by coincidence, have been modified for tourism consumption, notwithstanding their historic significance.

Historical sites have geopolitical interest through their uses to assert (or deny) territorial claims. Despite the international agreements that regulate the governance of Antarctica and Svalbard, and which settle or set aside the sovereignty discussion, issues of claims and jurisdiction are still part of regional politics (see e.g. Jacobsson, 2007; Pedersen, 2009) and rituals and symbols of possession (Seed, 1995) continue to be played out in both regions. While earlier ritual activities concerned the placement of infrastructure such as bases, huts, and mining works, some of today's historic sites, in the present rituals concern primarily the protection and management of these and other kinds of sites (Chapter 5). Material symbols of presence displayed at historic sites may include memorials, monuments, flags and flag masts; visitor management infrastructure such as huts, paths, and interpretation or regulatory signs; and added attractions such as souvenir shops or postal offices. These symbols project an image of presence and effective administration; and in some instances favour some narratives of historic events at the expense of others (e.g. Chapter 6). Tourists and other visitors provide an audience to these rituals and symbols and take them home as they record their visit with photographs, films, or post cards.

Rapid climate change is apparent in the Arctic and in parts of the Antarctic with consequential effects on historic sites (ICOMOS, 2008; Barr, 2008). Generally, the effects of climate change in the Polar Regions may affect both the landscape and the historic sites it contains. Elements designed for the cold or preserved in cold conditions deteriorate rapidly with a warming of the environment. Additional changes to historic sites may result indirectly through the warming of the polar landscape (ICOMOS, 2008). Effects may include the erosion of coastal sites, increased forms of biological decay, and increased tourism penetration (Barr, 2008). These pressures may in turn intensify the need for actions to ensure the long-term preservation of historic sites.

4.5 CONCLUSIONS

Historical sites in Antarctica and Svalbard contain the material remains of past activities of exploration and exploitation of these regions. These sites are subject to continuing transformation by cultural and non-cultural (natural) processes since their abandonment to the present. The sites are significant for the historical and other values they embody, and for research and management purposes it is important to assess and explain the changes they experience. In this research the status and change of seven historic features in Antarctica and Svalbard was monitored by means of repeat photography. Recent time-serial changes were observed in the seven photo couples examined here. Many minor and some major additive, reductive and *in situ* changes occurred at the seven historic features. In most cases the changes did not affect negatively the condition or integrity of individual features; rather, the deterioration of several features was stopped or reversed. Whether or not the changes affect the historical values of the sites is in some cases the subject of controversy. Some changes were interpreted to result from cultural processes, primarily conservation and other forms of active management; archaeological research; and organised tourism. Natural processes, particularly wind action, caused further changes. The ongoing conservation of the features discussed here may be an artefact of their relative accessibility and of their recognised historic significance. In contrast, many other historic sites in Antarctica and Svalbard are essentially unmanaged and evolving towards their eventual destruction (e.g., Barr *et al.*, 2005; Arenz and Blanchette ,2008).

Narratives about some historic sites in the Polar Regions emphasise the notion that these sites are largely “time capsules” that have remained untouched over the years since the last occupants left. This is reinforced by some interventions that aim to reproduce specific periods of the sites’ past. However, sites reflect traces of ongoing events over the years, and can be regarded as part of an evolving palimpsest. Historical sites participate in the behavioural system of contemporary activities in Antarctica and

Svalbard (in the systemic context of Schiffer, 1972) and are the focus of multiple contemporary interests and activities largely centred on the historical character and values of the sites. Local processes of change may be linked to regional developments, which are likely to have had an indirect effect on many of the sites discussed here, and in some instances have been significant drivers of site transformation. The evidence presented here suggests that historic sites in the Polar Regions are dynamic entities that not only reflect the past as it once was, but that are also a window onto the present.

4.6 ACKNOWLEDGMENTS

My thanks for their help to produce this article go, without implicating them in my interpretations, to the following people and institutions. Professor Louwrens Hacquebord and Dr. Maarten Loonen (University of Groningen, Arctic Centre) facilitated my fieldwork at the Netherlands Arctic Station and elsewhere in Svalbard, helped to develop this research, and reviewed this article. Some basic ideas about site monitoring emerged from discussions with Dr. Susan Barr (Norway's Directorate of Cultural Heritage) and Per Kyrre Reymert (Government of Svalbard). On Antarctic matters I benefited from discussions with Jim Barnes, Dr. Kees Bastmeijer, Ambassador Jorge Berguño, and Dr. Alan Hemmings.

I am grateful to the various institutions and individuals that invited, funded, transported, hosted or otherwise enabled me to travel to the Antarctic and Svalbard repeatedly and to conduct fieldwork there. These include the Instituto Antártico Argentino, Greenpeace, the Antarctic and Southern Ocean Coalition, and Gateway Antarctica, University of Canterbury. Fieldwork in Svalbard was funded by the Groningen Institute of Archaeology, University of Groningen. Particular thanks to Gary Evans, Per Kyrre Reymert, Dr. Michelle Rogan-Finnemore, Dr. Hans-Peter Ulrich, Base Decepción staff (2001–2002), Oleg Sakharov and Bellingshausen base staff (2005–2006), and Scott base's Antarctic field team (2007–2008). Reaching specific sites was made easier with the assistance or company of Wolfgang Dinter, Wojtek Moskal, and Steffen Vogt. Grete Marstein kindly acted as Norwegian interpreter when interviewing mining veterans.

Photographs supplementing my own collection were kindly provided by Dr. Antonio Bendala, the Governor of Svalbard, Professor Hacquebord, Dr. Loonen, Dr. Rubén Stehberg, and Christopher Weaver and Mike Salvarezza. Finally, I thank Janet Dalziell and two anonymous reviewers for their valuable criticism of earlier versions of this article.

4.7 TABLES CHAPTER 4

Table 4.1 – Case study sites

Site description	Site type and condition	Coordinates	Original occupation period	Original function	Current site function
Building foundations with cooking stove, Ny-London, Svalbard	Ruins of wood frame building	78° 57' 49"" N, 12° 03' 06" E	1911–1920	Marble quarrying	Recreational /organised tourism visits, field accommodation, research
Entrance to "Josefine" coal mine, Ny-Ålesund, Svalbard	Standing ruins of industrial building	78° 54' 56" N, 11° 55' 11"" E	1921–1924 (mining in area 1916–1963)	Coal mining	Sporadic recreational visits, conservation
Cross at Observation Hill, Ross Island, Antarctica	Memorial cross	77° 51' S, 166° 41' E (*)	1913	Ceremonial	Memorial, recreational/organised tourism visits
Scott's chart table at the *Terra Nova* Hut, Cape Evans, Ross Island, Antarctica	Group of assorted artefacts (not original assemblage) inside standing building	77° 38' S, 166° 24' E (*)	1910–1913; 1914–1917	Expedition, research	Recreational/organised tourism visits, conservation, research
Former cemetery at Whalers Bay, Deception Island, Antarctica	Group of mortuary artefacts (not original assemblage)	62° 58'38" S, 60° 33' 39" W	1912—1931; 1943–1969	Whaling, research	Memorial, recreational/ organised tourism visits, conservation, research
Sealer's shelter, Fildes Peninsula, King George Island, Antarctica	Temporary camp remains	62° 10' 00" S, 58° 58' 12" W	1820s	Sealing	Research, sporadic recreational visits
Boat house at Base A, Port Lockroy, Goudier Island, Antarctica	Standing wood frame building	64°49'S, 63°29'W (*)	1944–1962	Operation Tabarin, research	Organised tourism visits, "living museum", penguin monitoring

(*) Indicate that coordinates are from ATCM instruments designating the sites as Historic Sites and Monuments and give an approximate position.

Table 4.2 – Repeat photography characteristics

Feature	Date "before" photograph	Date "after" photograph	Years monitored	Comments
Building foundations with cooking stove, Ny-London, Svalbard	Late 1990s	2007	Approx. 10	Good match. "Before" photograph has low resolution.
Entrance to "Josefine" coal mine, Ny-Ålesund, Svalbard	1998	2008	10	Good match. "Before" photograph has relatively low resolution.
Cross at Observation Hill, Ross Island, Antarctica	1990	2007	17	Fair match.
Chart table at the *Terra Nova* Hut, Cape Evans, Ross Island, Antarctica	1990	2007	17	Fair match. "Before" photograph is dark.
Former cemetery at Whalers Bay, Deception Island, Antarctica	2002	2006	4	Good to fair match. There are photo-couples of individual features but not of the entire site.
Sealer's shelter, Fildes Peninsula, King George Island, Antarctica	1984	2006	22	Good match.
Boat house at Base A, Port Lockroy, Goudier Island, Antarctica	1993	2005	12	Good match.

Table 4.3 – Site characteristics, main changes, and interpretation

Feature	Site characteristics	Main changes during monitoring period	Interpretation of main causes of change
Building foundations with cooking stove, Ny-London, Svalbard	Building foundations containing an iron cooking stove, brick chimney remains, and mostly brick and wood fragments less than ca. 50 cm long. Several metal artefacts are on display above stove.	Artefacts removed from frame. Small to medium brick, wood, and metal objects moved within the frame.	Wind action, conservation, and visitation.
Entrance to "Josefine" coal mine, Ny-Ålesund, Svalbard	Wooden industrial remains with a scatter of wood boards less than ca. 200 cm long around them.	Partial reconstruction of main structure. Some boards moved within the frame. Location peg added to the site.	Conservation.
Cross at Observation Hill, Ross Island, Antarctica	Wooden memorial cross >3 m high.	Cross is shorter and faces a different direction than before. Ephemeral visitation marks.	Wind action, conservation, and visitation.
Scott's chart table at the *Terra Nova* Hut, Cape Evans, Ross Island, Antarctica	Chart table containing a display of documents, a stuffed Emperor penguin, and miscellaneous artefacts.	Objects rearranged and tidied up.	Conservation.
Former cemetery at Whalers Bay, Deception Island, Antarctica	Two crosses and coffin remains with associated stone cairns.	Two crosses brought from different locations and placed adjacent to coffin remains. Rock cairns and ephemeral visitation marks.	Conservation, management, and visitation.
Sealer's shelter, Fildes Peninsula, King George Island, Antarctica	Dry wall enclosure composed by stones less than ca. 50 cm long and containing some ceramic fragments and other artefacts.	Stones rearranged within site. Pottery remains removed. Exposure of objects at ground level.	Archaeological research, wind action, and animal activity.
Boat house at Base A, Port Lockroy, Goudier Island, Antarctica	Wood frame building surrounded by mostly 200 litre fuel drums.	Fuel drums removed. Building restored and plaque added.	Environmental clean up and conservation.

Table 4.4 – Summary of changes

Feature	Number of objects monitored	Objects that did not change (%)	Objects that changed in situ (%)	Objects removed from frame (%)	Objects added to frame (%)	Disorder
Building foundations with cooking stove, Ny-London, Svalbard	58	45	27	19	9	Decreased slightly
Entrance to "Josefine" coal mine, Ny-Ålesund, Svalbard	42	48	19	9	24	Decreased slightly
Cross at Observation Hill, Ross Island, Antarctica	1	-	100	-	-	No change
Scott's chart table at the *Terra Nova* Hut, Cape Evans, Ross Island, Antarctica	25	12	60	4	24	Decreased
Former cemetery at Whalers Bay, Deception Island, Antarctica	24	13	-	-	87	Decreased
Sealer's shelter, Fildes Peninsula, King George Island, Antarctica	105	52	14	18	16	Decreased slightly
Boat house at Base A, Port Lockroy, Goudier Island, Antarctica	12	8	17	58	17	Decreased

5 Antarctic scientific bases: Environmental and cultural heritage perspectives 1983-2008[†]

ABSTRACT

This Chapter examines the interplay of cultural and environment values and how they are reflected in the designation and management of historic scientific bases in Antarctica. In the context of negotiations leading up to the 1991 Protocol of Environmental Protection to the Antarctic Treaty, environmental NGOs denounced what were perceived as poor environmental practices of Antarctic Treaty parties. Whilst taking a hard line with regard to the need to clean up many sites and facilities, NGOs recommended that the historic merits of abandoned bases had to be assessed, and that they should either be protected as formally designated historic sites, and properly maintained, or otherwise removed. In this regard there was no conflict between environmental and cultural heritage perspectives with respect to protecting historic sites and the values they embody. After the entry into force of the Protocol, Antarctic Treaty parties conducting official inspections adopted a fairly similar position than NGOs did a decade earlier with regard to abandoned sites and facilities without historic designation. The fate of abandoned scientific bases in Antarctica under the Protocol has been diverse, and can be analysed from the perspective of Thompson's 1979 Rubbish Theory. Under this theory abandoned stations and waste dumps can be regarded mostly as "rubbish"; and Historic Sites and Monuments can be regarded as "durable". There is a basic tension between environmental and cultural heritage perspectives as ways of regarding human presence in Antarctica. However, there are convergences too, which result from the growing pressures on the intrinsic values of Antarctica and the legal requirement under the Protocol and its Annexes to protect them. The misuse of the historic designation process – assigning protected status to material remains that actually lack in significance – could result in an erosion of the wilderness, historic, and other values of Antarctica.

5.1 INTRODUCTION

Following the entry into force of the 1991 Protocol of Environmental Protection to the Antarctic Treaty (hereafter the Protocol), Antarctic Treaty parties have had to take ac-

[†] Published as:

Roura, R. (2008): Antarctic scientific bases: Cultural heritage and environmental perspectives 1983-2008. In: S. Barr and P. Chaplin (eds.): *Historical polar bases – preservation and management.* ICOMOS (International Council of Monuments and Sites). Monuments and Sites Series Volume XVII. (pp. 38-52). Oslo: ICOMOS International Polar Heritage Committee.

tion on their abandoned or underused scientific bases, while protecting those that have historic value. The Protocol and other Antarctic Treaty instruments protect sites of recognised historic value that have been designated as Historic Sites and Monuments (HSMs). Historic remains predating 1958 whose existence or present location is not known have also a degree of protection.[1] However, the remains of past activities that are not otherwise protected are subject to be removed under Protocol requirements to clean up past and present waste disposal sites on land and abandoned worksites.[2]

In 1983 a new wave of designations of historic sites began, following a hiatus of eleven years. The sites and monuments designated since then included some abandoned scientific stations that had at some point generated environmental concerns.

Non Governmental Organisations (NGOs) played a key role in the adoption of the Protocol by Antarctic Treaty parties. To promote its views the international environmental organizationGreenpeace International (hereinafter GPI) started an Antarctic campaign in 1983 (Roura, 2007). One of the approaches used was to conduct unofficial inspections of Antarctic sites and facilities.

This Chapter examines the designation of Antarctic scientific bases as HSMs from the evolving and at times contrasting perspectives of protecting environmental and cultural values. This analysis has been made on the bases of comparing and contrasting the findings of GPI's unofficial inspections[3] with Antarctic Treaty System (ATS) documentation including official inspection reports. These were used to track the interplay of cultural and environment values for over two decades and how they are reflected in the designation and management of historic scientific bases in Antarctica.

5.2 OVERVIEW: ANTARCTIC HISTORIC SITES AND MONUMENTS

The Protocol and other Antarctic Treaty instruments protect sites or monuments that have been designated as HSMs.[4] Any Party may propose a site or monument of recognised historic value for listing as an HSM. The proposal for listing may be approved by the Antarctic Treaty Consultative Parties (ATCPs) by a measure adopted at an Antarctic Treaty Consultative Meeting (ATCM).[5] By ATCM XXX (2007) there were 78 listed sites and monuments numbered 1-82 (some have been delisted or subsumed into other HSMs). Of these, approximately 70% are sites of past Antarctic activities and about 30% are memorials and monuments.

---

1 ATCM Resolution 5 (2001), *Guidelines for handling pre-1958 historic remains.*

2 Protocol's Annex III, Art. 1 (5).

3 The author led or co-led numerous inspections as a Greenpeace representative between 1990 and 1997. He has no current involvement with the organization, and all the information used here is based on publicly available sources.

4 For a review of Historic Sites and Monuments in Antarctica see for instance Bizarri (2006) and Chile (2007a).

5 Protocol Annex V, Art. 8 (2).

Fig. 5.1a-c – A diversity of historic sites and monuments in Antarctica. Upper: Borchgrevink's huts at Cape Adare erected in February 1899, the first buildings in the Antarctic continent, with remnants of Scott's Northern Party hut in the foreground (HSM 22). Middle: cross at Observation Hill, Ross Island, erected in January 1913 in memory of Capt. R.F. Scott and his companions (HSM 20). Below: replica of a bust of Capt. L.A. Pardo, commander of the Chilean expedition that rescued the survivors of the British vessel Endurance in 1916, located at Chile's Frei Station in King George Island (HSM 53). The original HSM 53 is in Elephant Island (Photos by author 1992, 2007, 2006).

Listed HSMs shall not be damaged, removed or destroyed. HSMs may be designated as protected areas under Annex V to the Protocol.[6] Past activity remains that are not otherwise protected are subject to removal under requirements of Annex III to the Protocol on Waste Disposal and Waste Management. Unlike the natural environment, historic values[7] are not normally subject to the previsions of Annex I to the Protocol on Environmental Impact Assessment (EIA).[8]

The list of HSMs is reviewed and updated periodically, but it is neither comprehensive nor representative of different periods or themes of Antarctic history. There is a certain inequality in the sites designated (Cochran and Collinge, 1994:15). For instance, the list includes "Heroic era" huts and other remains, graves of different ages, and monuments of different kinds, some of which represent national or religious figures unconnected to the Antarctic. Since 1995 sites and monuments must meet certain criteria prior to designation,[9] but this requirement post-dates the designation of many HSMs and is not enforced strictly. Finally, there are still no uniform criteria or accepted standards as to how an HSM should be maintained after designation (see Morrison, 2008).[10] This is of particular significance for the designation of scientific bases, which are likely to require costly and logistically complex maintenance.

5.3 PERSPECTIVES ON ANTARCTIC MATERIAL CULTURE

5.3.1 Scientific bases and the practice of inspection

The practice of inspection is established under Article VII of the Antarctic Treaty and Article 14 of the Protocol. The purpose of the inspections is to promote the objectives of, and ensure compliance with, the provisions of the Antarctic Treaty and its Proto-

---

6 The Protocol's Annex V on Protected Areas lists two categories of protected areas: Antarctic Specially Managed Areas (ASMAs) and Antarctic Specially Protected Areas (ASPAs). Both categories are subject to management plans approved by the ATCM. Entry to ASPAs requires a permit issued by competent authorities.

7 The Protocol is not internally consistent with regards to historic values. It sometimes refers to "historic values" and sometimes (in the context of areas) to "historic significance". Historic values are not listed in Art. 3 of the Protocol on Environmental Principles and are not usually regarded as part of the "Antarctic environment" in the meaning of domestic implementation (See Bastmeijer and Roura, 2008: 202, footnote 85).

8 *If the actual or potential presence of cultural heritage is not recognised during EIA, then that heritage is potentially at risk of being damaged or lost in activities subject to EIA, such as clean up operations. In these instances Antarctic EIA would be, in effect, "blind" to historic values. (Bastmeijer and Roura, 2008: 202).*

9 ATCM Resolution 8 (1995), *Guidelines for proposing new Historic Sites and Monuments.*

10 Update March 2011: ATCM Resolution 3 (2009) *Guidelines for the designation and protection of Historic Sites and Monuments* partly addresses this issue (see Appendix 6 of this thesis).

col. Inspectors designed under Article VII (1) (formally called "observers") "shall have complete freedom of access", including the conducting of aerial observations, "at any time to any or all areas of Antarctica." During inspections, observers should be given access including all parts of stations, installations, equipment, ships and aircraft and all records that are maintained at those facilities in accordance to Protocol requirements.

As part of its Antarctic activities GPI carried out unofficial inspections to over 160 sites or facilities from 1988 in the Antarctic Peninsula, Northern Victoria Land and the Ross Sea (ASOC and UNEP, 2003). Inspection procedures consisted of an interview with local authorities followed by visual site examination and documentation, and an official tour on invitation. Environmental sampling was carried out in some instances, and on-site protests of different kinds were occasionally conducted, although this latter approach was only used for the worst examples of preventable environmental damage. Inspection reports were produced after each season and were made available to Antarctic Treaty parties. Plainly these inspections had no formal status, but they provided information on the environmental performance of Antarctic scientific stations that was often unavailable at the time.[11]

ATCM Resolution 5 (1995) approved four official inspection checklists[12] that were produced as *aide memoires* for observers conducting an inspection. These checklists classify elements of the human built environment into four categories, which (probably unintentionally) constitute a useful classification of material culture in the Antarctic landscape:

- *Permanent Antarctic stations and associated installations, including remote field camps* (Checklist A);
- *Vessels within the Antarctic Treaty Area* (Checklist B);
- *Abandoned Antarctic stations and associated installations* (Checklist C); and
- *Waste disposal sites* (Checklist D).

For the purposes of this Chapter there are therefore permanent active stations and associated facilities, abandoned (or unused) stations, and waste disposal sites. Some of these may be, or may contain, designated HSMs or material remains of past activities that have heritage value and may one day be designated as HSMs. Any of these types of material culture can be examined from environmental or cultural heritage perspectives.

11 After the entry into force of the Protocol in January 1998 official inspections started to scrutinize environmental aspects of Antarctic operations much more closely than hitherto.

12 A fifth checklist for the inspection of protected areas was proposed by the United Kingdom and New Zealand at the XXVIII ATCM (2005). (Update March 2011: This was eventually approved as ATCM Resolution 4 (2008) *Checklist for inspections of Antarctic Specially Protected Areas and Antarctic Specially Managed Areas.*)

Fig. 5.2 – An abandoned Antarctic scientific base: Cape Hallett Station in 1992. In the foreground some of the artificial penguin mounds built by the base's operators to encourage the re-colonisation of the site by penguins, which had been displaced to establish the base (Photo by author).

5.3.2 Environmental perspectives: NGO position on Antarctic scientific bases

NGOs demanded higher environmental standards in the operation of active stations, and the removal of unused stations and waste disposal sites. A statement that summarises the environmentalists' position on abandoned stations appears in the concluding remarks of the 1990-91 Expedition Report:

> *The abandoned bases littering the Peninsula need to be dealt with as a matter of urgency.*
>
> *They are not looked after, and are deteriorating from the weather, and some cases, from visitors. Most of them have hazardous material lying around. They also have a visual impact which should not be ignored. Anything of genuine historic merit at these bases should be restored, designated, and maintained properly. Everything else should be removed, although the process should be subject first to environmental impact assessments, as wildlife is re-colonising many of the sites. (GPI, 1991: 23)*

GPI visited few designated HSMs in its Antarctic expeditions, and in those visits it did not follow its usual inspection and reporting procedures. For instance, when GPI ships were in Ross Island – where the organisation ran a year-round station, World Park Base, between 1987 and 1992 – personnel from Scott Base were made available to open the historic huts in the area so that the environmental activists could visit them. However, GPI did not report on the visits to these or other historic sites. Overall GPI accepted as valid the designation of historic sites as HSMs, even though some of these sites had obvious environmental problems. Some of those same historic

Fig. 5.3a-b – Southern stores dump near Scott's Terra Nova Hut, Cape Evans, Ross Island, part of HSM 16 (upper). The remains of a waste dump at Russia's Bellingshausen Station, King George Island, following a 2001 clean up (below). (Photos by author, 2007, 2006).

sites have since been recognized as contaminated and needing remediation, such as Scott's *Terra Nova* Hut at Cape Evans (e.g. Blanchette *et al.*, 2004). In contrast to historic *sites* the environmentalists did not openly criticize, but disapproved of, the proliferation of *monuments* in scientific bases that were designated as HSMs, on the suspicion that many such designations had a symbolic role that had little to do with science or even historic values.

A personal account of a former leader of GPI expeditions to the Antarctic region describes a 1993 encounter of environmentalists with the cruise ship *Kapitan Khlebnikov*. The account illustrates the tensions between environmental and cultural heritage perspectives with regard to the management of historic sites, and the different perspectives sometimes held by individuals with respect to GPI's official position. The encounter took place at Cape Evans, Ross Island, where a GPI team was monitoring environmental impacts at the former World Park Base site. From the site, which was 200 meters from the historic *Terra Nova* hut, the activists could see the landings:

> *After a brief pause, the Khlebnikov began disgorging inflatables to ferry passengers ashore...Many of them [the tourists] did seem to regard us with suspicion... One woman nearly wandered onto the site of World Park Base. Martin, one of our cooks, was there and asked if she could keep out because it was a scientific area. She responded that she was pleased the base had gone; it had been an eyesore next to such an historic monument. Martin pointed out that the historic monument was a wooden hut with trash all around it, and that the Antarctic surroundings were infinitely more historic and, indeed, monumental. She would have none of it. Yes, the hut might be surrounded by broken bottles, wood chips, wire, dead animals, food cans, and the kind of garbage that it is now illegal to even think of leaving on Antarctica, but it was historic garbage. (Mulvaney, 2003:170).*

To summarize, environmental NGOs demanded higher environmental standards of Antarctic scientific bases, and required the removal of abandoned facilities and waste materials that had no genuine historic value. However, they accepted as valid the designation of historic sites by the ATCM and did not scrutinize them despite the environmental risks some of them posed.

5.3.3 Cultural heritage perspectives: Historic site designation of Antarctic scientific bases

One of the characteristics of Antarctic science is (and has always been) its close involvement with politics, which results from the national scale support needed to conduct science in the Antarctic region (Fogg, 1992:2). Plainly, the reasons to operate an Antarctic base are not only scientific but concern a range of other reasons, including attaining consultative party status,[13] influence in the ATS, and geopolitical presence. Science, however, has a priority role in the ATS and should always be present, even if other factors are also at play.

Science may be defined as "...our attempts to understand the natural world – not the mere accumulation of facts about it but its assimilation into abstract concepts from which predictions can be made." (Fogg, 1992: 4). It follows that a scientific base is one where (or from where) scientific activities are carried out. However, the role of science in many Antarctic bases has never been and it still is not as high as it might be. An

13 Operating a national scientific base is no longer regarded as an essential requirement any longer to demonstrate active involvement in Antarctic research. For instance, The Netherlands attained Consultative status in 1990 without operating a scientific base.

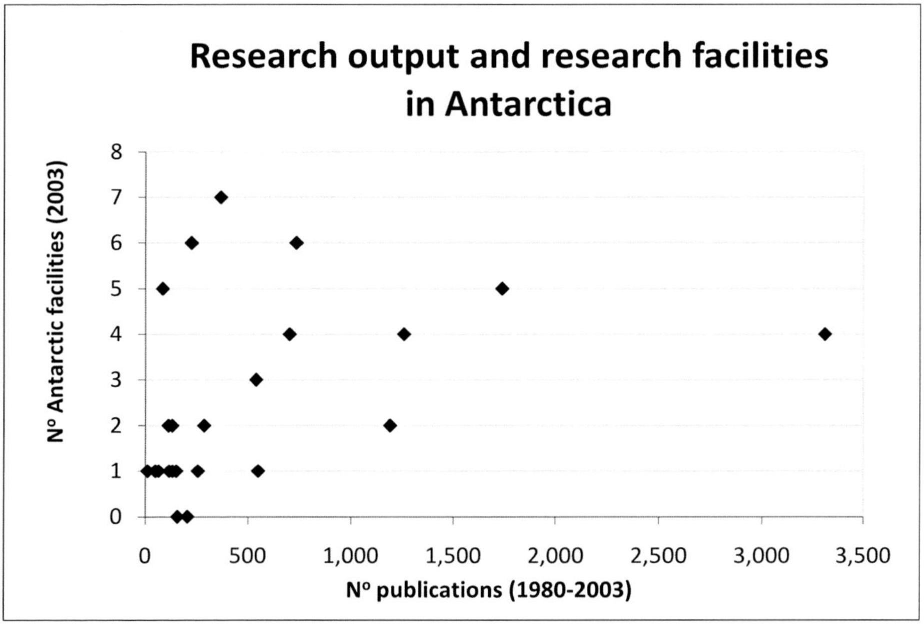

Fig. 5.4 – Research outputs and research facilities of Antarctic Treaty Consultative Parties. Research output data between 1980 and 2003 from Dastidar and Persson (2005). Facility data from COMNAP (available from www.comnap.aq, accessed June 2003).

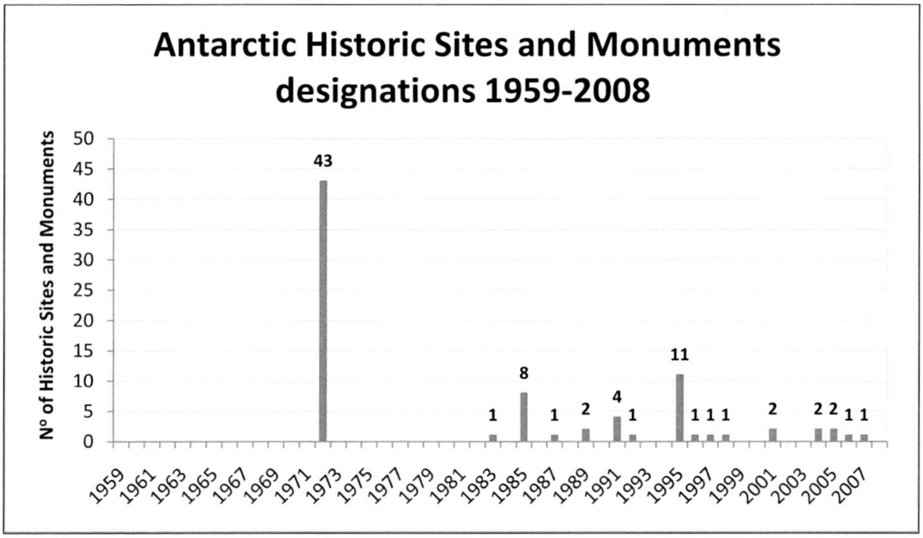

Fig. 5.5 – Designation of Antarctic Historic Sites and Monuments since the signature of the 1959 Antarctic Treaty. Data from the Antarctic Treaty Database (available from www.ats.aq, accessed May 2008).

official inspection conducted in the Antarctic Peninsula in 2005 made comments that resonate with those made by environmentalists a decade earlier:[14]

> *Although some stations were undertaking world-class scientific research into a wide variety of disciplines… a larger number of stations appeared to have relatively modest, or even rudimentary, science facilities. This was particularly so when viewed against the substantial size of some stations' infrastructure. (Australia et al., 2005: 6).*

This observation appears to be confirmed by comparing the output of research publications in peer reviewed journals between 1980 and 2003 and the number of extant Antarctic facilities in the early 2000s (Fig. 5.4). In that period, the most productive ATCPs ran three or four facilities, but so did other parties that are comparatively (and in some instances considerably) less productive. The production of scientific papers by parties with no Antarctic facilities is not much inferior to that of parties with five or more facilities.

Forty-three HSMs were designated in 1972, many of which are linked to early Antarctic exploration and/or science. Between 1983 (when a second wave of designations began) until May 2007, the ATCM approved 35 HSMs (Fig. 5.5). Of these, thirteen HSMs are the complete or partial remains of scientific bases or buildings, or structures used for scientific purposes; and seven are monuments located at or near to scientific bases. Among the HSMs designated since 1983, which include sites of historic significance such as the as yet unfound tent of Roald Amundsen near the South Pole, there are some of the sites and facilities that were once regarded as problematic from an environmental perspective. Some of these are now protected and remembered for the historic values they supposedly embody – including the role they played in Antarctic science.[15] In addition, some bases have been "memorialised" – meaning that a monument to or at the base becomes in itself an HSM, somehow legitimising the presence of Antarctic infrastructure (Fig. 5.6).

As a result of the rapid growth of Antarctic tourism since the 1990s, some stations have become (by chance or design) tourism attractions and regular destinations. Examples include Base A at Port Lockroy (HSM 61), and Chile's Gabriel González Videla, a station with two associated HSMs (30 and 56), which was refurbished to receive tourism visits (Chile, 2007b). Indeed, the 2006 US Inspection report noted:

> *… in light of the central role of science in Antarctica, the Team found it curious that some stations (including ones not visited by the Team) seemed to be going out of their way to attract tourist vessels. While there is nothing wrong with focusing attention on the historic aspects of stations and their locales, and visits by tourists*

14 GPI base inspection reports, while focusing on environmental issues, sometimes commented on excessive military presence for logistic support; presence for apparent political (territorial) reasons; and limited scientific activity at some stations.

15 ATCM Resolution 8 (1995), *Guidelines for proposing new Historic Sites and Monuments*, contains seven guidelines that should be addressed when proposing new HSMs, of which three refer explicitly to the history of science or the development of knowledge of Antarctica.

Fig. 5.6 – Monolith erected to commemorate the establishment of China's Great Wall Station in February 1985 (HSM 52). The initial construction of the station, carried out by a 600-strong construction crew, is alleged to have inflicted severe environmental impacts, and its subsequent operation was subject of criticism by environmental groups (e.g. GPI, 1990). A recent official inspection (USA, 2006) commended aspects of the station's operations and also recommended some improvements (Photo by author, 2006).

> *can increase general understanding of the role of stations in Antarctica, the Team felt that resources might be better focused on expanding science programs than attracting tourists. (USA, 2006:11).*[16]

With the entry into force of the Protocol in January 1998, Antarctic Treaty parties conducting inspections also required the removal of abandoned stations. A number of stations have indeed been fully or partly removed. In addition, a number of waste disposal sites have been fully or partly cleaned up since the entry into force of the Protocol – this being perhaps the requirement that Antarctic Treaty parties complied with more enthusiasm. However, a more recent official inspection made very similar recommendations to those made by environmental NGOs fifteen years earlier:

16 See also Uruguay (2005), Chile (2007b), and XXX ATCM Final Report, paragraphs 172-173: 24.

Fig. 5.7a-b – Port Lockroy, Base A, Goudier Island. Established in 1944 for Operation Tabarin and abandoned between 1962 and 1994 (upper). Designated as Historic Site and Monument 61 in 1995. Since 1997 it operates as a summer-only station (primarily a "living museum") for tourist visits. Upper photo – Boatshed building prior to restoration; photo below - Main base building, Bransfield House, after restoration (Upper photo by author, 1993; photo below by L. Hacquebord, 2005).

> *The Team observed a number of dilapidated huts and refuges...An effort should be made by those responsible ... to either repair them fully for some reasonable use (such as support for science) or remove them. There is no basis for old structures without historic designation to be left in such places simply on the basis that they might at some point provide a bit of safety to someone; if that were an appropriate bases for building and retaining huts, they would dot the landscape in contravention of basic environmental principles, if not the Environmental Protocol itself. (USA, 2006:11).*

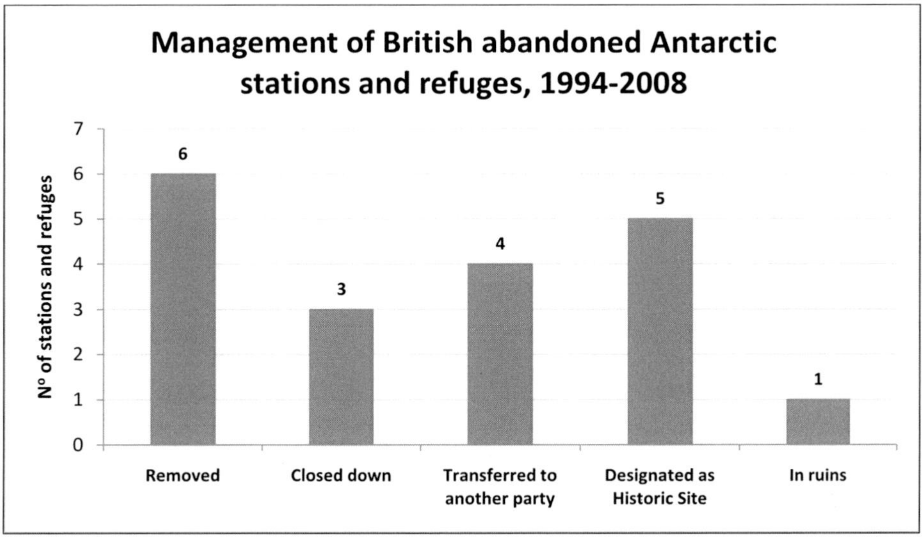

Fig. 5.8 – An example of the fate of abandoned Antarctic scientific bases. The United Kingdom, which as other claimant states established numerous Antarctic bases through the 1940s and 1950s, began to take action on its abandoned bases in the 1990s. (Data from Cochrane and Collinge (1994) and BAS (2008). Bases established outside the Antarctic Treaty Area were not included).

Science has always played a role in Antarctic endeavours, and since the signature of the 1959 Antarctic Treaty it has become the primary activity in the region. However, not all Antarctic bases have been primarily built or operated as scientific bases. In Antarctica, the placement of infrastructure on a certain place has long been played a practical, ritual and symbolic role in safeguarding (or countering) territorial claims and of keeping a foothold in the region.[17] Infrastructure has also been used to enhance a Party's influence in the ATS that derives from an active Antarctic presence. In the same vein, the process of designating an HSM is part of a ritual, and the HSM itself remains as a lasting symbol of presence that continues well after the original activity has been terminated. Each HSM reflects the wish of one or more parties to have an element of their own cultural heritage valued and protected. However, by designating Antarctic remains of past activities as an HSM an *international* status is given to elements that have primarily a *national* value. In what concerns Antarctic bases the risk is that an imaginary scientific past is created during the process of historic designation.

Overall, between 1983 and 2007 a number of Antarctic bases that had been active only some decades earlier were designated as HSMs, while others were removed. Some bases were memorialised or turned into tourism support facilities. Early

17 For a discussion of ritual and symbols of colonizing powers (albeit not specifically in the Polar Regions) see Seed (1995).

assessments that there were many redundant Antarctic facilities, and many active facilities that conducted limited science, appear to be as valid in 2008 as a decade earlier.

5.4 ANTARCTIC SCIENTIFIC BASES AND RUBBISH THEORY

Antarctic scientific bases and the process of designating HSMs can be examined from the perspective of Thompson's Rubbish Theory (Thompson, 1979). Thompson identified three categories of objects: the "durable" – an object that increases in value over time and has theoretically an infinite life span; the "transient"– an object that decreases in value and has a finite lifespan; and "rubbish" – that which has "zero and unchanging value" and usually does not disappear but "continues to exist in a timeless and valueless limbo" (Thompson, 1979: 7-10). Rubbish, however, may at a later day be rediscovered and categorized as cultural heritage (Thompson, 1979; Carman, 2002; Howard, 2003: 47). The process is thus:

Transient → Rubbish →Durable

Cultural remains on the polar landscape may belong to any of these broad categories. A certain cultural heritage item may go through all these stages during its life cycle. However, the distinction between items in one category and the other may not be clear-cut. The remains of past activities may blend with items from the contemporary present that are still in use or that have only recently stopped being used. In between the past and the present there is a grey zone where the "artefact or rubbish" question has not yet been resolved or addressed. Looking at the "things" in the Antarctic landscape (c.f. section 5.2.2) it is possible to broadly classify them thus, using Thompson's categories:

- Permanent stations can be regarded mostly as "transient";[18]
- Abandoned stations and waste dumps can be regarded mostly as "rubbish"; and
- Historic Sites and Monuments can be regarded as "durable".[19]

A scientific station may through its life cycle be in any or all three categories. Through the 1990s and 2000s, factors such as the continued existence of the ATS and the end of the Cold War had made the presence of Antarctic scientific bases no longer as necessary from a geopolitical perspective as they had been in the past. This result-

18 Some Antarctic stations have been active for decades and are likely to remain for the foreseeable future. However, the station buildings themselves may change over time. While the *location* of the station may be regarded as permanent, the *buildings* themselves can be considered transient using Thompson's categories. Some modern stations are being built with a projected lifetime and the ability to be dismantled with relative ease after their life cycle has been completed.

19 However, HSMs can be de-listed and there are a few instances in which this has happened.

Fig. 5.9 – Artefact or rubbish? Fuel and whale oil tanks at Hector Whaling Station in Whalers Bay, Deception Island, which are protected as part of a specially managed historic site (upper). Fuel tanks at King George Island formerly used to supply the Soviet fishing fleet, the removal of which has long been requested by both NGOs and Antarctic Treaty parties (below) (Upper photo by L. Hacquebord, ca. 2005, photo below by author, 2006).

ed in a number of scientific stations that were transferred from one Party to another, mothballed, removed, or abandoned (Fig. 5.8). This process was, in effect, a series of transfers from one of Thompson's categories to the next, with some bases remaining "transient" objects under a different operator, others becoming "rubbish" as abandoned stations, and yet others being designated as "durable" with a historic site designation. These decisions were made on the bases of national obligations (including obligations under international law) as well as national interests.

With the adoption of the Protocol environmental perspectives gained a dominant role in Antarctic affairs. Deserted stations and field huts as well as waste dumps had to be removed in compliance with the Protocol. However, abandoned bases were

Fig. 5.10 – Cape Hallett Station exhibit at the Christchurch Museum, Christchurch, New Zealand (Photo by author, 2006)

generally regarded as a waste management issue rather than as a cultural heritage issue. The debate was primarily about the environmental, economic and technical merits of keeping or removing buildings and other wastes from Antarctica, rather than about the potential heritage value these might have held. For instance, in discussing policy changes with regard to waste management in the Antarctic, Walton notes – almost as a side comment – "Let us also not forget that one man's rubbish, when given the acceptable patina of age, is another man's historical artefact!" (1990:103).

The focus on environmental perspectives was partly related to the spirit of the times (the period coincident with the Protocol negotiations), and partly to the fact that some past activity sites were relatively recent. Up until the late 1980s about half of the HSMs dated from the heroic era of Antarctic exploration or earlier. [20] More recent sites were not usually listed as HSMs – and therefore likely to be categorised as "rubbish". Nevertheless, some parties were already aware of the potential historic value of their own relatively recent remains. A 1991 GPI field report about Australia's Wilkes Station refers to an archaeological assessment of the station that had been carried out the previous year, which had recommended that "...nothing be removed from Wilkes [Station] as all the buildings plus surroundings refuse including fuel drums, should be viewed as artefacts." (Cited in the Annex to GPI, 1991: 6-7).

However, GPI noted that given the Australian government commitment to environmental protection it was unacceptable that Wilkes Station remained "in its appalling state" (Cited in the Annex to GPI, 1991: 7). It urged for the station to be cleaned up, except buildings deemed of historic value, and even recommended taking a compactor to the site to facilitate the removal of drums and other wastes. Today this may seem barbaric from a cultural heritage perspective, particularly since Wilkes Station was one of the original stations used in the 1957-1958 International Geophysical Year (IGY). However, the calls for an environmental cleanup should be placed in the context that

20 The early period of Antarctic exploration known as the "heroic era" is usually taken to span for the period between 1895 and 1917, ending with the departure from Ross Island of the *Aurora* relief expedition.

Wilkes Station (consisting of the base itself and a number of waste disposal sites) had at the time been abandoned for more than 30 years after a relatively short period of activity (1957-1969), and it was just one of a number of abandoned stations in the Antarctic at the time.[21] Many fairly noteworthy scientific bases had not yet, or had only recently, been designated as HSMs. Examples include East Base in Stonington Island (designated in 1987), Base A at Port Lockroy in Goudier Island (designated in 1995) and Base B and the ruins of Base Aguirre Cerda, both in Deception Island (designated in 1995 and 2001, respectively). All of these had been once abandoned scientific bases, and prior to designation as historic sites the environmental management of the sites had been a matter of concern for environmental groups.

Among the bases removed in compliance with environmental regulations there were some associated with the 1957-59 International Geophysical Year (IGY). A case in point is Cape Hallett Station, which was established during the summer of 1956/57 by New Zealand and the United States to provide weather data for American aircraft flying between New Zealand and Antarctica, as well as to undertake an IGY research programme. Cape Hallett Station was occupied year round between 1957 and 1964. After the IGY the station continued to undertake scientific research until a fire destroyed the station's main scientific laboratory in 1964. Subsequently the station was operated as a summer only research station until it was abandoned in 1973. Clean up operations started in 1984 and were completed in 2005 (New Zealand and USA, 2006; USA, 2008a: 3). Environmentalists had long regarded Cape Hallett Station as problematic not only because it had been abandoned by many years, but also because it was established in the midst of a colony of Adélie penguins (*Pygoscelis adeliae*) with more than 60,000 pairs, several thousand penguins of which were displaced to allow for the construction of the base. The base establishment and operation resulted the decline of the penguin colony (USA, 2008a:3).[22]

During the two decades that took removing the station some of its components acquired "durable" status. Many of the materials originating at Cape Hallett that were returned to New Zealand as refuse were collected by the Canterbury Museum in Christchurch. An exhibition was mounted on the bases that "the history of the station is both a human and an environmental account of an important era in Antarctic exploration and science" (New Zealand and USA, 2006:8). Paradoxically, after the removal of the station attention was drawn to the enduring historic and heritage values of the base site related to its former use (USA, 2008a:3). The revised management plan for Cape Hallett, which protects primarily the ecological and scientific values of

---

21 To date, Wilkes Station has not been proposed for designation as an HSM although it is protected by Australian legislation, which is applicable to Australian nationals.

22 A part of Cape Hallet was designated as a protected area in 1966 on account that it "... includes a small patch of particularly rich and diverse vegetation which supports a variety of terrestrial fauna and that the ecosystem, which includes a rich avifauna, is of outstanding scientific interest." (Recommendation ATCM IV-7, *Specially Protected Areas: Cape Hallett, Victoria Land*). This was expanded in 2002 to include the breeding area of a substantial colony of Adélie penguins. The area formerly occupied by the station has now been re-colonized (USA, 2008a) although population numbers in 2006-07 were about a third of those in 1959..

Fig. 5.11 – Greenpeace departs Cape Evans after five years of continued operations, February 10, 1992. Scott's Terra Nova hut is in the background (Photo by author, 1992).

the area, establishes that an appropriate authority should assess the historic value of any artefact that may appear at Cape Hallett as a result of e.g. erosion, before its removal and/or disposal. Management options are being discussed for the well preserved remains of a husky that died in 1964, which was not removed during the clean up (USA, 2008a:3). Visitors to the site should not to disturb or remove any artefact from the former station and to notify the find to the expedition leader and relevant authorities (USA, 2008b).

In the case of Cape Hallet Station environmental and cultural heritage perspectives reached a balance of sorts, although this took decades. The establishment and operation of Cape Hallett Station caused a significant impact on the local penguin population; and the base was then removed in compliance with environmental regulations requesting the cleanup of former worksites, but at the expense of historic values. The cultural heritage of Cape Hallett Station was eventually recognised as "durable", but only after the station had been categorised as "rubbish" and most material elements removed from the former station site.

GPI's own World Park Base can be examined from the perspective of Thompson's theory. Whilst not primarily a scientific base, some research was conducted there (Roura, 2007). With the signature of the Protocol in October 1991 – one of the main objectives of GPI's Antarctic campaign – a decision was made to remove it in order to free resources that could be used elsewhere in the region (GPI, 1992: 4; Roura, 2007). Conceptually all buildings, associated installations, and contents were transferred from the "transient" to the "rubbish" category.[23] World Park Base was entirely removed in 1992. A granite benchmark was placed at the site to aid environmental monitoring, but this too was eventually removed in 1996 (Roura, 2004: 55). There are now virtually no traces of past activity at the site (A.D. Hemmings, personal communication, 2 February 2005[24]; personal observation, 27 December 2007).

5.4 CONCLUSIONS

In the context of negotiations leading up to the Protocol, environmental NGOs denounced what were perceived as poor environmental practices of Antarctic Treaty parties. Whilst taking a hard line with regard to the need to clean up many sites and facilities, NGOs recommended that the historic merits of abandoned bases had to be assessed, and that they should either be protected as formally designated historic sites, and properly maintained, or otherwise removed. In this regard there was no conflict between environmental and cultural heritage perspectives with respect to protecting historic sites and the values they embody. After the entry into force of the Protocol, Antarctic Treaty parties conducting official inspections adopted a fairly similar position than NGOs did a decade earlier with regard to abandoned sites and facilities without historic designation. The fate during the past decades of abandoned scientific bases in Antarctica has been diverse, and can be analyzed from the perspective of Thompson's 1979 Rubbish Theory.

However, environmental and cultural heritage perspectives are fundamentally different ways of regarding human presence in Antarctica. From an environmental perspective historic values are not generally considered part of the Antarctic environment,[25] and the material remains of Antarctic activities mark the boundary of civilization and the encroachment upon the last great wilderness – conforming, in effect, an *Antarctic frontier*. Instead, from a cultural heritage perspective the same boundary conforms a cultural landscape that is a material record of human achievements in the region, a source of information about the past, and a source of inspiration for future generations. There is a basic tension between these two perspectives. However, there are convergences too, which result from the growing pressures on the intrinsic values of Antarctica[26] and the legal requirement under the Protocol and its Annexes to protect them.

Science is the primary activity in the Antarctic under the 1959 Antarctic Treaty and related instruments, but not all Antarctic bases were primarily built or operated for

23 In practice only a fraction of the base and its contents was dealt with as refuse – most materials removed from World Park Base, including buildings and other infrastructure, equipment, fuel, etc., were recycled or reused elsewhere.

24 *The site of Greenpeace's World Park Base, fourteen years after its removal.* Unpublished manuscript, 2 February 2005. (In file with author).

25 See e.g. Bastmeijer and Roura, 2008: 202.

26 As D. Lowenthal has noted, both the cultural and natural heritage are non-renewable and in limited supply; and both are endangered by the same forces (2005: 85-86).

scientific purposes. Safeguarding individual parties' territorial claims and influence in the Antarctic Treaty System were also factors.[27] Antarctic bases designated as historic sites should be remembered for the significance they may hold on scientific, historic, archaeological, or other grounds according to internationally accepted criteria, and not as symbols of an imaginary scientific past. The misuse of the historic designation process – assigning protected status to material remains that actually lack in significance – could result in an erosion of the wilderness, historic, and other values of Antarctica. Clarity of purpose in the designation of Antarctic historic sites and monuments will enhance the significance of those that are left as a legacy for the future.

5.5 ACKNOWLEDGMENTS

I acknowledge the support of Prof. Dr. Louwrens Hacquebord (Arctic Centre, University of Groningen, The Netherlands), which enabled my participation in the conference of the International Polar Heritage Committee in Barrow, Alaska, in September 2007, where the issues discussed here were first presented. Travel funding support from the Groningen Institute of Archaeology, University of Groningen, was much appreciated. Several entities and individuals enabled the fieldwork that informs this document, notably Greenpeace International and former colleagues of the Greenpeace Antarctic Expedition. Dr. Dag Avango (Arctic Centre, University of Groningen) usefully directed me to scholarly research on the rituals and symbols of occupation. I also thank Jim Barnes (ASOC), Janet Dalziell (Greenpeace International), and Dr. Alan Hemmings (University of Canterbury, New Zealand), for discussions on topics relevant to this article over many years, without implicating them in my interpretations.

27 Pursuing national interests, however, is also part of Antarctic history and cultural heritage, if not as universally valuable as scientific research, the preservation of international peace, or environmental protection.

6 The polar cultural heritage as a tourism attraction

A case study of the airship mooring mast at Ny-Ålesund, Svalbard†

ABSTRACT

This paper explores the use of narratives in the transformation of historic sites in the Polar Regions into attractions and consumable tourism products. The analysis is based on a case study of visitation to the airship mooring mast built at Ny-Ålesund, Svalbard, for the 1926 "Amundsen-Ellsworth-Nobile Transpolar Flight" of the airship *Norge*. The questions addressed in this paper are: How does cultural heritage in the Polar Regions operate as a tourist attraction? What is the role of tourism narratives in creating a tourism attraction? Direct observations constituted the main research method. Based on MacCannell (1976) and Leiper (1990), a tourism attraction is a system comprising a tourist or human element; a nucleus or central element; and a marker or informative element. Tourism narratives enable the different elements of the tourism attraction system to "click" together into a coherent whole. Through narratives, the mast becomes a place of significance and a symbolic marker of the North Pole and polar exploration. The application of this approach to other sites in Antarctica and Svalbard is discussed.

6.1 INTRODUCTION

How does cultural heritage in the Polar Regions operate as a tourist attraction? What is the role of tourism narratives in creating a tourism attraction? These are the questions addressed in this paper, which explores how historic sites in the Polar Regions are transformed in attractions and consumable tourism products through the use of narratives. The analysis is based on a case study of visitation to the airship mooring mast (the mast) built at Ny-Ålesund, Svalbard, for the 1926 "Amundsen-Ellsworth-Nobile Transpolar Flight" of the airship *Norge* (Fig. 6. 1). The mast is known locally as "Amundsen's mast."

†Published as:

Roura, R.M. (2009): Cultural heritage, tourism narratives, and tourism impacts: A case study of the airship mooring mast at Ny-Ålesund, Svalbard. *Special Issue on Polar Tourism, May 2009. Téoros 28*:1, 29-38. Reproduced with permission.

Fig. 6.1 – Tourists at the airship mooring mast at Ny-Ålesund, Svalbard. Photo by author, 2007.

Polar tourism combines two global processes: the long-established use of the Polar Regions as resource frontiers (Sugden, 1982), and the centrifugal tendency of the tourism system that constantly expands into new areas (Cohen, 1984: 382). Consequently, certain sites that have particular natural, cultural heritage or other attractions become tourism destinations where organized visits take place on a regular basis through the tourism season. According to G.J. Ashworth and B. Graham (2005: 7), heritage "is that part of the past which we select in the present for contemporary purposes, whether they be economic or cultural (including political and social factors) and choose to bequeath to a future." More succinctly, "Heritage denotes everything we suppose has been handed down to us from the past" (Lowenthal, 2005: 81). Historic sites – a tangible form of cultural heritage – are the subject of specialized polar tourism and also a component of mainstream tourism itineraries, complementing the "menu" of tourist attractions offered to customers. For the purposes of this article, a historic site is defined as any location that demonstrates past human activity, evidenced by the presence of artefacts, ecofacts, features, structures, or other material remains (Kipfer, 2007: 50).

In Antarctica and Svalbard, historic sites are among the most visited of all sites (Chapter 8). Historic sites are attractive to tour operators for several reasons. They may have originally been established because of logistic advantages such as having a safe anchorage or comparatively easy access. Unlike iconic Arctic fauna, for example polar bears or walruses, historic sites can reliably be found in the same place year after year. However, as noted by M.B. Schiffer (1987), sites are not static but subject to transformations resulting from both natural and cultural processes. Importantly, historic sites often have material remains (there is something to show) and a history (there is a story to tell).

In his analysis of the semiotic of attractions, D. MacCannell (1976: 109-110) suggested that tourism attractions are signs, that is, they represent something to someone. He defined a tourist attraction based on the empirical relationship between a tourist, a sight (the subject of sightseeing), and a marker (information about a sight). Significance is given to the sights by means of information, in turn giving sense to the tourism experience. Sights can become markers themselves, and certain sites interest tourists because of their markers rather than the intrinsic qualities of the sights. Based on D. MacCannell (1976) and C. Gunn (1972), N. Leiper (1990: 371) developed the concept of attraction further to include tourism activities other than sightseeing. He defined a tourism attraction as a system comprising three elements: "a tourist or human element; a nucleus or central element; and a marker or informative element. A tourist attraction comes into existence when the three elements are connected." N. Leiper (p. 382) further suggests that each of the elements is an important part of the attraction system; *e.g.*, a place with no visitors would not be regarded as a tourist attraction. This model will be applied with regard to the airship mooring mast at Ny-Ålesund, and to other historic sites in Svalbard and Antarctica. Let it be stressed that the cultural heritage discussed in this article consists of non-indigenous historic sites that are protected under applicable legislation. Plainly the polar cultural heritage can be more broadly defined and include intangible as well as tangible elements. The Archipelago of Spitsbergen is situated in the eastern part of the Arctic basin, at latitude 80° N on average, and approximately 1000 kilometres from the North Pole (Fig. 6. 2a). The archipelago is now usually known by its Norwegian name of Svalbard.

The islands had no indigenous populations and human presence began in the 17th century following discovery by Wilhem Barentz in 1596, which was followed by waves of activity, including exploration, natural resources exploitation, World War II confrontations, science, and tourism (Sugden, 1982: 283-285). Svalbard is subject to the 1920 Treaty Concerning the Archipelago of Spitsbergen, which grants the sovereignty of the islands to Norway while maintaining equal rights for all contracting parties, subject to Norwegian laws. Ny-Ålesund (78° 56' N, 11° 56' E) in Kings Bay (Kongsfjord) is a former Norwegian mining town turned into an international research centre (Fig. 6. 2b). Ny-Ålesund is owned and operated by Kings Bay AS, a Norwegian Crown company.

Between 1926 and 1928 three expeditions departed from Ny-Ålesund attempting to reach the North Pole by air: Richard E. Byrd's flight in 1926; the "Amundsen-Ellsworth- Nobile Transpolar Flight" in 1926; and Umberto Nobile's *Italia* expedition in 1928 (Amundsen and Ellsworth, 1927; Litchfield, 1927; Nobile, 1961; Nelson, 1993). The first attempt took place in a fixed-wing aircraft, and the latter two in airships, lighter-than-air crafts lifted by a gas and propelled by engines. These expeditions will be described briefly as narratives about them are central to the case study discussed here. Richard E. Byrd's was the first of these attempts. He and a co-pilot departed Ny-Ålesund on May 9, 1926, in a trimotor airplane named *Josephine Ford*. After a harrowing 19-hour flight, Byrd returned to Ny-Ålesund claiming to have reached the North Pole (1968: 233). However, Byrd's statement was brought into question within minutes

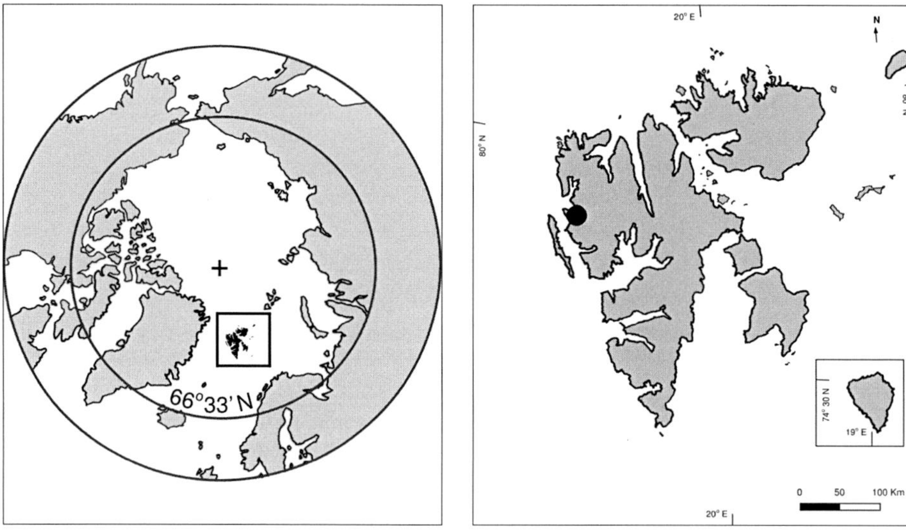

Fig. 6.2 – Left: location map of Svalbard. Right: location map of Ny-Ålesund. Maps by author.

of his return (Driveness and Jølle, 2006: 273), thus beginning a controversy that lasted for decades. A recent examination of his navigation notes, which had been lost for years, established beyond doubt that the *Josephine Ford* did not reach the North Pole in 1926 (Rawlins, 2000).

The "Amundsen-Ellsworth-Nobile Transpolar Flight," in the airship *Norge*, departed from Ny-Ålesund on May 11, 1926, two days after Byrd's attempt. The *Norge* reached the North Pole on May 12, 1926 and continued on to Alaska (Amundsen and Ellsworth, 1927; Nobile, 1961). The role and relative contribution of each of the expedition's leading members – Roald Amundsen, Lincoln Ellsworth, and Umberto Nobile – was the subject of some debate, not least among the participants themselves. It is not the intention here to contribute to that debate, simply to outline their roles succinctly. Roald Amundsen, who – among other polar exploration feats – had led in 1911 the first expedition to reach the South Pole, conceived the transpolar flight and provided his name and expertise in polar exploration. His official job was to look for new land as the airship flew over unknown parts of the Arctic Ocean. Lincoln Ellsworth, a US national, was a key expedition sponsor, and nominally the expedition's "leader of the scientific work and navigator." The airship's pilot was the Italian air force colonel Umberto Nobile, who had designed and built the *Norge*, and flown it from Rome to Kings Bay (Amundsen and Ellsworth, 1927: 128; Nobile, 1961; Driveness and Jølle, 2006: 267-276).

Relations between Roald Amundsen and Umberto Nobile were tense from the start. Roald Amundsen and Lincoln Ellsworth regarded themselves as the expedition leaders, whose job was to select the best people for the various posts and ensure that the "whole machinery" was in working order; an airship captain's role was to assist the leaders with flying the airship rather than to take command of the expedition (Amundsen and Ellsworth, 1927: 127-128). Umberto Nobile rejected being treated as a "hired pilot" (as Amundsen described him) and thought that he should be regarded as a joint leader, given his central role in the expedition and the fact that Roald Amundsen knew "absolutely nothing" about airships (1961: 92-93). After the trip, Amundsen and Nobile fell out with each other (Nelson, 1993; Driveness and Jølle, 2006).

Following the 1926 transpolar flight, Umberto Nobile organized a new airship expedition under his command on the airship *Italia*, which was an improved version of the *Norge* (Nelson, 1993). The *Italia* departed Ny-Ålesund on 23 May 1928 and successfully reached the North Pole 20 hours later. Weather conditions prevented the planned landing at the North Pole, and a decision was made to return to Svalbard. On May 25, the airship crashed on the sea ice at about 400 km NE of Kings Bay. Ten people survived the crash. Radio signals asking for rescue were picked up 12 days after the accident, triggering an international rescue effort that lasted several weeks. Roald Amundsen joined the rescue with tragic consequences: his chartered flying boat disappeared in the Barents Sea. Eventually, a rescue plane was able to land at the survivors' camp on 21 June. By then some survivors had started a trek to Svalbard. Umberto Nobile, who had sustained injuries and was expected to assist with the rescue operation from Kings Bay, was reluctantly the first to be evacuated. The rescue plane returned to collect more survivors but crashed on its second landing on the sea ice. The Soviet vessel *Krasin* rescued the remaining survivors almost three weeks after Nobile had been flown out, which time he spent as a virtual prisoner in an Italian warship at Kings Bay. The Italian fascist government demoted Nobile, who was forced into exile until 1943. The press, particularly in Norway, accused him of cowardice. His name was cleared before his death in 1978, and his polar achievements eventually recognized (Nobile, 1961; Nelson, 1993; Aas, 2005).

There are relatively few material remains at Ny-Ålesund of these dramatic events, of which the airship mooring mast is the most conspicuous.

6.2 METHODS

Several research methods were used to study tourism at historic sites in Svalbard and Antarctica between 2002 and 2008 (Chapter 2). These can broadly be categorized as unobtrusive research methods (Webb *et al.*, 1966; Lee, 2000). Fieldwork for this case study was conducted during June-July 2007 while based at the Netherlands Arctic Station at Ny-Ålesund. Direct observations (Martin and Bateson, 1983; Veal, 1992; Bruner, 2005; Pearce, 2005), complemented by photography (Collier and Collier, 1986 [1967]), constituted the main research method. Tourist landings at Ny-Ålesund take place almost daily during summer according to a pre-established schedule, from which it is possible to determine which groups of tourists will visit the mast. Overall, six groups of about 40-50 tourists each were observed closely (i.e. from within the group). Observations started just before the guided visit began, and ended as the group returned to the settlement 30-45 minutes later. Four different guides led the groups (two guides visited Ny-Ålesund twice during the observation period). Several more groups were observed opportunistically from the distance. Both approaches provided complementary perspectives of individual and group behaviour.

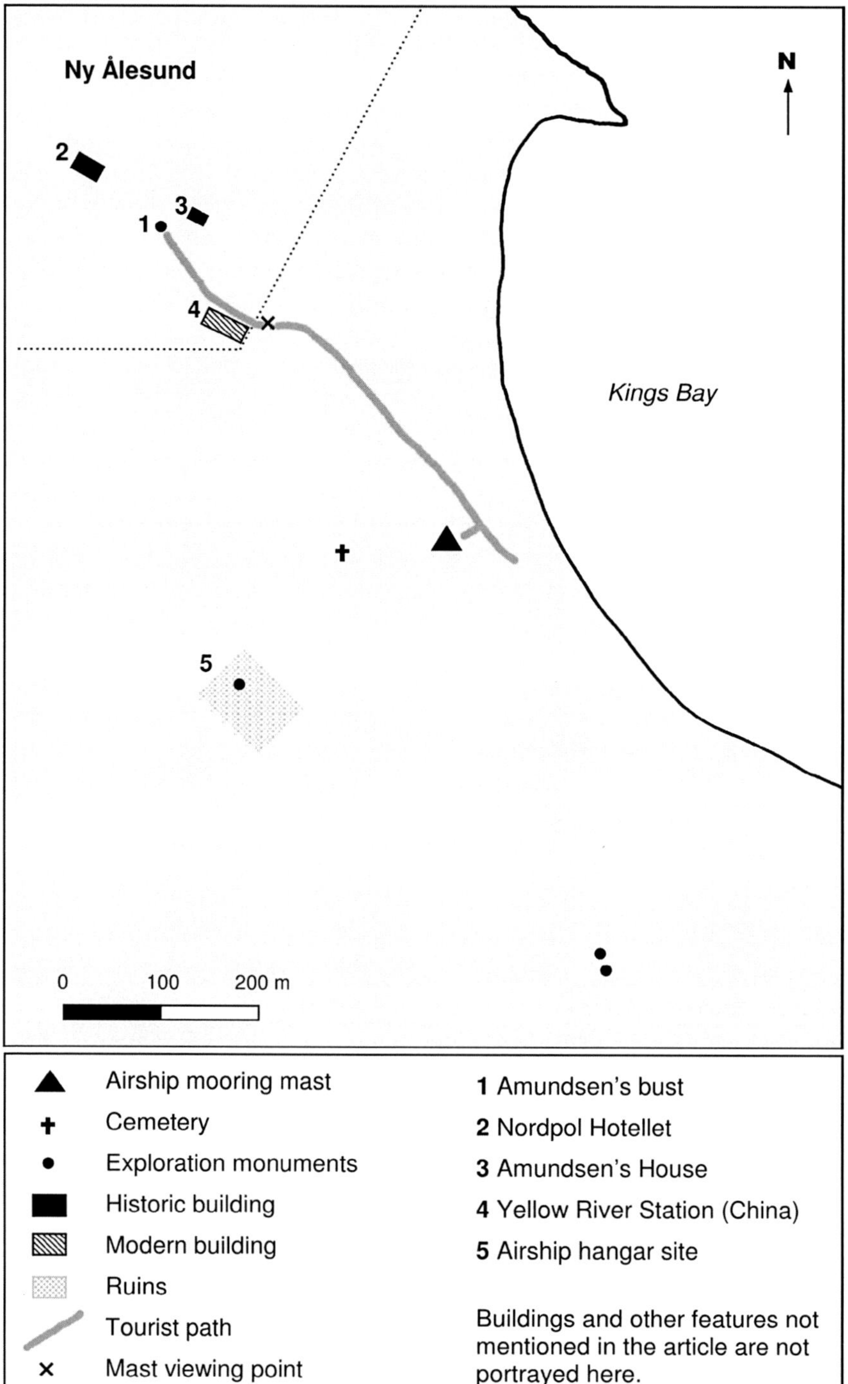

Fig. 6.3 – Sketch of Ny-Ålesund showing cultural heritage features associated with polar exploration by air and contemporary tourism. Sketch by author.

Observations were conducted from an etic perspective. This means that behaviour was described regardless of the apparent motivations or perspectives of the visitors (Pearce, 2005: 3-4). Observations followed an *ad libitum* sampling pattern in which whatever was visible and seemed relevant was recorded (Martin and Bateson, 1983: 84). Access to each group was negotiated on case-by-case bases with the expedition leaders, who in all instances allowed observation of the visit. The investigator's presence was not concealed from the tourists, although efforts were made to blend with the group rather than to stand out. For instance, observations were memorized and photographed rather than written down, thus outwardly the investigator's behaviour mimicked that of the tourists. Detailed field notes were written back at base immediately after each visit was completed.

Particular attention was paid to narratives about the historic events associated with the airship mooring mast as delivered by the tour guides in English. Following Dean MacCannell (1976: 110), narratives are regarded as markers – that is, information about a specific sight (or nucleus). Variations in the narratives shaped the version of historic events that was transmitted to different tourist groups. The content of each of the narratives, as transcribed in field notes, was analyzed to identify the individual components of the narrative – the various stories within the story (Bryan and Bernard, 2003). The key themes concerned the expeditions and actors that were included in each narrative. A different type of marker consists of the inscriptions of several memorials to the 1926-1928 transpolar flights, which were also documented and analyzed.

6.3 THE NUCLEI: THE AIRSHIP MOORING MAST AT NY-ÅLESUND AND ASSOCIATED CULTURAL HERITAGE

Heritage assets, both tangible and intangible, are the foundation of heritage/cultural tourism (Timothy and Boyd, 2003; McKercher and du Cros, 2002: 7). The main historic assets at Ny-Ålesund related to Arctic exploration by air are the airship mooring mast and associated cultural heritage (Fig. 6.3).

A mooring mast is a structure designed to anchor the air ship to the ground when it is not in flight for the purposes of personnel access, loading or replenishing the craft, or for standing down between operations. The mooring mast allows the airship to rotate on a horizontal plane, according to wind direction, just as a vessel at anchor. This enables an airship to tie up to the mast and wait for favourable wind conditions, for example to be moved into a hangar. The mooring system must be capable of making the craft secure under all likely weather conditions. That requirement influences both the technique adopted for mooring the aircraft and the determination of the mast's structural strength (Litchfield, 1927: 84; Howe, 1999: 301). Thus, an airship mooring mast is a substantially more complex engineering structure than it first appears to the layperson.

The mooring mast at Ny-Ålesund was one of three erected for the 1926 Transpolar flight along the path between Rome and Kings Bay (Nobile, 1961: 22). The mast is located some 300 metres to the southeast of Ny-Ålesund and just outside the settlement's limits. It consists of a triangular vertical metal tower approximately 14 tons in weight and 35 metres high. Each side of the mast is six metres wide at the base, and is

Fig. 6.4 – Tourists at the airship mooring mast. A history lecture was interrupted to watch a pod of beluga whales (Delphinapterus leucas) swimming in the fjord. Photo by author, 2007.

fastened to each corner to a 40-ton concrete block (Höver, 1927: 37, 43). A ladder on the inside of the mast reaches to a platform located near the top from where airships could be accessed. At the top of the mast there is a masthead, which is a rotating locking mechanism used to fasten the airship to the mooring mast.

Associated features include the remains of the airship hangar and some buildings that were in use in 1926, notably Amundsen's house and the "Nordpol Hotellet" where some expedition members and journalists stayed. These buildings, which are still in use, are legally protected, as are the mast itself and the hangar remains. Norway's 2001 Svalbard Environmental Protection Act protects structures, sites, and movable historical objects predating 1946, as well as some more recent cultural remains that are of particular historic or cultural value (Det Kongelige Miljøverndepartement, 2001). In addition, several memorials commemorate historic flights from Ny-Ålesund (Figs. 6.4 and 6.5).

In various parts of the world there is a scatter of material remains and memorials that are associated with the 1926-1928 transpolar flights. Some of these have become local tourist attractions themselves partly because of this association. These include an airship mooring mast at Vadsø, Norway; replicas of memorials at Ny-Ålesund located in Alaska and elsewhere; and the icebreaker *Krasin*, now a museum in St. Petersburg, which contains a permanent exhibit on the rescue of the 1928 *Italia* expedition. The recently opened Spitsbergen Airship Museum in Longyearbyen, Svalbard's largest settlement, contains a collection of items and documentation related to the airships that operated from Svalbard.

6.4 THE TOURISTS: ORGANIZED VISITS TO THE AIRSHIP MOORING MAST

Tourism in Svalbard began in the 19th century, with the first organized commercial cruise in 1871 (Conway, 1906: 302). Ny-Ålesund became a popular destination during the years immediately preceding World War II as the farthest north point of call (Polunin, 1945: 82). Tourism began to increase following the opening of an airport in Svalbard in 1975, and then with the Norwegian government designation of tourism as one of three priority areas for Svalbard in 1990 (Viken and Jørgensen, 1998; Norway, 2004: 3).

Currently cruise ships visit Ny-Ålesund from late May to mid-September. Visits are almost on a daily basis and, at the height of the season, there may be two or three passenger landings a day, with simultaneous landings occurring fairly frequently. Nearly 30,000 shipborne passengers landed at Ny-Ålesund in 2007, about three times the figure estimated in 1998 (Governor of Svalbard, 2006: 6; Kings Bay AS, 2007). At present, tourists visiting Ny-Ålesund are allowed to walk through a 1.5 km "nature, research and cultural heritage" trail around the settlement and visit the mining museum, the post office, and a shop. Outside the settlement they may visit the airship mooring mast, when accompanied by polar bear guards. Based on the passenger capacity of ships used by companies that organize guided visits to the mast, it is estimated that about 10,000 passengers could have visited the mast in 2007, which would make it one of the most visited historic sites in Svalbard.

The logistics of moving tourist groups of different sizes within a certain time determine whether or not tourists may visit the mast. The polar bear risk is low, yet armed guards are needed for the visit – for instance, a polar bear was reported to be swimming towards the mast just as a group of tourists was visiting it, hastening their return to the settlement (personal observation, 24 June 2007). The passengers of ships with smaller passenger capacity (*ca.* 50-100 passengers) are usually taken in guided visits to the mooring mast. However, not all passengers on any given cruise join the guided visit, particularly during poor weather conditions. Passengers from larger ships can view the mast from the edge of the built-up area, some 300 metres away (Fig. 6. 3). Sometimes groups from both large and small ships coalesce at the viewing point simultaneously, generating confusion among tour guides and tourists as to who may visit the mast and who is not allowed.

Notwithstanding the historic focus of the visit, the historic elements are only a component of visiting the mast. Judging from the tourists' behaviour – for instance, the choice of photographic subjects – , flora, fauna, and landscapes are attractions too (Fig. 6. 4). In most of the visits observed for this case study there were a few tourists who stood somewhat apart from the rest of the group and/or turned their back to the tour guides during lectures.

6.5 THE MARKERS: NARRATIVES OF THE TRANSPOLAR FLIGHTS

The airship mooring mast is a deceptively simple structure that, however, is difficult to interpret since most visitors would have never seen an airship. Understanding how such structure is linked to the operation of an airship may not be immediately obvious. For instance, a visitor apparently believed that the mast had been part of the airship and asked a guide (twice) if it had been returned to Ny-Ålesund after the North Pole flights (personal observation, 26 June 26 2007). Consequently, understanding the practical and historic role of the mast requires markers. On-site markers are provided as tour guide oral narratives and, more succinctly, as inscrip tions in various memorials.

There are several modalities of delivering verbal information to those who visit the mast. Usually the group gathers at Amundsen's bust in a crossroads of the settlement (Figs. 6.3 and 6.5), where the guide delivers an introductory lecture. Then the group walks to the mast where the story is completed (Fig. 4). This is usually punctuated by an interlude at China's Arctic Yellow River Station to examine the concrete dragon statues that guard its door. Sometimes the guides stop two or more times before reaching the mast, each time giving some additional information and slowly building up momentum to the story prior to a *grande finale* at the foot of the mast. Most groups gather on the south side of the mast, where a memorial plaque is located, but some gather on the north side and sometimes overlook the plaque altogether. In some instances, however, the guides do not deliver a lecture at all but simply respond to the questions of those who are interested.

Several approaches are available to transform a cultural asset into a consumable tourist product. These include mythologizing the asset, building a story around it, and showing a direct link between the past and the present (McKercher and du Cros, 2002: 128-132). Several of these tactics are at play at Ny-Ålesund. The airship mooring mast, with its dramatic history of success and tragedy, is easy to mythologize so that it becomes a place of significance. It may be regarded, as described by a guide, as "one of the ten, or perhaps five, most significant historic sites in the Arctic" (Chapter 8). In addition, some narratives contrasted present day tourism travel in the Polar Regions with the past days of polar exploration, with their obvious differences in purpose, comfort, and risks.

Narratives about the mooring mast are polyvocal – stories change and evolve as they are retold. Key differences in the narratives observed for this research included:

- All narratives described the *Norge* expedition, and most provided some background on the history of exploration of the Polar Regions, sometimes going as far back as the 1494 Treaty of Tordesillas. Some narratives also told the full story of the *Italia* expedition (Fig. 6. 6).
- The claims of Robert E. Peary (1906), Dr. Frederick A. Cook (1908), and Admiral Richard E. Byrd (1926) to have been the first to reach the North Pole by land or air were critically examined in most lectures, and sometimes these explorers were labelled as "liars."
- The first people to actually set foot on the North Pole were variously identified as one (or four) unnamed Soviet sailors; the US national Ralph Plaisted (1968); and

Briton Wally Herbert (1969). These had arrived at the North Pole in nuclear submarine, snowmobile, and dog sledge, respectively.

- Roald Amundsen was always portrayed as an experienced polar explorer, and in some narratives his polar career was reviewed from the beginning. While he was seldom criticized, he was not warmly described either. In other narratives his role in the transpolar flight was described as symbolic rather than practical. "He was sitting in the cabin looking at the [sea] ice and taking notes. What of?" (personal observation, 28 June 2007). Still other narratives suggested that Roald Amundsen's failed attempt to rescue the survivors of the *Italia* disaster, which cost him his life, resulted from a desire to protect his public image rather than from a willingness to help Umberto Nobile.
- Lincoln Ellsworth was a major sponsor of the transpolar flight and in some narratives this gave him a higher status than Umberto Nobile's. He was usually portrayed in passing, as a wealthy and somehow vain person who unnecessarily delayed the *Norge*'s departure date because he wanted to reach the North Pole on his birthday, May 12 – which he did, thus enabling Richard E. Byrd's attempt two days before.
- Umberto Nobile was sometimes portrayed as little more than a footnote – just an unnamed Italian pilot. Unsympathetic narratives disregarded his role in the expedition and sometimes subtly ridiculed him; his position as a military man in Italy's Mussolini made him suspected of chauvinism; and the fact that he had been the first to be evacuated from the *Italia* wreck implied weakness of character if not cowardice. That type of narrative appears to be inspired, in part, in Roald Amundsen's own opinion of Nobile (see, *e.g.*, Aas, 2005: 5-6). In contrast, sympathetic narratives emphasized Nobile's achievements of flying airships from Europe to Spitsbergen, let alone across the High Arctic, and justified his early rescue from the *Italia* as a rational decision.
- On closing the presentation some guides commented that Roald Amundsen's death fitted his life as an explorer, and speculated that he would have approved, or at least not minded, to end up his life in a risky Arctic mission: after his many achievements as a polar explorer, Amundsen had nothing left to accomplish. The guides emphasized that this was purely a speculation on their part; as the Italian proverb goes, *se non è vero è bene trovato* [even if it is not true, it makes a good story].
- Some narratives had minor theatrical touches. For instance, while standing at the foot of the mast, a guide described – accompanying the narrative with gestures – how Roald Amundsen and his companions climbed up the mast to board the *Norge*, waved farewell to the crowd below, and took off. That performance enlivened story telling; however, the mast was not in fact used for the 1926 flight, and period photographs show the *Norge* resting on the ground just before departure. This underscores that tourism lectures are more about the performance of storytelling than about accurate historic content.

The memorials that are most accessible to tourists are a Norwegian bust of Roald Amundsen at the centre of the village, where guided visits to the mooring mast begin, and an Italian plaque at the mast itself, where most visits end (Figs. 6.4 and 6.5). The

Fig. 6.5 – Roald Amundsen's bust in the centre of Ny-Ålesund, with Nordpol Hotellet in the background. Photo by author, 2007.

inscription in Amundsen's bust commemorates the first transpolar flight over the North Pole, "carried out by Roald Amundsen and his men, including an American, Lincoln Ellsworth, in the airship *Norge*," and bears the flags of Norway and USA. The inscription contains no references to Umberto Nobile or to any Italian involvement in the *Norge* flight. Such omission was noted in Italy when the bust was dedicated in 1976 (Aas, 2005: 9). The inscription in the 1983 Italian plaque portrays the flags of the three countries and commemorates the "glorious achievement of human endeavour" of the 1926 transpolar flight and names Amundsen, Ellsworth, and Nobile as well as, generically, "the crew of the airship *Norge*."

Markers act as the catalytic element, linking the human and nuclear elements of an attraction system (Leiper, 1990: 378). The mast is a plain, utilitarian structure, and in the absence of markers some people fail to realize its historic significance. The quantity and quality of information provided to the tourists may vary significantly depending on the different modalities in which the information is transmitted, as described above. For instance, a group of tourists looked briefly at the mast and went on to visit the small cemetery nearby (which, while interesting, lacks the iconic quality of the mast). The group's guide admitted not to know much about polar history (Roura, 2008a: 59).

Through tourists' narratives, the airship mooring mast becomes a tourist attraction system and a symbolic marker of the North Pole and of polar exploration. However, it is a matter of chance what version of the story any tourist group is told, or what inscriptions any individual tourist may notice. In some narratives the mast is primarily about Roald Amundsen and his achievements (hence "Amundsen's mast"), while in

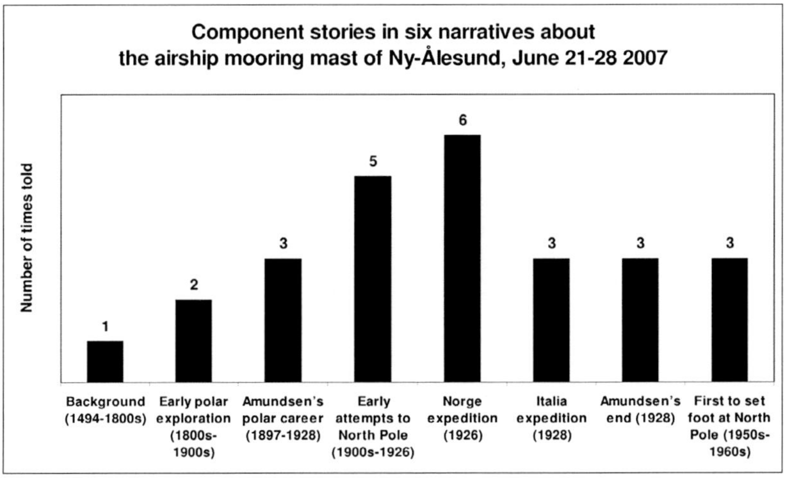

Fig. 6.6 – Component stories of six narratives about the airship mooring mast. Graph by author.

others it is about polar exploration and key polar explorers. Tourists may not notice a dissonance between oral narratives since they will normally listen to only one version of events (of course, tourists may have a previous knowledge of the mast's history or may pursue such knowledge after the visit). With regard to the memorials, it is likely that tourists will pay more attention to the vehicles that carry the inscriptions, which are sights (or nuclei) in themselves, rather than to the inscriptions they contain.

Regardless of the various versions of historic detail, as long as the narratives describe adequately the broad historic events, the mast emerges as a place of significance that, in tourism terms, translates as a place worth visiting. Arguably this is the main objective of the tourism narratives. Conversely, the absence of a narrative is likely to result in lesser appreciation of the monument's historic significance by the tourists and in a defective tourist attraction system.

6.6 POLAR CULTURAL HERITAGE AS TOURISM ATTRACTION SYSTEMS

By treating tourism attractions as systems, Leiper (1990: 371; see also 1979) suggests, they can be integrated conceptually within systemic models for tourism. The framework of tourism attraction systems is applicable to the tourism systems in the Polar Regions, including destinations that are significant because of their cultural heritage.

The tourism systems in the Polar Regions have characteristics of their own, which derive from the regions' remoteness, isolation, and harsh climate. Most tourists will travel in organized forms of tourism, and many nuclei in the Polar Regions would not necessarily become a tourist attraction by the choice of individual tourists that are "pushed" to visit those sites (see Leiper, 1990: 380-381). Rather, polar tourists in mainstream organized trips will be taken to wherever tour operators find suitable, given a range of significant practical constraints such as weather and ice conditions, and more

Fig. 6.7 – The wardroom of Robert F. Scott's Terra Nova Hut at Cape Evans, Ross Island, Antarctica, last inhabited in 1917. Photo by author, 2007.

prosaic concerns such as the time available between meals. Tour operators play a role in the transformation of historic assets of the Polar Regions into consumable tourist attractions, taking tourists to well-known sites as well as making the most of opportunities as they arise. For instance, the mast is visited partly because it is at Ny-Ålesund, where there is a range of other attractions and passenger ships call anyway, and partly because it has the essential elements – sights, stories, and tourist access – that enable it to be turned into a tourism attraction.

Historic sites (nuclei) characteristics will determine how important narratives are to create a tourist attraction system. Some historic sites are "open" in that their physical frame (such as site type, function, or condition) enables visitors to understand them even with minimal markers. At this type of places the nucleus may also act as a marker in the sense of providing information about itself. For instance, the base of Scott's ill-fated 1910-1913 British Antarctic Expedition was the *Terra Nova* hut in Ross Island (Fig. 6. 7). The hut, with its distinctive ambiance, is in itself a compelling sight and marker – an apparent "time capsule" of the early period of polar exploration or "Heroic era."

Other historic sites, instead, will be more "closed," and most tourists will need a framework about their history and uses to be able to grasp them. For example, some locations of the South Shetland Islands contain the remains of shelters used by sealers in the early 19th century. These shelters, which are among the oldest cultural remains in Antarctica, have a fascinating history, but usually consist of low dry-stone walls that are almost invisible to the untrained eye. Similarly, much of Svalbard's cultural heritage consists of ruins or standing ruins – of equipment used for long-abandoned industrial processes, for instance – and would require interpretation to make sense to the casual visitor. For example, Smeerenburg on Amsterdam Island contains the remains of furnaces that are fairly identifiable as such, but little obvious evidence that those furnaces were used in the 17th century for industrial whale processing.

Markers are important not only to understand historic sites but also, in a practical sense, to protect them (L. Hacquebord, personal communication, January 15, 2009). Just as historic remains transform the tourist experience, regular visitation by tourists has the potential to transform historic features and their immediate surroundings (Roura, 2008a: 60-64), either by inappropriate behaviour or simply by wear and tear resulting from usual visitor behaviour. In that context polar tourism can be regarded as a cultural process of historic site (trans)formation (in the sense of Schiffer, 1987). Most visitors will try not to damage historic elements they regard as important, but may unknowingly damage those elements if they are not aware of their significance.

In addition, markers may be essential to underscore the significance of some historic sites that contain material remains that would normally be regarded as rubbish rather than artefacts, particularly if those remains potentially constitute an environmental hazard. Some tourists may dislike the lasting presence of cultural remains on the largely pristine polar landscape. Without the knowledge that those remains are protected, tourists may not regard them as a sight worth seeing. (See Chapter 5, for examples of the categorization of Antarctic material remains as "artefact" or "rubbish.") For instance, the old mining area of Ny-Ålesund may look to the casual visitor like a vast industrial dumpsite, yet it is significant in a number of ways, including because it is a gravesite for the miners who died there and whose bodies were never recovered and, for that reason among others, it is legally protected.

The accessibility of nuclei to organized forms of tourism will determine whether historic sites become active tourism attraction systems. For instance, Scott's *Terra Nova* hut and other contemporary huts in the Ross Sea area of Antarctica are of obvious tourism interest, but only relatively small numbers of tourists are able to reach that remote area and, for conservation reasons, visitor numbers to the huts are capped at 2000 annually. Sealing sites in the South Shetland Islands are not currently active tourist attraction systems since they are for all practical purposes inaccessible to most tourists, and some are inside areas that are closed to non scientists. The old mining area of Ny-Ålesund is near to the settlement, but organized tourist groups are presently not allowed to visit it for practical and safety reasons.

6.7 CONCLUSIONS

Polar tourism, as an organized activity, is both systematic and opportunistic, and this influences both the selection of destinations (or sites visited) and that of attractions within a site. Tourism systematically visits places chosen because of their wildlife, history, scenery, or other attractions. It is opportunistic because it turns into an attraction anything that can be conceived as somewhat different, unique, and therefore worth seeing (Roura, 2008c).

How does cultural heritage in the Polar Regions operate as a tourist attraction? Historic sites are attractive to tour operators because they constitute an essential complement to natural attractions. In addition, historic sites have the sights and stories that can be turned into attractions and consumable tourism products, provided that the sites are accessible to organized tours. Ny-Ålesund has attractions of its own, but visiting the airship mooring mast makes for a better tourism product and a more meaningful landing.

What is the role of tourism narratives in creating a tour ism attraction? As John Urry (1992: 73) has noted, "the tourist experience involves something that is visually different and distinguished from otherwise mundane activities." One of the key skills of the tour guide is to interpret the (sometimes) minimalist polar landscape so that tourists can actually "see" what might otherwise escape their eyes (Roura, 2008: 115-116). Historic sites in the Polar Regions vary depending on factors such as site type, function, and condition. Some sites are easier to grasp than others in terms of what they were used for or what was their history. Particularly in the case of "closed" sites, such as the airship mooring mast at Ny-Ålesund, on-site tourism narratives enable the different elements of the tourism attraction system to "click" together into a coherent whole.

Tourism narratives recreate multiple and sometimes conflicting versions of polar mythologies of discovery and exploration. At Ny-Ålesund, as in other Arctic and Antarctic historic sites, tourists who listen to those narratives while standing around historic remains may be able to glimpse the Polar Regions as they once were for the early explorers: unknown, harsh, and dangerous, as well as alluring. Tourists may also realize that where they stand, however remote, is only the place from where journeys to yet more remote places began. Without markers the airship mooring mast may be regarded as a rusting metal structure standing outside an Arctic settlement, and tourists may simply walk by it.

6.8 ACKNOWLEDGMENTS

My thanks for their help to produce this article go, without implicating them in my interpretations, to the following people and institutions. Prof. Dr. Louwrens Hacquebord and Dr. Maarten Loonen, from of the Arctic Centre, University of Groningen, facilitated my fieldwork in Spitsbergen, provided me with interesting insights about my research, and reviewed an earlier version of this article. In particular, it was a pleasure to work with Dr. Loonen at the Netherlands Arctic Station in Ny-Ålesund. I had useful exchanges about issues relevant to this article, mostly while in Svalbard, with Dr. Elizabeth Cooper, Neil Cox, Ingunn Løyning, Bodil Paulsen, Per Kyrre Reymert, Monika Schillat, and Rolf Stange. Dr. Peter Capellotti, Janet Dalziell, Dr. Jan Wilhem Veluwenkamp, and two anonymous reviewers provided valuable editorial comments. The Groningen Institute of Archaeology, University of Groningen, kindly provided funding for fieldwork.

7 From extraction to exhibition: Tourism at a disused marble quarry at Ny-London, Svalbard

ABSTRACT

Historic sites in the Polar Regions contain the material remains of past activities of exploration and exploitation. These sites are subject to transformation by cultural and non-cultural (natural) processes from their abandonment to the present. For research and management purposes it is important to monitor and explain these changes. This chapter explores the broad themes of research technology, use of territory, and heritage at the former site of a marble quarry in Ny-London (Port Peirson) (78° 57'49" N, 12° 02'56.5" E), an abandoned marble quarry in the High Arctic archipelago of Svalbard that operated intermittently (and eventually unsuccessfully) between 1911 and 1920. The conceptual basis for this research is primarily drawn from the work of behavioural archaeologist MB Schiffer (1987, 1995) and tourism scholar N Leiper (1990). Currently most structures at Ny-London, except two wood frame buildings, are standing ruins or ruins. Organized tourist landings take place regularly through the summer. The site is automatically protected under the 2001 Svalbard Environmental Protection Act. Fieldwork was conducted at Ny-London during July 2007. Research methods were a pedestrian survey and documentation of the site; direct observations of visitor behaviour; and repeat photography. Management responses appear adequate although more clarity regarding expected visitor behaviour may be required. Cumulative impacts are potentially a cause of concern. To the extent that tourism is a generally accepted use of the polar historic environment – although not necessarily at all sites and at all times – it should be accepted too that some tourism-related changes to destinations are unavoidable and contributing factors to the overall site degradation brought about by other natural and cultural formation processes.

7.1 INTRODUCTION

> *"Spitsbergen bids fair to become not only a great mining country but the grandest [tourism] playground in Europe" (Brown, 1919)*

The economic potential of both mining and tourism in the High Arctic archipelago of Spitsbergen[1] has long been recognised (Brown, 1919; Conway, 1919) and both industries remain priority areas for the economic development of the region (Viken and Jørgensen, 1998; Norway, 2004: 3). The interface of these two industries is represented

1 The Archipelago of Spitsbergen (Spitzbergen) is now usually known by its Norwegian name of Svalbard.

Fig. 7.1 – Ny-London in its heyday. Photo: Trømso Polar Museum. Reproduced with permission.

Fig. 7.2 – Ny-London in 2007. 1 – Harbour; 2 – Fixed crane; 3 – Living quarters with standing buildings and building foundations; 4- Machinery hall; 5 – Rail bed; 6 – Lookout; 7 – Quarry area, mobile crane, workshop remains. Photo by author (2007).

by organised tourism at mining sites, both active and inactive. This chapter explores the broad themes of research technology, use of territory, and heritage at the former site of a marble quarry in Ny-London (78° 57'49" N, 12° 02'56.5" E), also known as London. Ny-London is located on the Blomstrand Island[2] in the Kongsfjord in NW Svalbard. The site contains the remains of a marble quarry and associated facilities established in 1911 by Ernest Mansfield of the Northern Exploration Company Ltd. of England (Fig. 7.1).

The chapter aims to answer the following questions with respect to the assessment, use, and management of this historic site: What opportunistic, low technology research methods can be used to assess the status and change of polar historic sites and the effects of tourism at those sites? How does mining and tourism compare with respect to the use of territory at Ny-London? And what are the processes of site formation affecting Ny-London, and their implication for site management? The conceptual basis for this research is primarily drawn from the work of behavioural archaeologist

2 Formerly believed to be a peninsula and known as Blomstrand Peninsula.

Fig. 7.3 – Surface of marble outcrops at Ny-London. The white vein is about 2 cm wide. Photo by author (2007).

M.B. Schiffer (1987, 1995) and tourism scholar N. Leiper (1990 and refs.). Polar tourism is conceived as a behavioural system, i.e. as a set of patterned behaviours that link specific activities with the environment – in this instance, the historic environment. There are different ways in which visitors may interact with the historic environment, such as visually, physically and through the use of narratives. In a physical sense tourism results in specific traces – usually ephemeral but sometimes lasting – at the sites it visits, and from that perspective tourism is a cultural process of historic site (trans) formation. Narratives about a specific site or features and associated historic events enable the creation of tourism attractions. The example of tourism at a former mining site illustrates in a practical sense these theoretical considerations.

The Northern Exploration Company was primarily interested in prospecting and claiming resource-rich areas and then selling them to third parties for a profit (Avango and others, 2010). However, a substantive camp was established at Ny-London containing machinery, transport infrastructure, workshops, stores, and living quarters (Figs. 7.1 and 7.2). Ny-London appeared to be promising due to the quantity and quality of marble. Marble – metamorphic crystallized limestone – occurs in different locations of West Spitsbergen, and the bedrock of Blomstrand Island consists predominantly of medium to high-grade metamorphic marbles of Palaeozoic age belonging to the Generalfjella Formation (Hjelle, 1979) – so much so that it has been referred to as "Marble Island" (Brown, 1915: 16). In 1915 an observer noted:

Fig. 7.4 – One of the "London houses" at Ny-Ålesund, with tourists photographing an Arctic fox (Vulpes lagopus) (Photo by author (2007).

> *During the last three or four years a great deal of marble-cutting plant has been erected and much marble exported. Besides real marble a great deal of breccia and conglomerate are quarried, but all seem to have a high value. The range of colour is quite extraordinary.* (Brown, 1915: 19)

However, quarried marble blocks crumbled during storage due to the effect of frost on micro-fractures (Fig. 7. 3) (Governor of Svalbard, undated). Activities were suspended during World War I and the venture closed in 1920. Before World War II Ny-London was used briefly to experiment with the drying of cod. The site was eventually sold to the Norwegian Government in 1932, that is, after the entry into force in 1925 of the 1920 Treaty Concerning the Archipelago of Spitsbergen, which grants the sovereignty of the archipelago to Norway.[3] Some of the wood frame buildings were relocated to Ny-Ålesund in the 1950s, where they are still in use – including the building used as the Netherlands Arctic Station – and known as the "London houses" (Fig. 7. 4).

---

3 Now usually referred to as "Svalbard Treaty". As with other British companies operating in Svalbard, the Northern Exploration Company lobbied for Britain to claim the sovereignty of Svalbard but it did not receive any support from its national government (Avango *et al.*, 2010).

The first tourism cruise to Spitsbergen was organised in 1871 (Conway, 1906: 302). It was a popular activity in the first part of the 20th century with regular visits by crew ships, the establishment of some hotels, and great expectations for future developments (Conway, 1919; Brown, 1919). Tourism in Svalbard came to a halt during World War I and II, but it experienced a new boom in between the wars. More recently, tourism began to increase following the opening of an airport in Svalbard in 1975, and then with the Norwegian government designation of tourism as one of three priority areas for Svalbard in 1990, along with mining, and education and science (Viken and Jørgensen, 1998; Norway, 2004: 3; Kolltveit, 2006). The Norwegian government enforces a restrictive preservation policy complemented by industry self-regulation (Viken and Jørgensen, 1998; Norway, 2004). Restrictions on tourism traffic have now been introduced on the eastern part of the archipelago, including a general traffic prohibition at eight protected cultural heritage sites (Governor of Svalbard, 2009).

Organised tourist groups call into Ny-London periodically. This reflects the growth and expansion of polar tourism in the past decades, and the interest that polar historic sites generate as tourism destinations and attractions (discussed in Chapters 6 and 8). Other visitors, including Ny-Ålesund residents and Svalbard government officials, also visit Ny-London for recreation, accommodation, and other purposes. Overall, in recent years annual visitor numbers have increased from less than 100 visitors in 1996 to more than 1,500 in 2006 (Geitz, 2004; Governor of Svalbard, 2006).

7.2 METHODS

Fieldwork was conducted at Ny-London during 2-7 July 2007 in the context of doctoral research with the Arctic Centre, University of Groningen, The Netherlands. The Governor of Svalbard granted the use of one of the remaining buildings in Ny-London as field accommodation. Data collection methods were a pedestrian survey and documentation of the site; direct observations; and repeat photography (Chapter 2). An integration of these methods enabled a rapid appraisal of the status and recent change of key historic features, the main pressures (or site formation processes) affecting the site, and management responses.

Key historic features were surveyed and documented during a pedestrian survey of the site. It was not the purpose of this work to obtain information about historic activities, but rather to describe the site as the "scenario" in which contemporary tourism takes place. Notes were made of evidence of site formation processes affecting specific features, including contemporary traces of behaviour and activity. Four small groups of visitors landed at Ny-London during fieldwork (Table 7.1). Observations were made of the behaviour of individuals and groups and of their spatial and functional use of the site. Finally, the changes experienced by specific features in the recent past (ca. 10 years) were assessed by means of repeat photography. The earlier or "before" photographs were collected from a brochure about Ny-London that had been produced by the Government of Svalbard in the late 1990s (Governor of Svalbard, n.d.). Each of these photographs was reproduced as accurately as possible to obtain a later or "after" photograph. "Before" and "after" photographs were then compared, based on Roura (2008) (see also Chapter 4; Roura and Loonen, 2009).

7.3 RESULTS AND DISCUSSION

7.3.1 Rapid site appraisal using low-technology methods: Status and change

There are numerous features of different sizes and functions at and around Ny-London, which illustrate the historic process of resource extraction, the establishment of a mining settlement, and the staking of mining claims (Fig. 7.2). The main features include two standing wood frame buildings, three standing ruins, eight foundations, and the remains of a machinery hall. In addition, there are two quarries, a spoil dump, a fixed crane and a movable crane. A gravel rail bed connects the quarry area with the machinery hall and with a crane by the shore, which was used to transfer cargo to and from ships. There is also a water supply system and a number of collapsed claim boards extending away from the camp. Several dumps are located around the site, as well as artefact scatters, and the remains of a radio antenna. The standing buildings are maintained and used periodically. Government field officers use the building known as "Camp Mansfield" as field accommodation, while the welfare club of Ny-Ålesund uses the second building for recreational purposes. The buildings are kept locked so that only authorised visitors can use them.

The condition of most features is poor, but it should be noted that already in the 1930s and 1940s London had been described as a "ghost town" (Polunin, 1941). Natural site formation processes affecting the site include wind, fluvial action and animal activity (e.g. reindeer and polar bears). Landslide erosion is apparent along a low cliff on the southern part of the site. Coastal processes affect some remains by the beach. Snow accumulation and meltdown are likely to be influential too, although during fieldwork the ground was snow free.

Recent changes to several features were assessed through repeat photography (Figs. 7.5 and 7.6). The features re-photographed between the late 1990s and 2007 were too few and random to assess changes to the site as a whole, but they illustrate several key historic features, including those that attract visitors. The most common changes affected small metal and wood items. Most changes occurred *in situ* (i.e. within the location that appears in the photographic frame), followed by the removal of objects from the frame, and the addition of objects to the frame. Changes of position were the most common *in situ* change, although some objects had also fragmented. Overall, none of the features monitored appeared to have changed substantially in a period of eight to ten years although numerous small changes have taken place (Roura, 2008).

These changes were interpreted in view of the likely site formation affecting specific processes (Table 7. 2). However, in most instances it was not possible to ascertain unambiguously the cause of the changes (Roura, 2008). As observed in other polar historic sites (Chapter 4) some of the changes had decreased, rather than increased, the disorder (the formal, spatial and relational arrangement of material cultural elements), suggesting active preservation.

Fig. 7.5 – An example of repeat photography. The white rectangle in the 2007 photograph shows the framing of the late 1990s photograph. From the late 1990s to 2007 the changes have included the removal of a table and a wood rack, which in 2007 was lying nearby broken into pieces; the addition of an a metal antennae fragment and a wood board, which were stored near the dog hutch; and a reindeer antler that has been hung above the door. In addition, there have been a number of minor changes in features that have been removed from or added to the frame, or that have changed in situ. Upper photo Governor of Svalbard, photo below by author.

7.3.2 Territory: Use patterns by visitors

Ny-London is a coastal site with a relatively sheltered harbour, Port Peirson – described as "...an ideal harbour...perhaps the best in all Spitsbergen" (Brown, 1912) –; a nearby water supply; and a gentle topography in coastal areas. The marble outcrops cover

Fig. 7.6 – Visitors at one of the marble quarries. Photo by author (2007).

much of the area and suitable quarrying sites can be found a short distance from the shore. These characteristics enabled the establishment of a substantial mining camp adjacent to the quarries and the point of embarking and disembarking cargo and quarry products.

Many of the characteristics that made London a suitable site for the establishment of a mining settlement are also suitable for its development as a tourism destination. The site is located at about 7 km across the fjord from Ny-Ålesund, which is a key tourism destination in Svalbard, so that it can be used as a complement or as an alternative to visits to that site. Port Peirson is sheltered from the predominant along-fjord winds. The gently sloping beach enables landings by organised tourism groups. Furthermore, the key historic elements are a short, easy walk from the landing site. Had the mining remains been located further inland or uphill they would not have been so easily accessible, for instance, and it is likely that there would be fewer visitors.[4] While the site is largely derelict, the extant historic remains serve to illustrate the quarrying

4 The distances walked by different groups of organised tourists observed in Svalbard during July and August 2007, both at Ny-London and elsewhere, usually were below 2 km – even those advertised as "extended walks" – and in most cases below 1 km. This suggests that visitor numbers to a historic site located more than a kilometre away from the landing site would drop considerably.

process (Figs. 7.2 and 7.6). In a protected wilderness like Svalbard, the attraction of a former mining site lies in part on the juxtaposition of images between the industrial and non-industrial landscapes (Cole, 2004). The interest generated by this contrast has been apparent in some of the visits observed here. Narratives about the site's history can be used to complement visual attractions – for instance, a detailed account of the process of extraction and processing of marble was the subject of tourism narratives in one of the visits observed. In addition, the site has many attractive natural features.

7.3.3 Site management

Ny-London is automatically protected under the 2001 Svalbard Environmental Protection Act, which protects all sites pre-dating 1946. According to the Act, "No person may damage, dig up, move, remove, alter, cover up, conceal or disfigure protected structures and sites or movable historical objects, including any security zone, or initiate measures that may entail a risk of this happening."[5] Furthermore, in the security zone "…it is not permitted to erect tents, light fires or undertake similar activities." [6] Ny-London has been identified as a priority site in the cultural heritage Strategic Plan 2000-2010 and the action plan for the site includes the maintenance of machinery and equipment, and site monitoring (Governor of Svalbard, 2000: 57-58).

The Svalbard Cruising Handbook, produced by the Norwegian Polar Institute in consultation with the Governor of Svalbard, Norway's Directorate for Cultural Heritage and Directorate for Nature Management, highlights the vulnerable elements of key landing sites and contains general and site-specific behavioural guidelines for visitors. It is worth examining how these regulations are put into action by the four visitor landings observed during fieldwork (Table 7. 3). These visits will be described briefly below:

The first group landed from a commercial sailing vessel at the promontory situated to the W of Port Peirson, where there are some remains from the transport system of the mine. The group of eleven tourists and one guide moved slowly and systematically, visiting one feature after the other. The guide was obviously very knowledgeable and described to the tourists the industrial process of marble extraction and cutting. In turn, the tourists asked many questions about the site and how the quarrying process had worked. After visiting the main structures in the camp the group walked to a tarn located some 500 m to the E of the settlement, making detours along the way to look at wildlife, soil polygons, a collapsed claim board, and a waterfall where the old water supply pipelines are connected.

The second group, with six passengers and one guide, landed at the E end of the beach while kayaking around the island where Ny-London is located. The group visited the most emblematic elements of the site – the two buildings and steam boiler room, and the quarries and crane – before heading to a nearby hilltop. Overall, the group looked cursorily at historic features and took most interest in natural features – for

5 Det Kongelige Miljøverndepartement (2001): Chapter V. The cultural heritage, § 42 (the substance of protection).

6 Ibid.

Fig. 7.7 – Visitors standing around artifacts. Photo by author (2007).

instance, they spent some time watching a fox being chased by Arctic terns. In contrast to other groups observed in Svalbard, this group spread quite widely over the area.[7]

Then, three biologists landed briefly at Ny-London. They were not conducting research specifically at that site but simply "being tourists" – as one of them said – while waiting from some scientific equipment to arrive to Ny-Ålesund. They wanted photographs of a long-tailed skua (*Stercorarius longicaudus*), which they knew was nesting among the cultural remains. In their visit they briefly examined the standing buildings, located and photographed the long-tailed skua nest, and departed.

Finally, one man was briefly seen running along the beach and onto the hills to the west of the harbour. It is not known how this person came to be there, what he was doing or where he was going.

The visits were generally unremarkable, although some observations of visitor behaviour are relevant to site management:

- One of the guides instructed the tourists where to walk so as to minimise their impact. Another guide did not appear to have instructed the tourists beforehand, but

7 Observations of other tourist groups elsewhere in Svalbard suggest that guides prefer keep tourists in tight groups for greater protection against polar bears. For instance, a guide instructed his group not to walk more than twenty to thirty meters from the armed polar bear guards, a point he repeated frequently during a weeklong observation period.

Fig. 7.8 – Visitor picking up an artifact (upper); artifact that has just been broken (below). Photos by author (2007).

three times during the visit requested them to modify their behaviour when some of them walked on the path leading to the houses, approached the nest of an eider duck, or walked on a building foundation.[8]

- Visitors tended to concentrate around features of interest, some of which were surrounded by patches of bare ground, suggesting erosion (to which repeated trampling may be a contributor).

---

8 Here there may be issues of reactivity – that is, people altering their behaviour in the knowledge that they are being observed.

Fig. 7.9 – Access to the Ny-London buildings from the beach. Photo by author (2007).

- While approaching historic features visitors need to make a judgment as to the appropriate distance from where to stand from a historic feature. However, the sensitivity boundaries – the boundary beyond which behaviour may result in some disturbance to cultural remains – are not necessarily obvious to casual visitors. This is particularly the case since many artefacts at Ny-London are composed of fragments of wood, metal, brick and other materials resulting from the decay of the site (Fig. 7. 7). What distinguishes them from other natural or cultural substrates is that that these cultural features are legally protected. As a result, visitors sometimes were seen to step on the artefacts lying around historic features.
- One of the guides picked up a small ceramic bottle that was lying on a streambed in order to photograph it; when he did the same with an adjacent glass vial it broke (Fig. 7. 8). "It was already busted" – he remarked – but it was apparent that the bottle broke while being lifted.
- The instructions as to where to walk differed. One of the guides recommended walking on existing paths and to avoid stepping on vegetation, and another advised to do the opposite so as to not to widen existing paths, as recommended in the Svalbard Cruise Handbook. This is somewhat counterintuitive – the view of the buildings from the beach does not immediately suggest that walking off paths is the best option (Fig. 7. 9).
- There are some minor traces of contemporary activity, including traces of visitor trampling around parts of Ny-London. For instance, the rail bed connecting vari-

Fig. 7.10 – Footprint erosion. Visitors walking across the historic rail bed at Ny-London have caused localized erosion on its sides. Photo by author (2007).

ous features of the sites has evidence of visitors walking up or down its sides as visitors move to and from the most visited parts of the site i.e. the standing buildings, the quarries, and the machinery hall (Fig. 7. 10).

- To the extent that the original and contemporary uses of the site are different (i.e. mining and tourism), the areas of activity of the original and contemporary activities overlap only partly (Fig. 7. 11). The tourism activity area, as assessed here, was less extensive than the original mining activity area, that is, not all historic mining remains appear to be used as tourism attractions.
- As observed at other polar sites (and discussed elsewhere in this thesis), tourism visits to polar historic sites focus not solely on historic elements alone but also on the various natural and cultural elements at the site. These include flowering plants, opportunistic sightings of wildlife, and elevated points that serve as lookouts. In addition, they may engage on a range of activities aside sightseeing.
- The direct impact of these few visits on the condition and integrity of the historic site could be regarded as negligible. However, a repetition of low-intensity direct impacts could result in cumulative effects on specific elements or parts of the sites.

These limited examples and observations illustrate the range and frequency of visitors at Ny-London, some common forms of behaviour, and some of the resulting traces of visitation. However, other visits to the site may have a different pattern than those described here. For instance, shortly after the brief stay at Ny-London described here a group of 25 Ny-Ålesund residents held a barbeque at the site,[9] and subsequently a small group spent a weekend there – a recurrent activity during summer.

7.4 CONCLUSIONS

Many sites of past activities in the Polar Regions look like abandoned places. Some sites are in ruins; others are well preserved and are described in some narratives as "time capsules" (Chapter 4). Yet this appearance is deceptive: many historic sites have become an arena where contemporary activities take place, notably tourism. These activities contribute to site transformation and add to the changes brought about by other natural and cultural site formation processes. The outcome is a dynamic balance between decay and preservation. This chapter has examined three broad themes relevant to the assessment, contemporary use and management of a former marble quarry in NW Svalbard. The questions posed at the outset will be reviewed below.

What opportunistic, low technology research methods can be used to assess the status and change of polar historic sites and the effects of tourism at those sites? Assessing the status and change of polar historic sites may be a complex and expensive task (e.g. Governor of Svalbard, 1999). The methods used in this research, while simple and low technology, allowed producing a rapid and reasonably informed appraisal of site's status, recent change, tourism modalities, and management needs. In its conception, albeit not necessarily in specific methods, the approach used here is based on the tradition of rapid appraisal used in some disciplines of applied social sciences (e.g. Kumar, 1993; Beebe, 1995). This approach is appropriate to the Polar Regions, where site visits may be infrequent, opportunistic, and brief. Naturally, more comprehensive assessment methods may be required for different (or more specific) research or management needs.

How does mining and tourism compare with respect to the use of territory at Ny-London? Plainly marble quarrying and tourism are substantially different activities – mining is an extractive industry while tourism is not. Rather, mining tourism " ... reflects a change in the perception of what constitutes a resource – from a productive raw material to a consumptive viewing of the past." (Pretes, 2002: 447). In addition, both are industries based on the sustained use of particular sites. However, tourism tends to focus on specific parts of the site, "cherry-picking" certain features that could be conceived (and narrated) as attractions. This includes not only historic features but also natural features such as flowering plants, passing wildlife and lookouts from where to examine the surrounding landscape. As a consequence, the activity areas of historic mining and contemporary tourism overlap but are not identical. It should be noted that narratives (or the absence of narratives) are related to the use of territory. Tourism narratives influence behaviour, in that they direct visitors towards (or away from) specific

9 Curiously, the fireplace is located at 62 m from the buildings, closer than the 100 m stated in the Svalbard Environmental Protection Law as a mandatory distance to keep from cultural heritage. It is speculated that the fireplace itself pre-dates the law. The content of the ashes included tins and wood fragments, none of which appeared to be "historic".

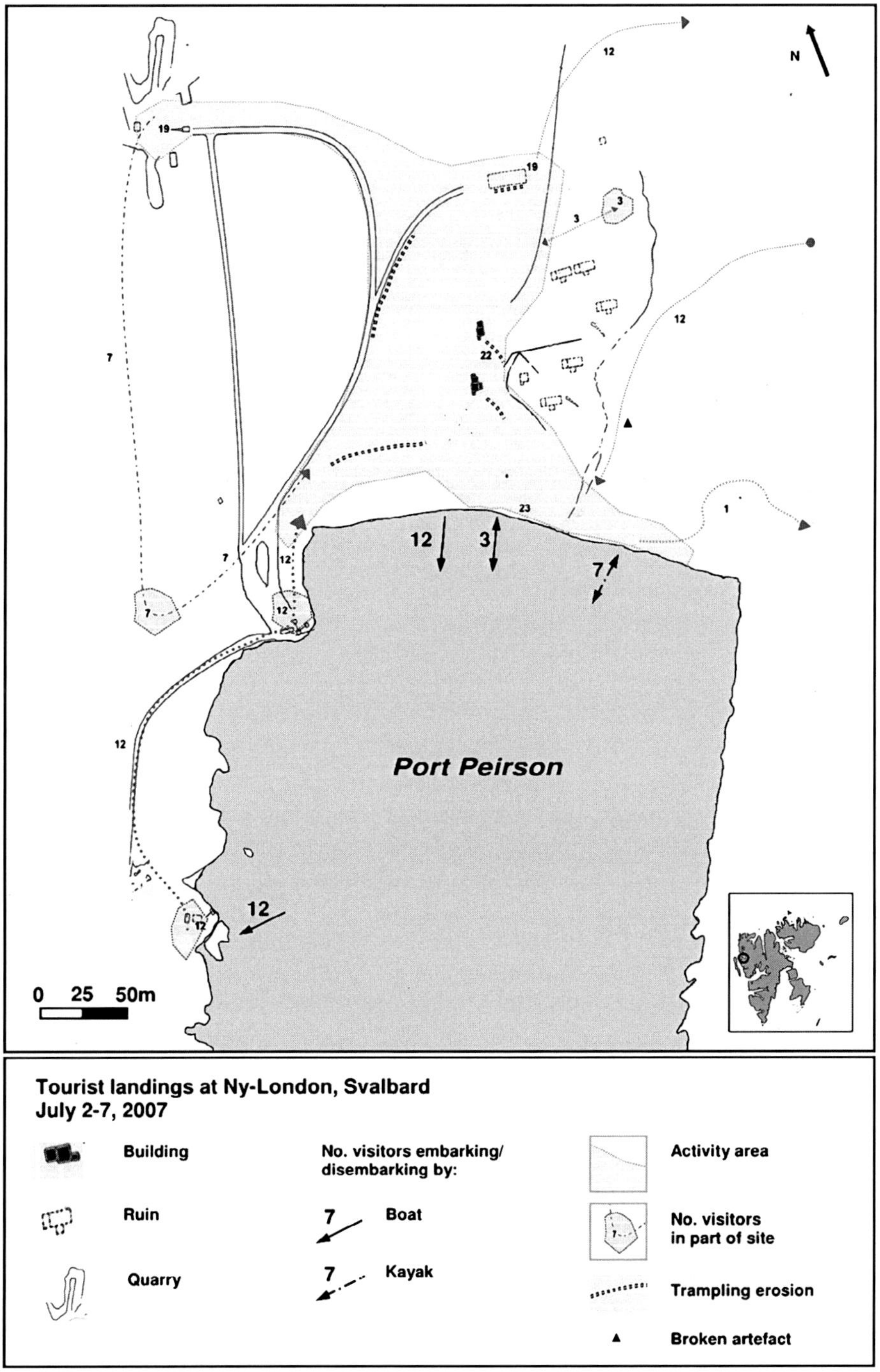

Fig. 7.11 – Sketch showing tourism landings at Ny-London, Svalbard, 2-7 July 2007, outlining the movements of visitors and the estimated activity area. Sketch by R. Roura. Base map by Nash and Løkken (1990).

features or parts of a site, and encourage (or discourage) some forms of interaction with the site's attractions. In addition, narratives about a site's history and uses are important to create "tourism attraction systems" in certain sites that may appear unremarkable to the casual visitor (Chapter 6). Conversely, an absence of narratives may result in tourists missing out on important elements of a site, visiting areas that are off-limits, or behaving in ways that are not recommended.

And what are the processes of site formation affecting Ny-London, and their implication for site management? Judging from the monitoring of a limited number of features, there has not been substantive change in a period of about a decade, but this assessment does not hold for the site as a whole. Tourism has the potential to modify the cultural remains that resulted from mining as well as the surrounding environment – for instance, through the establishment of paths or localised footprint erosion. It should be noted, however, that some traces of visitation would be comparable to those of other site formation processes – a concept described as "equifinality" (Lyman, 2004; see also Chapter 4). In some instances it may only be possible to attribute an impact to visitation by direct observation. While regular visitation may accelerate the degradation of historic sites (or particular features therein), industrial sites are on the whole less vulnerable than other kinds of historic sites, on account of their large scale and hard-wearing nature. In addition, it is apparent that Ny-London's accessible location not only enables frequent visitation but also facilitates active preservation.

Current site management of Ny-London appears to be adequate. Compliance with guidelines, as assessed in a limited set of observations, was generally adequate. However, guidelines concerning where to walk/not to walk are not immediately clear, and it is not easy to avoid stepping in artefacts in a place that is littered with them. A casual visitor would not immediately recognize many artefacts as such since many of them are fragments that originate from the ongoing decay of movable and unmovable site features. In addition, many artefacts are obscured by vegetation or half-buried in the ground. Consideration should be given to identify and set aside some areas with a high density of artefacts as off limits for visitors so as to preserve the information and heritage value of the site, and to establish routes joining the most visited features.

The commercial exploitation by private interests of a common natural and cultural heritage, which may result on the degradation of that heritage, is an issue of increasing concern across the Polar Regions. Nevertheless, if polar historic sites that are legally protected are not actively preserved and consistently managed they will eventually decay, gracefully or otherwise, regardless of whether or not they are tourism destinations. To the extent that tourism is a generally accepted use of the polar historic environment – although not necessarily at all sites and at all times – it should be accepted too that some tourism-related changes to destination are unavoidable, and a contributing factor to site decay caused by other site formation processes. This is a more realistic assessment than assuming that tourism, if regulated and supervised, will cause no changes at the sites it visits. The challenge – and the alternative to closing off the sites entirely to visitation – is how to manage tourism so that it does not accelerate the decay process and in doing so diminishes the information, heritage and other intrinsic values of polar historic sites. The example of Ny-London suggests that the exhibition of industrial cultural remains is compatible with their protection and preservation, but that there are no perfect management solutions to regulate the inherently flawed human behaviour.

7.4 ACKNOWLEDGEMENTS

I am grateful to Professor Dr. Louwrens Hacquebord (University of Groningen, Arctic Centre), Dr. Susan Barr (Norway's Directorate of Cultural Heritage), and Per Kyrre Reymert (Government of Svalbard) for sharing their knowledge about Svalbard´s history and cultural heritage. Frigga Kruse (University of Groningen, Arctic Centre) kindly shared her photographs of Ny-London and her insights about the Northern Exploration Company. Of course, none of these people should necessarily be implicated in the positions taken or interpretations made by the author here.

The Trømso Polar Museum and the Governor of Svalbard kindly allowed the use of their photographs (Figs. 7.1 and 7.5a). The logistic support provided by Dr. Maarten Loonen (University of Groningen, Arctic Centre and Netherlands Arctic Station), Per Kyrre Reymert, and Wojtek Moskal (Norwegian Polar Institute) is much appreciated. Funding for fieldwork was provided by the Groningen Institute of Archaeology, University of Groningen, The Netherlands.

7.5 TABLES

Table 7.1 – Visits observed at Ny-London, Svalbard, 2-7 July 2007

Date	Transport	Landing	No. visitors	Visit duration (AM/PM)	Itinerary	Distance walked
2 July 2007	Sailing vessel; inflatable boat	Headland W of Port Peirson; central part of beach	11 tourists 1 guide	<2 hr (PM)	Crane, road on E of site, buildings, quarries, boiler room, waterfall, tarn, claim board, polygons,	2300m
3 July 2007	4 kayaks	E end of beach	6 tourists 1 guide	1 hr (PM)	Two buildings, boiler room, quarries and crane	1300m
3 July 2007	Rigid hull boat	Beach, central part	3 scientists	<1 hr (PM)	Two buildings, area to W of site	400m
3 July 2007	?	?	1 visitor	<1 minute (PM) (observed)	Running eastwards along beach, climbing over headland.	300m (observed)

Table 7.2 – Repeat photography: Feature characteristics, main changes, and interpretation

Feature description (as in late 1990s photograph)	Main changes during monitoring period	Interpretation
Machinery shed ruins – S aspect Detail showing metal cylinder in upright position [b]	A few heavy items e.g. segment of iron rail added or removed from frame or changed position within frame.	Handling
Machinery shed ruins – S aspect Three metal cylinders in upright position and one in a horizontal position. Some scrap metal.	Large metal box, use unknown, has apparently changed position.	Handling
"Camp Mansfield" hut - S aspect Pre-fabricated woodframe building with a south-facing porch and extensions.	Some heavy items e.g. metal antennae fragment stored next to building; reindeer horns hung above door. Several large objects (e.g. garden table) removed from frame; saw rack is broken, located outside the photograph frame. Metal lid added to frame.	Site management Wind action
Machinery shed ruins – NE aspect Detail of standing ruins containing wood fragments and scrap metal of different sizes and shapes; and drum barrel-shaped block (chalk?), with the drum itself no longer present. Larger items partly stored orderly inside and outside the feature, and partly lying about on the ground.	Pieces of wood, metal, ca. 40-50 cm long, added to frame. Some pieces of metal and a wooden board have been removed from frame or changed position. Half broken box containing metal scrap collapsed	Handling, trampling; Wind, snow accumulation (?)

Table 7.2 continued

Feature description (as in late 1990s photograph)	Main changes during monitoring period	Interpretation
Building foundations Consisting mostly of unmovable wooden beams and some loose wood fragments.	Fragment of wood ca. 30-40 cm now visible; some beams of the foundation changed position slightly.	Handling; trampling?
Building foundations Containing an iron cooking stove, brick chimney remains, and brick and wood fragments less than ca. 50 cm long. Several metal artefacts are on display above stove.	Artefacts on display on top of stove removed from frame. Small to medium brick, wood, and metal objects moved within the frame.	Site management, visitation, trampling. Wind.
Waste dump adjacent to the "Welfare" hut Containing an assortment of metal and glass artefacts from primary deposition.	Metal plate added to dump; iron bar, fragment of metal coil removed from frame. Several small metal objects have changed position.	Cultural deposition/removal; wind action; gravity/slope processes.
Half barrel partly buried in the ground	Circular piece of metal added to frame; four staves broken, remaining fragments *in situ.*	Wind and weathering; cultural deposition?
Wooden wheelbarrow with a metal wheel	Large nail stored inside wheelbarrow. Wheelbarrow moved to different location.	Site management

b This structure was formerly described to be a shed for "steam engines" but later research suggests that it consists of the remains of so-called channelers, which are the machines used for mechanically cutting the marble prior to extracting the blocks (Aalders *et al.*, 2008).

Table 7.3 – Observations of compliance with guidelines from the Svalbard Visitor Handbook [c]

✓ - General compliance with guidelines
× - Instances of non-compliance

Guideline	Group 1	Group 2	Group 3	Comments
Cultural remains must not be disfigured or destroyed.	×	✓	✓	Accidental breakage of artifact
Avoid walking and standing in the cultural remains (does not apply to the cabins).	×	✓	✓	Numerous artefact fragments scattered through area, many may not be recognised as "artefacts" and be stood on.
Do not remove any cultural remains from the site.	✓	✓	✓	
Leave the objects in place. Do not move the items.	×	✓	✓	Artefact held by visitor to be photographed
Avoid walking where you may enhance erosion of the ground in the area between the beach and the houses.	✓	×	×	It is not immediately clear how to apply the rule

c Note guidelines appear in narrative form as a listing of vulnerable elements, and as a listing of expected behaviour.

8 Cultural heritage tourism in Antarctica and Svalbard: Patterns, impacts, and policies[†]

ABSTRACT

This chapter examines the patterns and impacts of organized tourism in Antarctica and Svalbard (Spitsbergen) in the context of existing tourism policies for each region. Protection mechanisms in both regions are compared and contrasted. Explorers, whalers, seal hunters, scientists and others have left many material remains that are significant because they tell the history of exploration and exploitation of the Polar Regions. Following on these earlier activities, contemporary polar tourism represents a new phase in the "exploration" and exploitation of the Polar Regions. Polar tourism has increased significantly over the past few decades and is characterised by growth, diversification, and geographic expansion. Key cultural heritage sites are regularly included in standard tourist itineraries and are also the subject of specialized tourism. The patterns of tourism developments in both Antarctica and Svalbard have been broadly similar. Both regions are subject to international treaties – the 1959 Antarctic Treaty, and the 1920 Treaty of Spitsbergen, respectively. However, in practice the regime of governance is international for the Antarctic and national for Svalbard. This difference has implications for the protection of cultural heritage sites, which are examined in this article.

8.1 INTRODUCTION

For several centuries the Polar Regions have been the focus of activities of people originating from outside these regions, including explorers, whalers, seal hunters, and scientists. Their activities have left many material remains in the polar landscape such as supply depots, huts and shelters, industrial remains, research stations, and crosses, graves and other memorials. These and similar cultural remains are significant, among other reasons, because they tell the history of polar exploration and exploitation of the Polar Regions. Following on those earlier activities, contemporary polar tourism represents a new phase in the 'exploration' and exploitation of the Polar Regions. Many significant historic sites in Antarctica and Svalbard have now become tourist 'destinations' where tourism landings are a seasonal feature (e.g. Hughes, 1994; Hughes and Davis, 1995; Geitz, 2004; Roura, 2009).

This chapter compares and contrasts organized cruise tourism at protected historic sites in Antarctica and Svalbard (Prior to 1925 the archipelago was known as Spitsbergen, it is now usually referred to by its Norwegian name of Svalbard) (Figure 8.1). The

† Published as:

Roura R (2010): "Cultural heritage tourism in Antarctica and Svalbard: Patterns, impacts, and policies." In: Michael Hall and Jarkko Saarinen (eds.): *Tourism and change in the Polar Regions: Climate, environment and experiences*. London and New York: Routledge. pp.180-203.

purpose is to identify the patterns and impacts of organised tourism at historic sites in the context of existing tourism policies for each region. In this context an historic site (any location that demonstrates past human activity, evidenced by the presence of artefacts, ecofacts, features, structures, or other material remains (Kipfer, 2007) is distinguished from a landing site (places where people set foot ashore). A landing site may or may not also be a historic site – and vice versa. A further distinction exists between sites of past human activity depending on whether or not they are legally protected. The protected sites and remains of past activities can be regarded, in a legal sense, part of the regions' cultural heritage. Heritage, however, may be more broadly defined (e.g. Ashworth and Graham, 2005; Lowenthal, 2005).

Cultural heritage tourism in Antarctica and Svalbard consists primarily of organized visits, mostly by ship, to historic sites. Visits focus on the places, material remains, and associated historic events. The following research questions are addressed in this chapter: What are the patterns of tourism at historic sites, both regionally and locally? How do visitors behave at historic sites, and what are the traces of their behaviour? And how effective is the protection of historic sites in the context of existing tourism policies?

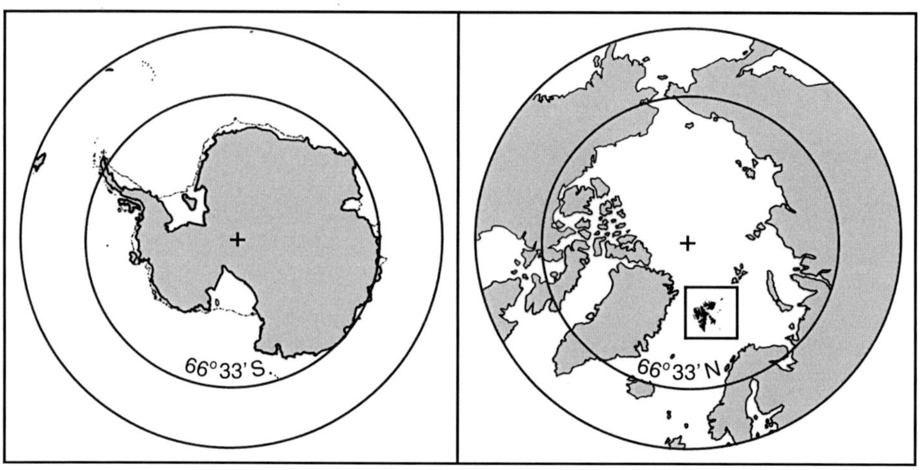

Fig. 8.1 – Antarctica and Svalbard (Spitsbergen).

8.2 OVERVIEW: CONTRASTING ANTARCTICA AND SVALBARD

8.2.1 Background

The Dutch pilot Wilhem Barentsz is generally acknowledged to have discovered the High Arctic archipelago of Spitsbergen in 1596. The earliest sighting of Antarctica might have been in 1599 by the Dutchman Dirk Gerritsz Pomp (Ligtendag, 2002), but others suggest 1820 as a likely date (Headland, 1989). At any event, neither of these regions had indigenous populations, and human presence on the land began in the 17th in Spitsbergen and in the early 19th in Antarctica. Following their discovery both regions have experienced broadly analogous historic developments, including exploration, natural resources exploitation, science, tourism, and the establishment of comprehensive environmental protection regimes. However, the timing and tempo of developments were different in each case, and a range of events in Spitsbergen's history – hunting and trapping, large-scale industrial developments, and warfare during World War II – have no or a limited Antarctic equivalent.

Both Antarctica and Spitsbergen have been regarded as *terra nullius* and originated territorial ambitions. Resulting tensions were in both instances resolved through international treaties (e.g. Mathisen, 1954; Beck, 1986; Triggs, 1987; Ulfstein, 1995). Svalbard is subject to the 1920 *Treaty Concerning the Archipelago of Spitsbergen*, which comprises all the islands situated between 74° to 81° N and 10° to 35° E, and grants the 'full and absolute' sovereignty of Spitsbergen to Norway (Article 1) while maintaining equal rights for all contracting parties, subject to Norwegian laws. These include rights to hunt, fish, or engage in all maritime, industrial, mining and commercial operations (Article 3). Norway's *Svalbard Act* of 17 June 1925 establishes that Svalbard is part of Norway. A Norwegian-appointed Governor of Svalbard is responsible for ensuring compliance with the provisions of the Treaty and current Norwegian legislation (Ulfstein, 1995; Norway, 2004).

The Antarctic Treaty Area, which includes the Antarctic continent, islands, and seas south of 60° south of latitude, is governed by the 1959 *Antarctic Treaty* and related instruments, notably the 1991 *Protocol of Environmental Protection to the Antarctic Treaty* (the Protocol). This set of instruments comprises what is known as the Antarctic Treaty System. Seven states claim parts of Antarctica and two others reserve the right to place future claims. There are overlapping claims over the Antarctic Peninsula area as well as one unclaimed sector. Article IV of the Antarctic Treaty 'freezes' the status quo regarding territorial claims. The Antarctic Treaty Consultative Meeting (ATCM), which currently meets annually, makes decisions about the governance of the region (e.g. Beck, 1986; Triggs, 1987; Bastmeijer and Roura, 2004).

8.2.2 Cultural heritage

Cultural heritage protection is a subset of environmental protection laws in both regions. Many more cultural heritage remains have been inventoried and listed as protected in Svalbard than in Antarctica. Cultural heritage in Svalbard has a broader range of types and functions than in Antarctica (Figures 8.2 a-d).

The Protocol and other Antarctic Treaty instruments protect the Antarctic cultural heritage, which is represented by designated Historic Sites and Monuments (HSMs). Any Party may propose a site or monument of recognised historic value for listing as a HSM. The proposal for listing may be approved by a Measure adopted at an ATCM (Protocol Annex V, Article 8(2)). By ATCM XXXII (2009) there were 79 listed sites and monuments numbered 1–84 (some have been de-listed or subsumed into other HSMs). Of these, approximately 70% are the sites and remains of past Antarctic activities and about 30% are memorials and monuments. The list of HSMs is subject to periodic review and updates, but it is neither comprehensive nor representative of different periods or themes of Antarctic history. HSM designation gives an international status to elements that have primarily a national value (Roura 2008).

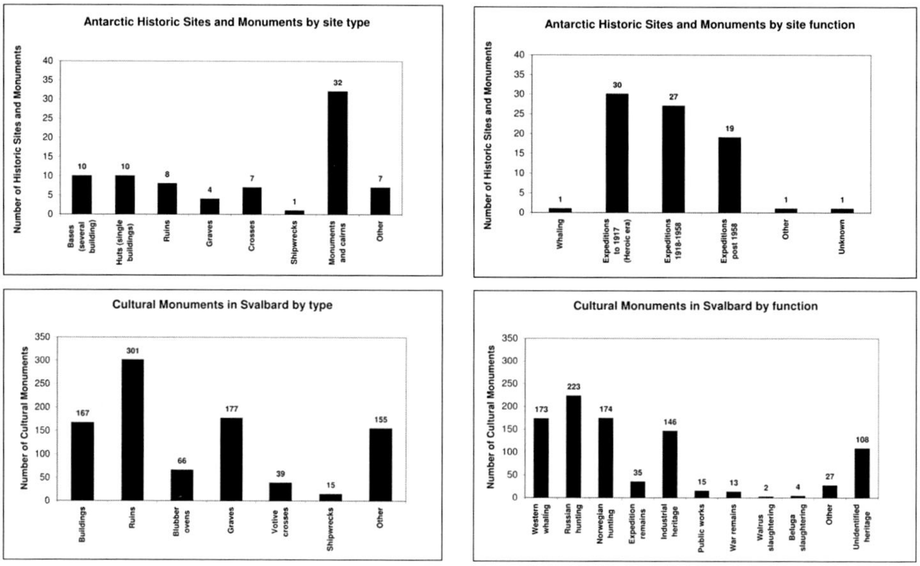

Fig. 8.2 a-d – Type and function of historic sites *in Antarctica*[1] *and Svalbard.*[2]

Listed HSMs shall not be damaged, removed or destroyed. Historic remains predating 1958 whose existence or present location is not known also have a degree of protection (ATCM Resolution 5, 2001). HSMs may be designated as Protected Areas. Past activity remains that are not otherwise protected are subject to removal under Protocol requirements (Annex III, Article 1(5)) to clean up past and present waste disposal sites on land and abandoned worksites. Existing provisions pertaining the designation and management of historic sites have now been consolidated into a set of guidelines (ATCM Resolution 3, 2009).

Unlike in the Antarctic – and indeed, mainland Norway – from a Norwegian legal perspective the natural and cultural heritage of Svalbard are considered together as part of the environment (miljø) (P.K. Reymert, personal communication, September 22, 2006). The Treaty Concerning the Archipelago of Spitsbergen allows Norway to set measures to ensure the preservation of fauna and flora (Article 2). In Norway this has been interpreted as a requirement to protect the archipelago's environment. Norway's 2001 Svalbard Environmental Protection Act protects Svalbard's cultural monuments (*kulturminner*) and states that Svalbard's cultural heritage shall be protected 'as an

1 Based on official descriptions of Historic Sites and Monuments as per designation instruments (www.ats.aq).

2 Based on Marstrander 1999:134; and PK Reymert pers.comm. 10 July 2007.

element of a coherent system of environmental management' (section 38). Approximately 920 cultural monuments in Svalbard have been identified to date, covering all periods of the archipelago's history (see also Marstrander 1999). The law protects structures, sites, and movable historical objects predating 1946, as well as more recent cultural remains that are of particular historic or cultural value and that are protected by a decision of Norway's Directorate of Cultural Heritage. Cultural features such as human graves and certain animal slaughtering sites are also protected irrespective of their age. Camping, lighting a fire or leaving other traces is prohibited within 100 metres from a cultural monument (Det Kongelige Miljøverndepartement, 2001). Over 50 sites containing cultural monuments have been identified as having high priority for protection (Governor of Svalbard, 2000).

8.2.3 Organised tourism and tourism policies

Antarctica and Svalbard present fairly different conditions for cruise tourism, which is the main form of tourist access to historic sites. The landmass of Antarctica (14 x 10^6 km^2) is much larger in surface area than Svalbard (61,022 km^2). Less than one percent of Antarctica is ice free, compared to approximately 40 percent in Svalbard (Statistics Norway, 2005). With the exception of the Antarctic Peninsula, the Antarctic coastline is located 3,000 kms or more from the closest continent. Instead, Svalbard is about 1,000 km from Norway. Svalbard has a comparatively mild climate for its latitude as a result of the influence of the Gulf Stream. Unlike Antarctica, Svalbard is accessible by air year round and tourism is promoted for both summer and winter. The regions' flora and fauna are substantially different and the ecology of species such as penguins, polar bears and walruses influence tourism patterns. In terms of basic characteristics such as access, density of landing sites, and contemporary cruise tourism patterns, Svalbard bears a closer resemblance to the northwest Antarctic Peninsula than to the rest of Antarctica.

The first tourism cruise to Spitsbergen was organised in 1871 (Conway, 1906). Tourism began to increase following the opening of an airport in Svalbard in 1975, and then with the Norwegian government designation of tourism as one of three priority areas for Svalbard in 1990 along with mining and education and science (Viken and Jørgensen, 1998; Norway, 2004; Kolltveit, 2006). In Antarctica the first organised cruise was in 1958, and while it continued for the next three decades it did not begin to increase substantially until the 1990s (e.g. Bastmeijer and Roura, 2004; see also Hall and Saarinen, 2010). Cruise tourism is largely characterized by 'expedition cruising', in which vessels are comparatively small and thus able to land people ashore despite a lack of infrastructure (Geitz, 2004: 7). Periodic landings are complemented by excursions on inflatable boats and by cruises to scenic areas, and by on-board lectures. Some cruises visit either region as part of more extended itineraries including adjacent regions such as mainland Norway or South America. In Svalbard there are also day cruises and other forms of local tourism. The cruise tourism season usually extends from November to March in Antarctica and from June to September in Svalbard.

Cruise tourism in these regions is currently characterised by growth (an increase on ships, tour operators, and tourists); diversification (a increasingly broad range of transport modalities and activities on offer), and geographic expansion (an increase in the number of sites used for tourism) (e.g. Bastmeijer and Roura, 2004; Geitz, 2004; see also Hall and Saarinen, 2010). In 2006 more than 28,000 tourists arrived in Svalbard in overseas cruise ships, and more than 12,000 tourists boarded 'expedition' cruises and day cruises (Governor of Svalbard, 2006). More than 35,000 shipborne tourists visited Antarctica in the summer of 2006-2007, of which more than 28,000 landed ashore (IAATO, 2008).

Tour operators in Antarctica and Svalbard are organised in industry associations, respectively the International Association of Antarctic Tour Operators (IAATO) and the Association of Arctic Expedition Cruise Operators (AECO). Among other functions, these associations agree on self-regulation measures. More than 70 percent of AECO members are also IAATO members and operate in both regions. Some guides work regularly in both regions, and some tourists visit both.

Antarctic Treaty Consultative Parties generally have accepted that tourism will continue to operate in the region, and that it needs to be managed to preserve the intrinsic values of Antarctica (New Zealand and Australia, 2006: 4). However, how tourism is or should be managed is the subject of intense debate, the detail of which is beyond the scope of this chapter. Antarctic tourism is primarily ruled by generic regulation applicable to all Antarctic activities, by specific regulation on insurance and contingency planning and landing requirements (ATCM Measures 4 (2004) and 15 (2009) are the only tourism instruments that will be legally binding once they become effective), and by non-legally binding guidelines emanating from the ATCM and from the tourism industry. There is generic guidance for visitation to historic sites (ATCM Recommendation XVIII-1) and site-specific guidelines for some frequently visited sites. Voluntary site-specific guidelines have been developed for 25 of the most visited sites in Antarctica including seven designated HSMs. Visits to some 'Heroic era' huts require a permit. Overall, Antarctic tourism must meet certain conditions but it is not subject to restrictions (Bastmeijer and Roura, 2004). However, recent shipping accidents involving cruise vessels may result in the future adoption of more restrictive tourism regulation. For instance, the recently adopted Measure 15 (2009) seeks to prevent landings from vessels carrying more than 500 passengers.

Tourism is one of the activities prioritized to support Norwegian presence in Svalbard but it should take place in the context of environmental protection (Vistad *and others* 2008). The Norwegian government policy against tourism in Spitsbergen changed officially in 1990. Tourism is managed by state regulation, which enforces a restrictive preservation policy, complemented by industry self-regulation (Viken and Jørgensen 1998; Norway 2004). The 1991 "Regulations relating to tourism and other travel in Svalbard" cover matters such as insurance, accountability and notification for trips organised in Svalbard. The Governor has the authority to alter or prohibit trips. Some natural or historic sites have permanent or seasonal restricted access (e.g. Geitz 2004). However, one of the goals for Svalbard's cultural heritage is protecting and sustaining it 'for the enjoyment of present and future generations' (Governor of Svalbard 2000: 8). Tourism is therefore acceptable under current cultural heritage policy.

In 2009 the Governor of Svalbard recommended that a general ban on traffic be introduced at or near several cultural heritage sites that are automatically protected by law. The objective of the proposal is to preserve several relatively unaffected and significant cultural heritage sites from wear and tear and other negative effects caused by traffic (Governor of Svalbard 2009). This proposal was under discussion at the time of writing.

8.3 ELEMENTS OF VISITATION: METHODS

An integrated approach was used to assess the *behavioural* elements and *material traces* elements of visitation at historic sites (Rathje 2001). These provided a mutually complemented perspective of tourism use of historic sites. Direct observations and site surveys of different kinds constituted the main research methods. Tourism statistics (Governor of Svalbard 2006; IAATO 2008) were analysed to assess landing patterns. Semi-structured interviews with key informants complemented other research methods.

Cultural heritage tourism in Antarctica and Svalbard is largely ship-based; and tourists are regularly taken ashore to visit places of interest. Observers can be part of a tourism contingent and follow its itinerary (which may be a form of participant observation). Alternatively observers can be based at a landing site and observe landings as they take place. Both approaches were used for this study. In Antarctica direct observations were conducted on opportunistic bases during stays at Deception Island (2002), King George Island (2006), and Ross Island (2008). In Spitsbergen observations were conducted on day cruises in the Ice Fjord (2006); on a weeklong cruise to NW Spitsbergen (2007); and during stays at Ny-Ålesund and Ny-London in Kings Bay (2007). Observations were made of both group and individual behaviour. Observations were conducted from an *etic* perspective, that is, tourist behaviour was described regardless of the apparent motivations or perspectives of the individuals (Pearce 2005). Observations followed an *ad libitum* sampling pattern i.e. whatever was visible and seemed relevant at the time was recorded (Martin and Bateson 1993). The traces of behaviour and activity visible on the terrain were surveyed (Schiffer 1987; Bernard 2002). The focus here was not on the structure of behaviour itself – as identified in direct observations – but on the consequences of that behaviour.

8.4 RESULTS AND DISCUSSION

8.4.1 Tourism patterns and cultural heritage

Cruise tourism in Antarctica and Svalbard relies on the long-term seasonal use of certain sites that have been established as destinations. Historic sites are attractive destinations because they have material remains (there is something to show) and a history (there is a story to tell) (Roura 2009). Key historic sites are regularly included in mainstream tourist itineraries both in Antarctica and Svalbard, adding to sites that are visited for their natural or scenic values or for other reasons. Tourism landings at historic sites therefore reflect general tourism patterns. Cultural heritage is also the subject of more specialized tourism.

Antarctic tourism visits are concentrated at less than 35 sites (IAATO 2008). An examination of Antarctic tourism statistics shows that tourists have visited approximately

200 sites in the Antarctic Peninsula region since 1989, including 20 research stations. About 50 of these sites have received 100 or more visitors in any one season. In contrast, about the same number of sites have been visited only once. At least about 100 sites have been visited elsewhere in Antarctica, mostly in the Ross Sea area. Tourists have visited almost 300 sites in Svalbard over the years (Geitz 2004: 10–11). Between 100 and 200 landing sites are currently visited each season in both regions (Figure 8.3). Visitor landings currently exceed 80,000 for Svalbard and 170,000 for Antarctica (Figure 8.4).

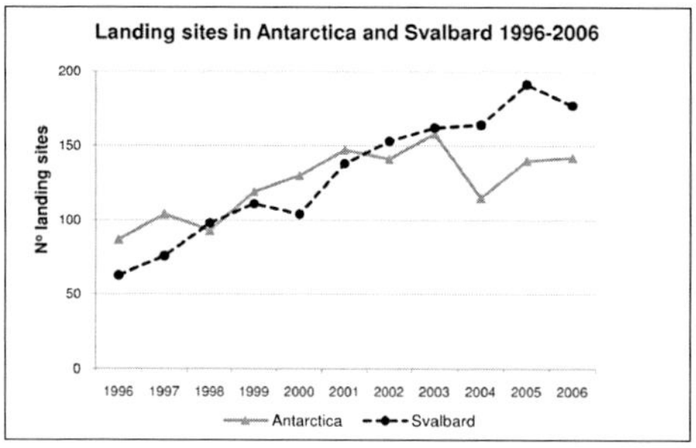

Fig. 8.3 – Number of landing sites in Antarctica and Svalbard.[3]

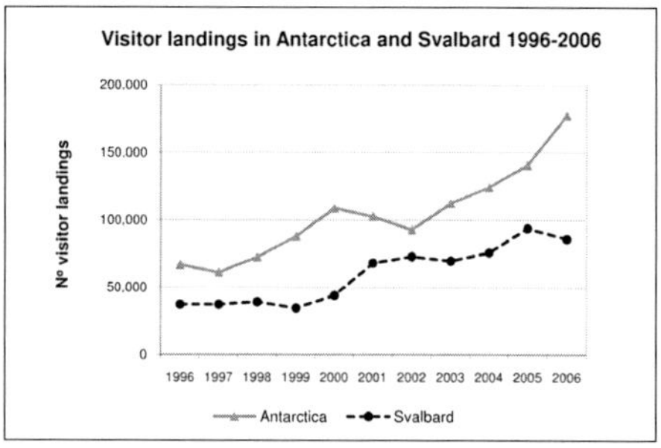

Fig. 8.4 – Visitor landings in Antarctica and Svalbard. Visitors include tourists, staff, and crew.[4]

Key historic sites are among the most visited sites in both regions (Figures 10.5 a-c). The most visited historic sites are situated in the NW Antarctic Peninsula and the Ross Sea area of Antarctica (Figure 8.6) and in NW Spitsbergen (Figure 8.7). In 2006–07 nearly 50,000 visitor landings took place at 15 significant historic sites in Antarctica (about 20 per cent of all landings). Some historic sites have become what can be regarded, by Antarctic standards, as mass tourism destinations. These sites receive upward of 10,000 visitors annually and include Port Lockroy in Goudier Island and Whalers Bay in Deception Island. In 2006 more than 25,000 visitors landed at 30 high-priority historic sites in Svalbard (more than 30 percent of all landings). These figures

3 Data from Governor of Svalbard (2006a, 2006c) and IAATO (2007). Antarctic visitor figures based on tourist numbers with an estimate of one guide every 20 tourists. If a passenger lands in two different locations in one day this is registered as two landings, and therefore the number of landings (instances when people set foot ashore) is higher than the number of visitors.

4 Data from Governor of Svalbard (2006a, 2006c) and IAATO (2007). Svalbard landing data prior to 2001 are uncertain due to the various methods of reporting and unavailability of data (Government of Svalbard 2006a:6). Antarctic visitor figures based on tourist numbers with an estimate of one guide every 20 tourists. If a passenger lands in two different locations in one day this is registered as two landings, and therefore the number of landings (instances when people set foot ashore) is higher than the number of visitors.

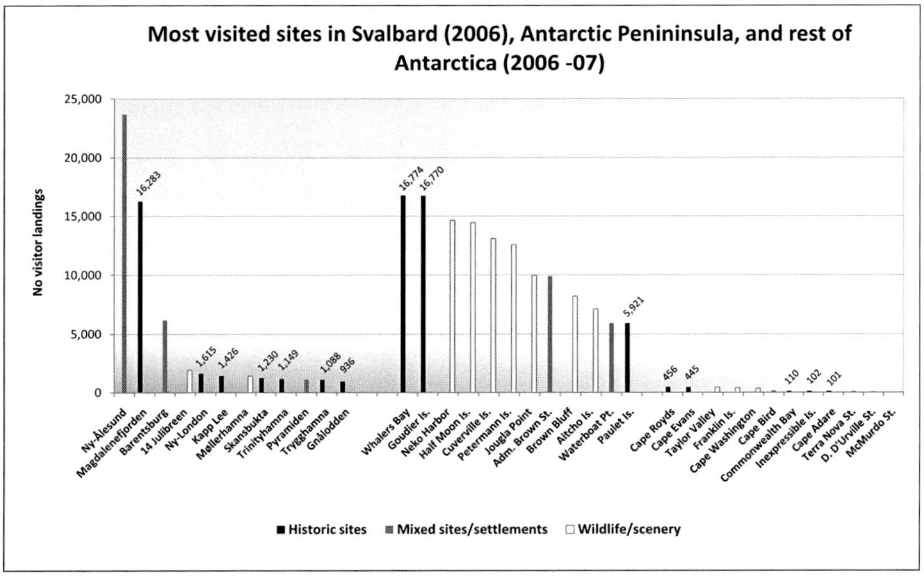

Fig. 8.5 – The twelve most visited historic sites in the Antarctic Peninsula; in continental Antarctica excluding the Antarctic Peninsula; and in Svalbard, classified according to main attractions.[5]

exclude sites that contain a range of attractions in addition to historic sites, such as some Antarctic research stations and Svalbard settlements.

Each landing is choreographed differently depending on site characteristics, site-specific regulations, and logistic or other factors. Generally tourists visit the most significant (or prominent) features of a historic site, where guides may deliver a lecture or interpret the site. At a landing groups may remain together, disperse, or combine both behaviours. Although these aspects are not developed here, it is apparent that the dominant nationality of a tourist group and the ability of the tour guides and tourists to communicate with each other influence visitation and behavioural patterns. Free roaming is allowed in some Antarctic sites, but is generally not allowed in Svalbard because of polar bear risk. For instance, all tourists that land at Ny-Ålesund are allowed to walk around the settlement's 1.5 km 'nature, research and cultural heritage' trail. However, only the passengers of ships carrying ca. 50–100 passengers are taken to visit the airship mooring mast of the 1926 "Amundsen-Ellesworth-Nobile Transpolar Flight" ("Amundsen's mast"), which is only 300m from Ny-Ålesund but requires the presence of polar bear guards (Figures 10.8 a-b).

5 Data cover the 2006 Boreal summer and the 2006-2007 Austral summer. Data from IAATO (2007); Government of Svalbard 2000, 2006a; and data provided by Government of Svalbard officials, 4 October 2006.

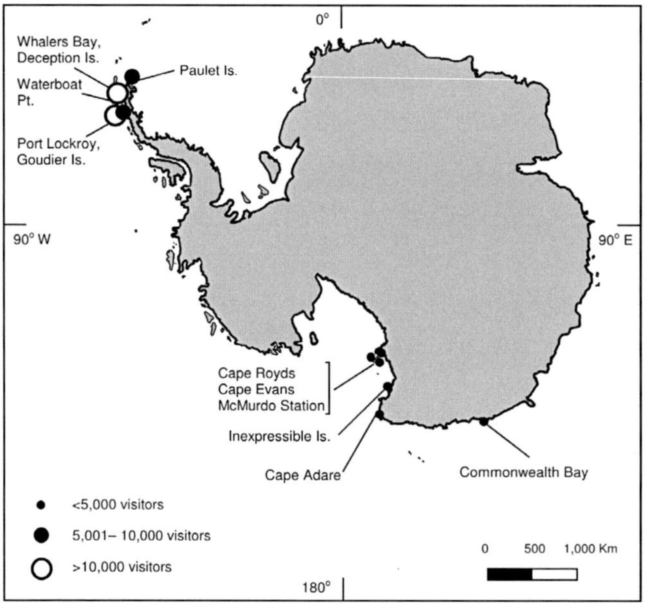

Fig. 8.6 – Most visited cultural heritage sites in Antarctica, 2006-2007 Austral summer.[6]

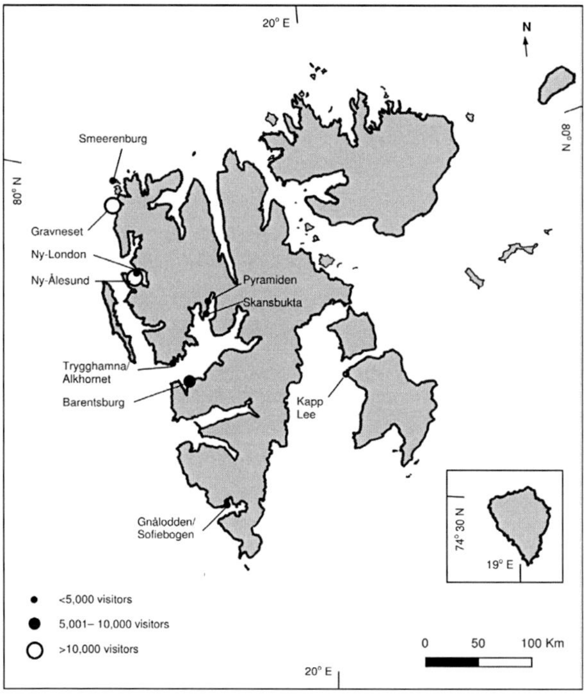

Fig. 8.7 – Most visited cultural heritage sites in Svalbard, *2006 Boreal summer.*[7]

---

6 Data from IAATO (2007).

7 Data from Government of Svalbard (2000, 2006a).

Figs. 8.8 a-b – Different choreographies of visitation. A group of tourists approaches the airship mooring mast at Ny-Ålesund (upper) while another group watches it from the distance (below). Photos by R. Roura 2007.

Site specific knowledge influences visitation patterns. The more visitors know about a site and its history, the more thorough and extended the visits can be. For instance, a group of tourists visiting Ny-Ålesund looked briefly at Amundsen's mast – which one guide described as 'one of the ten, or perhaps five, most significant sites in the Arctic' – and instead visited the small cemetery nearby. This group's guides reported not to know much about the mooring mast's history (pers. obs. 29 June 2007). Most tourist groups visiting Whalers Bay in Deception Island spend two or more hours visiting whaling and other remains spread around the bay. Eight tourists travelling in a commercial yacht visited only conspicuous parts of the site and returned to their vessel 75 minutes later. The yacht's skipper was ashore supervising the visit but did not lead the tourists through the site (pers. obs. 15 February 2002).

Fig. 8.9 – Basic tourist behavioural repertoire at a historic site. Photo by author 2002. Key: 1 - Moving around; 2 - Gazing upon the site (note group standing inside derelict building); 3 - Documenting the visit; and 4 - Gathering/receiving information (exchanges with a tour guide).

Fig. 8.10 – A lecture around a blubber oven at Smeerenburg, Svalbard. Photo by author 2007

Antarctic tourism statistics (IAATO 2008) suggest an increase in activities at or near historic sites that are unrelated to the sites' historic values. These include kayaking, scuba diving, and hiking at landing sites such as Cape Royds in Ross Island and Snow Hill Island, which have significant remains of the 'Heroic era' of exploration. At least one marathon took place at Whalers Bay in Deception Island, in which participants run among whaling remains (Racing the Planet nd). A comparable Svalbard example is the holding of beach buffets and parties next to the XVI whalers' graveyard at Gravneset in the Magdalenafjord.

8.4.2 Tourism impacts and cultural heritage

8.4.2.1 Tourism behaviour

Observations of tourist visits to historic sites enabled to develop a basic behavioural repertoire. Basic visitor behaviour includes moving around the sites (walking, standing); gazing upon historic features and surrounding flora, fauna, and landscapes; gathering and receiving information (reading, listening, asking); and documenting the visit, whether by taking something away (e.g. photographs) or leaving something behind (e.g. a signature in a visitors' book) (Figure 8.9). These behavioural events often take place simultaneously e.g. people moving objects in order to take photographs (e.g. Hughes 1994). Naturally, these are not the only possible forms of visitor behaviour at an historic site, but the broad patterns of usual behaviour. There might be unacceptable (or illegal) forms of basic behaviour e.g. looting an artefact may be regarded as a way of documenting the visit.

For individual tourists impact-causing behaviour may result from a single action or a concatenation of actions exposing cultural remains to physical contact and possible damage. For instance, behaviour observed at Amundsen's mast included people standing, leaning or sitting on parts of the mast to take photographs, gain a higher viewing point, rest, and tie their bootlaces. The airship mooring mast may not be particularly vulnerable to these forms of behaviour, but other cultural remains might be.

Group behaviour that may generate impacts may include spreading out or crowding over or around sensitive areas. For instance, in Smeerenburg, a 17th century whaling station, visitors must stay away from the remains of the furnaces, which are built of sand cemented with whale blubber (Figure 8.10). However, the whole area is an archaeological site (Figure 8.11). Even if visitors stay away from such fragile features, they tend to stand in the area immediately around the furnaces, which may contain vulnerable surface or subsurface remains.

Visitation to a historic site does not necessarily imply full immersion in history. Tourists may focus on a site's historic features, but also on the views from the site, and on plants, wildlife, scenery, or people. For instance, tourists leaned on Amundsen's mast to photograph saxifrage flowers inside the mast's structure, which protected the plants from trampling, or stood on the mast's footings to have a picture taken (Figures 10–12 a-b). Some tourists looked away as lectures on the mast's history were delivered (Roura 2009). The highlights of a visit to Smeerenburg were the natural events unfolding during the visit, rather than the site's historic features (Figure 8.11). The visitors' focus on something other than historic features when visiting a historic site is not inherently problematic, but it may lead to inadvertent damage to those features.

Tour guides are primarily responsible for enforcing regulations and supervising tourists. During observations some guides delivered enthusiastic lectures on historic events, others lectured mechanically, and others simply answered questions when asked. Some guides informed tourists about the rules that applied to a site before a visit, and others let the tourists know as some behaviour seen as reproachable was happening or immediately after. Some guides reminded tourists not to touch or stand on cultural remains, keep a certain distance, or not to walk in some areas, but other guides did not seem to issue any instructions while on site. Most guides took

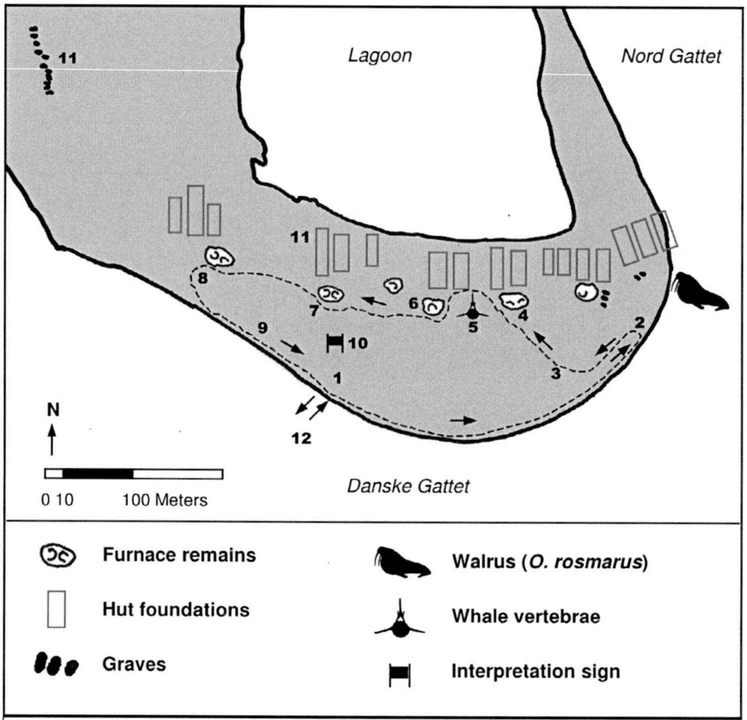

A landing at Smeerenburg, Svalbard, 17 June 2007

Map by R. Roura.

Base map source: Hacquebord L, F Steenhuisen and H Waterbolk (2003), 'English and Dutch whaling trade and whaling stations in Spitsbergen (Svalbard) before 1660,' *International Journal of Maritime History* XV:2, 117-134.

1 Three guides and about 50 tourists land at Smeerenburg, which the expedition leader describes as "one of the ten most significant historic sites in the Arctic".

2 The group approaches a walrus hauled out at the beach keeping a 30m distance. The walrus is badly wounded and the expedition leader orders to retreat. The easternmost part of the site is not visited.

3 The expedition leader calls the group's attention to three birds from different species fighting for food in full flight.

4 The group gathers around a furnace. The expedition leader talks about the two "natural Arctic dramas" just witnessed – the walrus and the birds. Lecture on cultural heritage protection in Svalbard.

5 The group gathers around whalebone. Lecture on bowhead whales ecology and whale processing.

6 The group gathers around a furnace. Lecture on Smeerenburg's history.

7 The group stops briefly around a double furnace.

8 The group gathers around a double furnace. The history lecture continues. A seal swims along the shore past the group. About half of the tourists turn around to look at it, interrupting the lecture.

9 After the lecture the group return to the landing site. Some tourists pick up plastic marine debris along the way.

10 Some tourists examine the Government of Svalbard plaque describing Smeerenburg before returning to the ship. Most tourists overlook or ignore it.

11 The hut remains and the cemetery are covered by snow and are not visited.

12 On return to the ship a whale is seen. The inflatable boats follow it.

Fig. 8.11 – A landing at Smeerenburg, Svalbard, 17 June 2007. Base map from Hacquebord et al., 2003. Map by R. Roura.

Fig. 8.12 – Tourist behaviour at Amundsen's Mast, Ny Alesund: Leaning on the mooring mast (upper); standing on the mast's footing (below, note broken edge). Photos by author 2007.

photographs during landings, and for a small minority this constituted a distraction from guiding. In turn, most tourists broadly followed the instructions given to them, but some tourists did not, whether inadvertently or otherwise.

8.4.2.2 Tourism behaviour traces and impacts

Based on Schiffer (1987) polar tourism can be regarded as a cultural process of historic site (trans)formation (Roura 2009). From a management perspective, an activity-output-exposure-impact framework, usually applied to the natural environment (ATCPs 2005 ((EIA guidelines)), is useful to differentiate conceptually the activity itself from its consequences. Each activity at a historic site has specific outputs to which cultural

Fig. 8.13 – Effect of trampling at Amundsen's mast. Photo by author 2007.

remains may or may not be exposed. Exposure to a particular output may or may not result in an impact. Thus a footfall may leave a footprint on the ground but not in itself cause an impact on historical values. However, a footfall on an artefact may break it. In addition, the cumulative effect of successive footprints may damage the subsurface archaeological archive and contribute to overall site degradation (e.g. Hofman and Jatko 2000; Hacquebord 2004).

Direct visitation impacts may include trampling damage, the dismantling of sites, and the removal of artefacts (Table 8.1). However, it can be difficult to differentiate unambiguously actual traces of tourism behaviour from those caused by other contemporary or historic cultural processes, by nature, or by a combination of processes. In some instances it may only be possible to attribute an impact to visitation by direct observation. Nevertheless, it is possible to overlay observed patterns of visitation with behaviour traces and identify locations within a historic site where activity traces can reasonably be attributed to recurring tourism landings. These can be defined as contemporary 'tourist activity areas' partly overlaying the area where earlier activities took place.

The most obvious evidence of visitation identified in this study was trampling disturbance on the ground and vegetation cover at some frequently visited sites, including path development and erosion on or around prominent features at Whalers Bay in Deception Island and some 'Heroic era' remains at Ross Island. In Svalbard trampling disturbance was observed at parts of the 17th century whaling station in Smeerenburg; around Amundsen's mast (Figure 8.13); and on and around some historic remains

Fig. 8.14 – Effect of trampling at historic earthworks in Ny-London. Photo by author 2007.

in the mining settlement at Ny-London (Figure 8.14). Smaller scale behaviour traces included footprints on top of a 17th/18th century mass grave; a broken glass artefact damaged during handling; urine stains at a historic cross; and the occasional presence of lost or discarded items. To the extent that it was possible to differentiate the effects of visitation from those caused by other processes, these could be characterised as minor but fairly widespread over historic sites, as well as more significant but localised effects, e.g. in a single artefact.

An additional effect of tourism is the transformation of the original site function (e.g. industry, exploration) and its replacement by a new function (tourism). This transformation may entail not only changes to the site through e.g. wear and tear, but also through the installation of infrastructure to manage (or attract) tourism (Table 8.2). In some instances historic site adaptation to tourism uses, which may involve intrusive restoration and the establishment of tourism management structures, may be regarded as an indirect tourism impact on historic values.

8.5 CONCLUSIONS

The patterns of organised cruise tourism in Svalbard and Antarctica are broadly similar in terms of modus operandi, scale, and characteristics of landing sites. Key historic sites are among the most visited sites in both regions. However, more visitors land at fewer historic sites in Antarctica than in Svalbard. Consequently, current tourism pressure on popular Antarctic historic sites (in terms of the number of visitors) is greater than on equivalent sites in Svalbard. Historic sites in both regions may be affected by wear and tear resulting from normal behaviour, or by acts of misbehaviour.

During this study, tourist behaviour that *could* cause an impact in vulnerable historic sites – but not necessarily *did* – was noted in most visits. Visitation traces were evident at most landing sites that contained historic remains, underscoring the role of tourism as an agent of site transformation. However, 'impact' is a management term that implies a value judgement. Opinions will vary widely as to what traces of visitation constitute an impact, because of varying perceptions on their scale, duration, intensity, and overall significance. However, some behaviour traces may detract from the condition or integrity of a historic site or its components, and should be regarded as impacts.

The potential of tourism to cause a significant impact on historic remains depends on spatial and functional patterns of site use and on tourist behaviour. Minimising the exposure of cultural remains to visitors depends on separate mechanisms to manage tourism and manage historic sites themselves. Under current policies tourism is likely to continue to increase in both regions, with an ensuing increase on visitation pressures at key historic sites. Tourism in Svalbard is promoted as a key economic activity, and regulated by Norwegian authorities, which aim to make the archipelago one of the best-managed wilderness areas in the world. In contrast, there is no Antarctic Treaty wide policy on tourism at present, although XXXII ATCM (2009) approved Resolution 7 (2009) 'General Principles of Antarctic Tourism', which may be the basis of a future Antarctic tourism policy. So far the preferred approach has been to manage tourism using existing generic legally binding instruments, such as the Protocol, complemented with specific non-binding guidelines. Following the rapid growth of tourism and several accidents involving cruise vessels Antarctic tourism may in the future be subject to additional legally binding regulation. However, Antarctic decision-making on tourism issues has so far been slower than the pace of tourism developments.

The cultural heritage protection regime in Svalbard is *inclusive* (all cultural remains predating 1946 are automatically legally protected, plus some more recent remains), whereas in Antarctica it is *exclusive* (with some exceptions only designated HSMs are protected). In addition, the regime of governance is *national* for Svalbard and *international* for the Antarctic region as a whole. Norwegian authorities manage all cultural remains in Svalbard regardless of their origin. This brings greater consistency and integration with regards to the management of both tourism and historic sites. It also brings the advantages of an administrative, monitoring and enforcement structure associated with an acting government authority. However, this approach takes the direct control of non-Norwegian historic sites away from the country where that heritage originated, resulting in potential issues of site access, ownership, and management approaches. Instead, individual historic sites in Antarctica are usually managed by the Party or Parties that originated the sites and promoted their designation as HSMs. Consequently, the degree to which Antarctic historic sites are managed actively, and to which tourism at those sites is restricted or promoted, varies considerably from site to site. Overall, there appear to be greater consistency in the management of historic sites in Svalbard than in Antarctica, even though many key Antarctic historic sites are effectively managed.

Historic sites are sometimes time capsules of past times and lives, but are not in themselves static. Historic site transformation takes place in the context of broad changes across the Polar Regions, in which tourism is both a cause and a consequence of change. Notwithstanding the legal and practical framework in place to manage both tourism and historic sites in Antarctica and Svalbard, and the diverse circumstances of each region and each site, the potential exists for the cultural remains of these remote regions to be degraded, destroyed or removed from context as a result of tourism – and with them the history they tell.

8.6 ACKNOWLEDGMENTS

I thank Prof. Dr. Louwrens Hacquebord and Dr. Maarten Loonen (University of Groningen and Netherlands Arctic Station), Dr. Kees Bastmeijer (University of Tilburg), Dr. Alan Hemmings (University of Canterbury), and Per Kyrre Reymert (Government of Svalbard) for discussions on issues relevant to this article, without implicating them in my interpretations. Fieldwork in Svalbard was funded by the Groningen Institute of Archaeology, University of Groningen. Opportunistic fieldwork in Antarctica was enabled by the Antarctic and Southern Ocean Coalition and Gateway Antarctica. Finally, I thank Dr. Dag Avango, Dr. Peter Capelotti, Janet Dalziell and Dr. Jessica O'Reilly for helpful criticism of an earlier draft.

8.7 TABLES CHAPTER 8

Table 8.1 – Examples of known impacts on historic and past activity sites in Antarctica and Svalbard from tourism and other visitors

Impact	Type of remains and location	Reference
Erosion of footpaths	Several popular landing sites in the Antarctic Peninsula, including Penguin Island, Barrientos (Aitcho) Island, and Hannah Point. Sites are visited primarily by their natural attractions but contain artefacts/ecofacts from whaling times.	United Kingdom (2006a).-
Erosion of footpaths, trampled vegetation	Material remains from World War II at Signehamna, Svalbard	Geitz 2004
Wear and tear - ground erosion around prominent features	Airship mooring mast at Ny-Ålesund, Svalbard	Roura 2009.
Wear and tear	Whaling graveyard dating from the early 17th to the late 18th century, Gravneset, Magdalenafjord, Svalbard	Viken and Jorgensen 1998; Geitz 2004
Recent graffiti on buildings, oil tanks	Early- to mid-20th century whaling and scientific base remains at Whalers Bay, Deception Island, Antarctica.	Argentina, Chile, Norway, Spain, the United Kingdom and the United States (2006)
Damage to outer patina of historic hay bales	Shackleton's Nimrod Hut, Cape Royds, Ross Island, Antarctica	Hughes and Davis 1995:244
Removal of artifacts - Souveniring (by National Program expedition members)	Wilkes Station, Antarctica, an abandoned IGY station protected by Australian domestic law	Hughes and Davis 1995:247
Removal of artifacts - Souveniring (ca. 1950s)	'Heroic era' huts in the Ross Sea region, Antarctica.	Harrowfield (2005)
Removal of artifacts - 'Almost certain' souveniring (ca. 1960s-1970s)	Sealers' refuges in Byers Peninsula, Livingston Island, Antarctica	Lewis Smith and Simpson (1987)
Unspecified acts of vandalism.	Nordenskjöld Hut, Snow Hill Island, Antarctica	Argentina and Sweden (2007)

Table 8.1 continued

Impact	Type of remains and location	Reference
Dismantling of site - Bricks removed from historic tramping hut ruins, used to write the word "OPEN" on the ground	Russetuft in Trygghamna, Svalbard	Governor of Svalbard (1999)
Dismantling of site - Logs, boards from historic hut ruins used as firewood; stones from historic site used to hold tents	Laegerneset, Belsund; Beautnest, Barenstsoya, Svalbard	Governor of Svalbard (1999); Per Kyrre Reymert, Cultural Heritage Adviser, Sysselmannen på Svalbard, personal communication, September 22, 2006, Longyearbyen
Dismantling of site - Wood from historic site used to line a thermal spring in volcanic beach	Whalers Bay, Deception Island, Antarctica	http://mikehorn.com/index.php/site/page/dairy_archive_october_2008
Damage to inactive station - Breaking into locked buildings. Doors left open resulted in significant wind and snow damage to buildings	Leningradskaya, a Russian scientific station unoccupied since the early 1990s, Antarctica. While not currently protected as a historic site, the base is a 'time capsule' of Soviet times.	Russian Federation (2008)

Table 8.2 – Examples of changes to historic sites in Antarctica and Svalbard resulting from the management of tourism and other visitors

Changes	Location	Reference
Building of observation hut and fence around graveyard	Graveneset, Svalbard	Geitz (2004)
Use of a historic hut as a manned "living museum", which includes post office and souvenir shop	British Base A at Port Lockroy, Goudier Island, Antarctica	United Kingdom (2006)
Additions to historic hut ruins (remains of an upturned boat) include placing a monolith with a plaque and metal structure with an interpretation sign; digging around the hut's foundation so that its outline became more visible; and erecting a fence around remains	Waterboat Point hut, which is adjacent to Chilean station Gabriel González Videla, Danco Coast, Antarctica	Chile (2008)
Stains caused on historic huts by interpretation plaques	(Unspecified) 'Heroic era' huts in the Ross Sea region, Antarctica.	Hughes and Davis (1995)
Planned private salvage of a historic aircraft forced authorities to remove it from site. [8]	Mid-20th century scientific base remains at Whalers Bay, Deception Island, Antarctica.	United Kingdom (2004)[9]
Reconstruction, fencing of historic stone hut	Nordenskjöld hut at Hope Bay, Antarctica, which is adjacent to Argentina's Esperanza Base.	Described as "...a *recreation* of an original Swedish stone hut..." (emphasis added) in United States Department of State (2006: 29).

8 While not strictly a tourism-related impact, this is a dramatic example of the actions taken to manage private activities in Antarctica.

9 See also http://www.dehavillandmuseum.co.uk/dhc3otter/dhc3otter.html, accessed 5 February 2007.

9 Overview and synthesis

9.1 INTRODUCTION

Tourism can be examined on different spatial, temporal and behavioural scales (LaMotta and Schiffer, 2001; Heilen *et al.*, 2008).This chapter provides an overview of how polar tourism functions on these different scales, linking it with both the main theoretical framework outlined at the outset of this thesis, and the empirical findings of this research (Figs. 9.1-9.4). In addition, it summarises the management issues that result from the interaction of tourism and historic sites. The focus of this overview is tourism in Antarctica and Svalbard, although the broad considerations may also be applicable elsewhere in the Polar Regions.

9.2 TOURISM AS BEHAVIOUR

9.2.1 Destination selection

The main systemic process of polar tourism is the procurement, establishment and consolidation of tourism "destinations", often conceived as "landing sites" where "shore visits" by tourists take place (see e.g. Naveen *et al.*, 2006; Splettstoesser *et al.*, 2004). As noted by the tourism scholar E. Cohen, tourism regional development is essentially a centrifugal phenomenon: it usually starts around core locations in a certain region and then expands towards the periphery of that region. Intensification of tourism at the core causes a range of impacts on the natural and cultural environment, while tourism expansion brings to the periphery the same impacts that had previously occurred in the core area (Cohen, 1979). This model aptly summarises tourism developments in Antarctica and Svalbard. Up until the late 1980s-early 1990s tourism in Antarctica and Svalbard was a niche industry involving a few companies and a small number of customers. Since then polar tourism has consolidated as a major presence in these regions. A network of sites where tourism activity occurs has been established. Some of these sites have emerged as regular tourism destinations used by most tour operators where a high concentration of passenger landings occurs, while other sites are visited less frequently. Historic sites are included in many cruises, but this is only one aspect of a larger menu of tourism destinations and activities.

Polar tourism is both systematic and opportunistic toward the selection of sites visited. This dual approach results in the establishment of certain sites as tourism destinations, and the ongoing recruitment of new sites as opportunistic visits take place. Tourism operations take place regularly at certain locations endowed with attractions such as wildlife, landscapes, cultural heritage, and whose characteristics made them suitable to land tourists. Sites that have a certain unique or novel feature are more likely to attract visitors than areas where such novelty is absent (IAATO, 2002). However, landing site selection is a dynamic process, and the number of visitors at specific sites fluctuates over the years (e.g. Argentina, 2006; Serrano, 2007; ASOC, 2008;

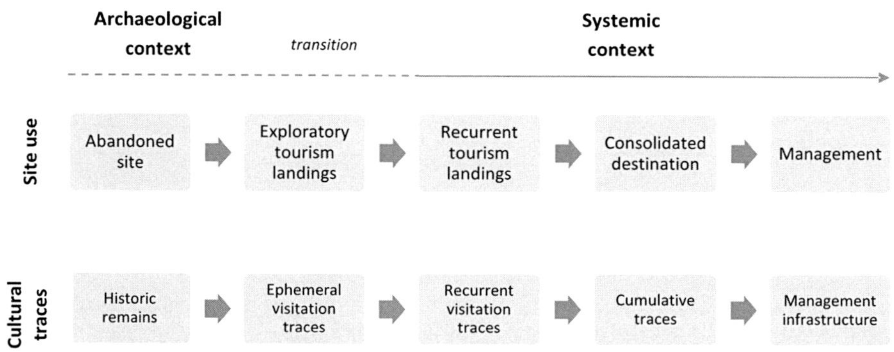

Fig. 9.1 – A simplified evolution of an abandoned site into a tourism destination, with increasing direct, cumulative and indirect effects of tourism. Graph by author.

Lynch *et al.*, 2009). Diachronic change is influenced by changes in factors such as ice conditions, site attractiveness, and regulation directing tourism towards or away from specific sites (e.g. Lynch *et al.*, 2009).

9.2.2 Behavioural patterns of shore visits

Once ashore, tourists engage on sightseeing and a range of other activities. Tourism shore visits usually follow a consistent pattern that is often tailored to specific sites. According to Splettstoesser *et al.* (2004), most shore visits last 2-3 hours; usually not all passengers on board actually land; and the majority of those who do land stay ashore a fraction of the allocated time, with the rest staying either less than one hour or the entire time. However, most of the visits observed in this research lasted under two hours. In Antarctica, and in most of Svalbard, no more than 100 tourists are allowed to land simultaneously. During landings, tourist groups are often divided into two or more smaller subgroups.

The selection of attractions at each landing site also combines systematic and opportunistic approaches. Certain features within a site are visited regularly, and other features (both permanent and ephemeral) are opportunistically turned into tourist attractions, often through mediation by the tour guides. Anything that can be represented as something either archetypical of the Polar Regions, or unique, or otherwise remarkable in some way can be turned into an attraction. This may include representative or well preserved historic features; the highest point of a landing site; passing wildlife; and encounters with non-tourists. Plainly visits to historic sites in the Polar Regions tend to focus on the historic components of the site, but there may be other (non-historic) elements to which visitors may be taken or be attracted to, sometimes in preference to historic features.

Site-specific behaviour of groups and individuals depends on the type and location of attractions, and on how the visit is choreographed. Most commonly visitation will include circulation to and from an attraction (A=>B) or following a circuit (A=>B=>C=>D=>A),

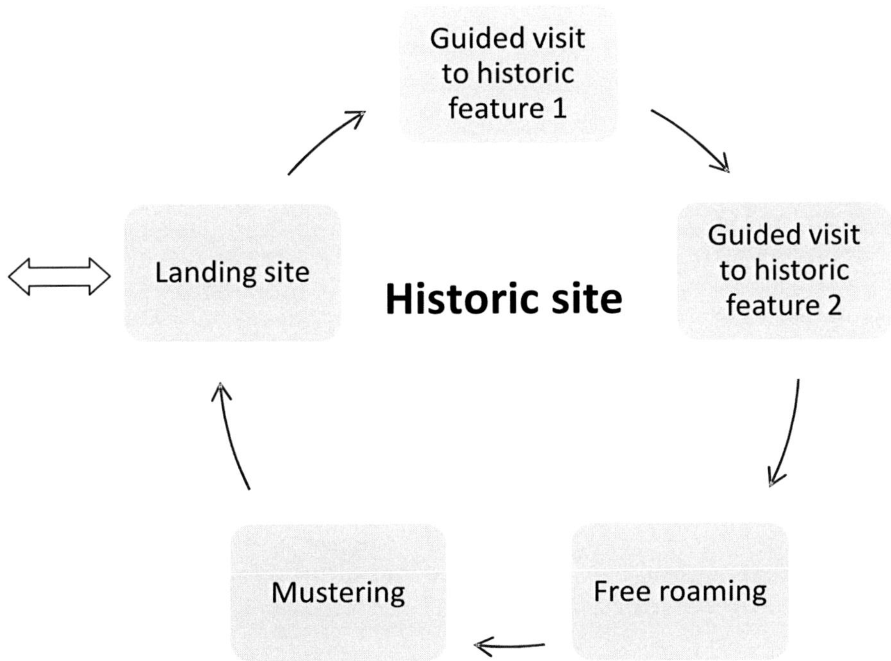

Fig. 9.2 – A typical landing at a historic site. Graph by author.

although there are alternative patterns. Tourist groups would change in composition and density as they move, whether spontaneously or following directions by guides. Free roaming is allowed in some sites, but not in others, and furthermore some sites may have zones that are off limits to visitors for safety or environmental reasons. In Svalbard tourists groups tend to stay close to armed guides due to polar bear risk. Visitation may include periodic stops at focal points such as historic features, lookouts, or places with suitable "photo opportunities". Tour guides may deliver lectures and/or respond to questions from tourists. Synchronic variation and diachronic change may result from e.g. changing weather or ice conditions.

9.2.3 Individual behaviour

As described in Chapter 8, visitors display a basic behavioural repertoire that combines moving through the sites; examining visually the site for sightseeing or navigation purposes; gathering and receiving information; and producing a record of the visit. Plainly visitors may display many other forms of behaviour including potentially eccentric forms. Visits consist of a continuous repetition of these individual actions, which take place through the processes of arriving, circulating through the site, and departing. During the visits people interact with cultural and natural features historic sites mechanically, visually, and in other ways (see LaMotta and Schiffer, 2001; Schiffer, 2010; Chapter 1).

Fig. 9.3 – Basic forms of behaviour, common types of interaction between people and historic sites, and resulting effects. Graph by author.

A pattern of visitor behaviour may result in a corresponding "archaeology" of visitation traces. Mechanical interactions between visitors and elements of the site may result in "tourism artefacts" of some kind. Fig. 9.3 illustrates common forms of visitor behaviour, different types of interaction with historic features, and resulting traces, some of which are commonly regarded as "impacts". However, visitor behaviour would normally have to be in compliance with existing regulations, to the extent that these exist and are known and enforced, so that some impacts that are conceptually possible are not supposed to occur in practice, such as for instance the removal of artefacts as souvenirs.

A behavioural archaeology theory of communication suggests that any form of human behaviour may be regarded as a response to information, which may include non-verbal communication modes such as tactile, visual, and chemical (through smell and taste), and may also include the acquisition of information from non-living entities (Schiffer and Miller 1999a-1999b; quoted in LaMotta and Schiffer, 2001: 33-36).

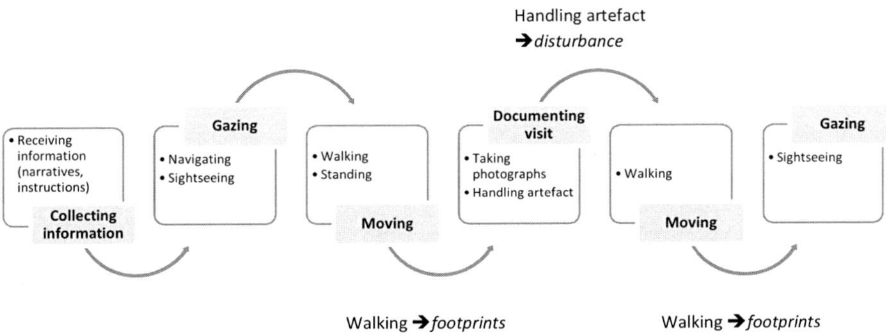

Fig. 9.4 – A hypothetical example of tourist behaviour as a visitor moves through a site responding to information cues and documenting the visit, and resulting traces of behaviour. Graph by author.

The conduct of tourism involves constant information exchanges so that the tourist identifies, for instance, what the sights are, where they are located, and what is their meaning. This is equivalent to the concept of tourism attraction systems (Leiper, 1990), by which an attraction is created by combining a person, a sight (e.g. a historic feature), and information about that sight (Chapters 1 and 6). Tourist behaviour may be influenced by information not only in terms of constructing meaning (*i.e.* creating the attraction itself) but also in terms of directing visitors towards (or away from) particular features or parts of a site, as it might be with respect to areas that are off-limits to visitors or as a cue to features of interest that might otherwise be overlooked. The significance of information is highlighted by the fact that visual cues about historic sites are not always self-explanatory (Chapter 6). In addition to verbal communication – such as instructions and narratives – non-verbal communication also influences visitor behaviour. This may include the legend in an interpretation sign; prominent features that stand out in the landscape; and particular smells as, for instance, the peculiar smell inside Scott's *Terra Nova* hut.

9.2.4 The footprint of tourism

Archaeological sites, according to P. Drewett, "consist essentially of activity areas and rubbish. That is where people have done things in the past and left some residue of having done something." (1999:17 quoted in Green and Moore, 2002: 52). The same concept broadly applies to the sites where contemporary polar tourism takes place. Regulations in place prevent visitors from littering, and this obligation is generally adhered to, but other features resulting from tourism may be identified.

In behavioural terms, a pattern of visitor activity may result in a corresponding "archaeology" of visitation traces. The choreography of visits to specific sites tends to be repeated visit after visit, as a result of the way individual sites are "packaged" as attractions. Consequently, there is a pattern to individual visits and to the resulting direct and cumulative traces of visitation. These define a *tourist activity area* superimposed over a historic activity area. These two areas may overlap but are not necessarily the same. The tourism area may include a landing site; the site's "must see" features; and the circulation network to get to/from any location.

The outputs or direct traces of tourism and other forms of visitation at historic sites may be of two kinds: additive and reductive (or accumulation traces and erosion traces). Additive traces, which result from deposition, include painted graffiti, discarded items (litter), and lost items. Items that are discarded or lost may include cigarette ashes or buts, paper handkerchiefs, gloves, hats, water bottles, walking sticks, and plastic bags. Rock cairns built gradually by visitors are another type of additive trace. Chemical traces such as hydrocarbon traces may also potentially result from tourism (e.g. Roura *et al.*, 2008) although these would not normally be identifiable with the research methods used here. Reductive traces, which result from erosion, include individual footprints, trampled areas or large areas of overlapping footprints, and paths.[1] Trampled surfaces may develop around prominent historic features where visitors tend to gather. Plainly paths develop between two or more locations that are frequently visited, and become particularly obvious on sloping ground. Reductive traces may also include abrasion marks (striations) on surfaces like wooden floors or walls; broken artefacts or parts of immobile features; and carved graffiti. Individual traces result from different forms of behaviour.

As noted by Weiner (2010: 46), "...the archaeological record is missing a huge amount of information when compared to the same site when occupied." This requires assessing what may be missing and, in particular, differentiating "...what is missing because it was never present or because of degradation (*diagenesis*)." (2010: 46). Unless there are material traces suggesting absence – such as surface decolouration or indentations – identifying what is missing requires previous knowledge of the existence of an item that is no longer present. This knowledge may well be lacking at many historic sites. The same concept applies to items whose condition or integrity has changed as a result of disturbance. For instance, up until the 1970s the collection of artefacts from historic sites by casual visitors was common in parts of Antarctica[2] and Svalbard (AHT-NZ, 2004; Harrowfield, 2005; Governor of Svalbard, 2000). This type of looting is now illegal and has subsided over the past decade. While looting was regarded to be more of a problem associated with recreational visitors than organised tourism, the problem potentially remains; for instance, some "very iconic" items had to be removed from Scott's *Terra Nova* hut as a precautionary measure (N. Watson, personal communication, 7 August 2006).[3] Barring detailed knowledge about the artefact inventory of specific sites (e.g. Splettstoesser, 2001:75; see also Chapter 4), it may

1 Technically a footprint is formed by a combination of erosion (where sediments have been removed by a footfall) and accumulation (where those sediments have deposited).

2 "The historic huts were often plundered in the early days. Richard Pape visited Cape Evans in 1959...he records quite candidly that he pocketed "a glass inkwell on which 'R.F. Scott' had been painted, also a bottle of Indian ink marked 'Wilson'" (Wheeler, 1999: 55).

3 The reversal of looting is also taking place: some people are returning objects looted in the past, and some of them are in better condition than similar objects that had remained in the huts (N. Watson, personal communication, 7 August 2006).

Fig. 9.5 a-b – Destruction of an emergency hut, possibly as an indirect effect of tourism. The Argentine "Naval Refuge Capt. Fliess", Neko Harbour, Danco Coast, Antarctica, established in 1949, was not a protected historic hut but it pre-dated 1958, the age by which temporary protection may be afforded to historic remains. Similar behaviour may result in similar effects elsewhere. Photos by R. Werner 2006, 2008.

only be possible to attribute changes to the presence, condition or integrity of an item by direct observation.

The traces listed above constitute direct consequences of visitation, which may also be cumulative. The footprint of tourism includes also a range of traces caused indirectly, for instance damage caused by visitors who leave a door open and expose a building to the elements (Russia, 2008; Argentina, 2010; Fig. 9.5), or melt water

Fig. 9.6 –Melt water erosion along a tourist path to the airship mooring mast at Ny-Ålesund. The path is used occasionally by tracked vehicles. (Photos © R. Roura 2007).

erosion along pedestrian paths (Fig. 9.6). Tourism management infrastructure of different kinds may also be regarded as an indirect impact and result in additive and reductive traces such as temporary flags or traffic cones used to demarcate specific areas, or permanent infrastructure such as walking boards, boundary fences, interpretation signs, monitoring huts, tourism reception centres, and souvenir shops and post offices inside historic buildings (Fig. 9.7). In contested areas a constantly renewed audience of tourists provides an incentive to display material symbols of territorial presence such as flags and flag posts, monuments, and plaques, or to use tourism management infrastructure for that purpose (e.g. Timothy, 2010; see also Chapter 4). Other reductive traces may include items that are removed from a historic site because they pose a risk to visitors, such as items that contain asbestos or that may be blown into the air in strong winds.

Tourism traces at historic sites are most noticeable, and arguably more significant, as an *assemblage* – that is, as an associated set of contemporary artefacts related to tourism (*i.e.* things made by or for tourists) that have a cumulative (and obvious) effect on landing sites. Broadly speaking, cumulative effects such as paths and extended areas covered in footprints are likely to occur at sites that have consolidated as tourism destinations. As tourism traces increase, minor historic traces may decrease by effect of regular disturbance. These may include loose artefacts, artefact scatters, rubbish dumps, including artefacts that are half-buried or below the ground surface. Larger or

Fig. 9.7 a-b – Impacts of activities linked to tourism: Chilean interpretation signs placed at Whalers Bay, Deception Island, in the 1990s, reflecting a version of historic events that was regarded as controversial by opposing claimant states (UK et al., 1999). They were subsequently removed under international management of the island.[4] *Photos S. Barr (2002).*

unmovable historic features may remain unscathed, depending on their characteristics and the use to which they are subject. The ground surface itself – such as tundra vegetation or volcanic terraces – may also be affected regular visitation as well as, potentially, the fauna present at a site. However, the overall footprint of tourism depends, in part, on the context of other site formation processes affecting the sites, and on the sensitivity of specific sites and features (Table 9.2).

4 In the late 1990s both the United Kingdom and Chile, independently from each other, mounted interpretative signs at Whalers Bay describing the history, geology, clean up, and individual components of the site (UK and Germany, 1999a: 106; Splettstoesser, 2001:75). An official inspection at the time noted that a number of signs had been erected at the historic site in Whalers Bay and that these "...appeared to contain misleading information and could be construed as visually obtrusive" (UK and Germany, 1999b: 10).

9.2.5 *Tourism in the context of other site formation processes*

In 1961 R. Ascher noted with respect to the impact of archaeological research:

> *What the archaeologist disturbs is not the remains of a once living community, stopped as it were, at a point in time; what he does interrupt is the process of decomposition.* (324; quoted in Gamble 2008:97).

Conceptually, these comments are equally valid for the conduct of tourism at polar historic sites. Tourism does not take place at unchanging sites that are "frozen in time", but rather at sites that are undergoing decay. The underlying question is whether or not tourism causes less, the same, or more changes than ongoing processes of decay. The relative effect of tourist visitation at cultural heritage sites as compared with the effects caused by natural or other cultural processes is relevant to site managers but difficult to assess. A Svalbard official described this as "the question of a million dollars." (P.K. Reymert, personal communication, 22 September 2006).

There are no definitive answers to this question but some empirical comments can be offered, based on observations conducted through this research at polar historic sites that are regular tourism destinations, and at "control sites" with no tourism and/or no historic remains:

- Tourist visits to natural sites leave an assemblage of visitation traces comparable to those left at historic sites – footprints, paths, lost or discarded items, and so on. These reflect the unfolding of the visits and the location of key attractions.[5] However, these traces may be ephemeral, depending on the site characteristics and the frequency of visitation. For instance, no obvious traces of visitation were found at one large penguin colony known at the time to have about twenty landings each season (or about 1,000-1,500 visitors), and which had already received about half that number of visitors when it was inspected..
- At historic sites that receive few or no visitors the traces of visitation (such as footprints) will eventually disappear and be replaced by natural features such as a lag pavement. Sites that receive regular visits will continue to exhibit traces that are inherently ephemeral (such as footprints) but that are recurrent and that may not disappear entirely from the end of one season to the beginning of the next. However, this is in itself a complex process that is dependent on the substrate characteristics, and environmental factors such as wind speed and snow precipitation (e.g. Campbell *et al.*, 1998).
- Visitors who were not tourists, as observed in this research, often behaved in comparable ways to tourists. However, a limited set of observations conducted in this research suggest that non-organised visits to historic sites tend to be briefer, cover smaller spatial areas, and be less focused on historic features than for organised tourist groups. Sometimes these groups would miss key historic features

5 The interaction of visitors with wildlife would also cause specific effects, but these have not been studied here.

altogether. Information about the sites helps to inform behaviour towards sensitive features (e.g. Pearson *et al.*, 2009: 59) and simultaneously may also make visitors aware of the existence and location of particular sights.

- Natural processes affect historic sites in the Polar Regions continuously, comprehensively, and sometimes with great intensity. As a result cultural materials affected by wind, flowing water, freeze and thaw cycles, waves, wildlife, etc. are eroded, transported and deposited along with natural sediments and become embedded in local landforms. In contrast, tourism and most other cultural processes tend to focus on specific site features, and are usually seasonal. To that extent, the decay of historic polar sites by effect of natural processes tends to be more substantial than the direct disturbances caused by tourist behaviour. However, tourism tends to focus on significant elements of historic sites so that disturbances to these elements have a comparatively greater potential to diminish the archaeological value of a site.
- A range of cultural processes such as conservation or environmental management may cause major changes to historic remains, either in terms of increasing protection (and decreasing disorder in the arrangement of items) or in terms of diminishing protection (and increasing disorder). For instance, it has been suggested that a "political urgency to 'clean up the environment'" and the implementation of waste management regulations under the 1991 Protocol pose the greatest threat to the Antarctic cultural heritage (Evans, 2007: 1). Equally, some conservation approaches used at some historic huts have been regarded as exceedingly intrusive in their attempts to recreate the past as it was at a particular point in time (Save the Huts, n.d.). To the extent that tourism may indirectly trigger some of these changes, these could be regarded too as part of the footprint of tourism.

9.3 MANAGEMENT PERSPECTIVES

9.3.1 The effects of tourism on historic sites: interactions and exposures

Some forms of behaviour may lead to disturbances at historic sites, or "impacts" in management terms. Behaviour always result in impacts, however not all behaviour causes *substantive* impacts. As noted at the outset (Chapter 1) an activity-output-exposure-impact framework can be used to differentiate conceptually the activity itself from the consequences of the activity, or impacts. Each activity has specific outputs to which the cultural remains may or may not be exposed; the exposure of a certain feature to a particular output may or may not result in an impact. The key word here is *exposure* (or interactions in behavioural terms): certain forms of visitor behaviour may result in an impact if a historic feature is exposed to visitors' actions.

Exposure of cultural materials to visitation would normally result from interactions such as trampling over these materials, handling them, and other forms of physical contact. As a result of these actions cultural materials may experience breakages, striations, displacements, and so on, depending on their characteristics. Many impacts result from a transfer of kinetic or other forms of energy from people to an object, as a result of interactions, with the potential effect on the condition, integrity or other characteristics of the objects.

Major changes such as the removal of "iconic" objects or the destruction of particular features or generally the fabric of the site would affect the site substantially across a range of assigned values. Minor changes of this sort may not affect the aesthetic qualities of a site (which would still *look* "historic") but may change the information content by altering the archaeological record – particularly if there is no awareness that some changes are the result of tourism and not of other site formation processes.

Some energy transfers are naturally thresholded, in that there is a limit to the disturbances that an individual tourist carrying usual artefacts such as boots, walking sticks, and tripods could cause to a historic site. An individual visitor may cause damage to minor traces of past activities such as an artefact or small feature, but not normally to large features (other than perhaps by careless behaviour or vandalism). However, a site with many visitors may experience widespread minor disturbances. Tourism would not easily cause damage to a large industrial site as a whole, but may damage specific site components. Mechanical interactions between people and historic artefacts can be minimised or prevented using a range of management tools. These may include keeping some areas off limits, discouraging or banning some types of behaviour such as touching or handling artefacts, or capping the number of visitors at any given time or during any given period of time. Historic site conservation then becomes an issue of compliance with these regulations, which highlights the empirical role of tour guides in site conservation.

Nevertheless, mechanical interactions may be accidental/inadvertent (the visitor cannot see or recognise historic remains) or may also be intentional (the visitor walks on those remains, for instance, to reach somewhere else; to look at them closely; or to take a photograph). A historic feature may be the main subject of attention (the visitor stands on a feature to photograph it from nearby), but may also be used in an action directed towards some other subject (the visitor stands on a feature to photograph something else). Historic items designed to perform a certain technological function may be particularly exposed to the actions of visitors. These may include, for instance, entering a building, climbing a ladder,[6] or lying on a bed.[7] However, these actions may not be permitted for conservation or other reasons, which in itself does not ensure compliance.

Contemporary visitors are usually well behaved and fairly tightly controlled, and the most common evidence of visitation is the "wear and tear" resulting from cumulative

---

6 Climbing to the top of the airship mooring mast at Ny-Ålesund (Chapter 6) is not allowed now, but this was not the case in the past: "We climbed the high mooring mast, which the airships *Norge* and *Italia* had previously used…" (Strijbos 1957:45; author's translation).

7 "Then I laid my head on his pillow, and went to sleep". Thus finishes Sarah Wheeler's account of visiting Antarctica (Wheeler 1999: 334). The pillow was Scott's pillow in the *Terra Nova* hut, and Wheeler was traveling with the American National Science Foundation's Antarctic Artists' and Writers' Program in the 1990s, during which she spent some days at Cape Evans. This behavior would be unlikely to happen with an organised tourist group, but it illustrates the endless possibilities of visitor behavior.

Fig. 9.8a – Visitation traces. Footprints on edge of 17th century Dutch whalers' grave at Ny-Ålesund, Svalbard.[8] (Photo © R. Roura 2007).

visits. This does not necessarily imply that unintentional or intentional damage might not occur. Unintentional damage may result from accidents, such as people stumbling on a fragile object or causing unintentional damage. For instance, a visitor at the New York Metropolitan Museum of Art fell accidentally into a rare Rose Period painting by Picasso, causing a 30 cm tear of the canvas (Vogel, 24 January 2010). Deliberate attacks to art pieces by visitors have taken place in museums worldwide, including slashing, vomiting, spraying with ink or acid, hammering, punching, painting or engraving with graffiti, and dynamiting (Hind, 27 January 2008). In Svalbard visitors have been prosecuted for entering protected areas, painting rocks, and opening a 17th century whalers' grave (Governor of Svalbard, 1998: 21).

In this research physical evidence was found of visitors walking on a mass grave (Fig. 9.5), and urinating on a memorial cross (Fig. 9.6). This behaviour did not cause lasting damage but suggests, at a minimum, a lack of awareness about the presence or value of these remains, and would be regarded as inappropriate anywhere. For instance, in 2009 and 2010 there were four cases in Britain of people being sentenced to fines for urinating on war memorials (Wainwright, 20 August 2010). Historic sites are protected because their values are recognised (or defined) by legal criteria, but not all

8 A whalers' cemetery containing over 50 individual graves dating from the 17th century was relocated into a mass grave at Ny-Ålesund's cemetery in 1958 to allow for the construction of a coal washing plant (Roura, 2008). The original remains are automatically protected under the Svalbard 2001 act because of their age even though the monument post-dates the 1946 cut-off date. Under the management plan for Ny-Ålesund the graveyard is included in a cultural heritage area (Kings Bay, 2009).

Fig. 9.9b – Detail of urination marks at the memorial cross in Observation Hill, Antarctica (Photo © R. Roura 2007).

visitors may share those values. Polar tourism is the regional manifestation of a global phenomenon, and what is exceptional is where it occurs rather than the activity itself. Thus, there is nothing particularly unique about polar tourism that would exclude this type of behaviour, and it is conceivable that some visitors would take action to remove, damage, or desecrate polar historic remains such as national or religious symbols, if they could, for what they represent.

Regulations in place and voluntary codes of conduct explicitly prevent most of these actions. In Svalbard, government field staff may be present at some sites, some of which have observation huts. In Antarctica the presence of "official" guides (staff from heritage institutions or representatives of National Antarctic Programs) is required for visits to some sites, some of which function as "living museums" with resident staff. However, in both areas there are many instances in which compliance with regulations is left to tour guides or to the tourists themselves.

9.3.2 Historic site sensitivity factors

Common or likely forms of exposures that may cause disturbance or damage to historic sites and their components include handling movable objects; handling (touching) non-movable objects; standing on or close to historic features; entering off-limit areas; and disposing of waste.[9] Some of these forms of behaviour appear to be inherently non-damaging, however they are not allowed at most historic sites in Antarctica and Svalbard and furthermore they are commonly restricted at many museums and historic sites worldwide. These forms of behaviour will affect different sites and features differently depending on their sensitivity.

Site sensitivity is a function of several intrinsic characteristics of historic sites:

- **Site type** – In broad terms, historic *sites* (or specific components of historic sites) are more sensitive to visitation than historic *monuments*. However, some monuments, such as plaques, are susceptible to being taken away as souvenirs.
- **Site function** – Industrial sites are generally less susceptible to tourism impacts than other kinds of sites (e.g. a small hut containing a collection of artefact) because of their scale and hard wearing nature. Contemporary tourism is unlikely to cause substantive changes to massive industrial structures, even though it may cause a number of minor visible changes.[10]
- **Scale and complexity** – Impact assessment is multi-scalar, both in terms of space and time. Sites that have a small scale and/or a large complexity (including e.g. artefact scatters, collections of artefacts, and a diversity of features of different characteristics) are more susceptible to damage than sites of a large scale and/or low complexity.
- **Site condition and integrity** – The "impact potential" might be higher for sites in good condition/high integrity than those in poor condition/low integrity. However, sites in poor condition may be structurally more vulnerable to some forms of behaviour.
- **Visibility/obtrusiveness** – Historic remains having low visibility and poor contrast with the environment may be inadvertently damaged if people do not identify them as artefacts.[11]
- **Age** – Older sites may be more sensitive than more recent sites.
- **Relative values** – Historic sites are significant *tout ensemble*, however not all elements of a historic site may be equally significant. For instance, the body managing the four historic huts in the Ross Sea Area has prepared a conservation

---

9 At natural sites other forms of exposure include approaching wildlife too closely, or entering sensitive environmental areas that are off limits.

10 This depends on the type of pressure. For instance, a portable barbeque was found inside a wooden building of one Svalbard mine, which could potentially have caused a fire (T. Eliassen, personal communication, 22 September 2006).

11 For instance, Stehberg *et al.* report damage caused to a sealing site in Byers Peninsula code-named "Vietor 1" by scientists taking rock samples from an outcrop adjacent to the site (2008: 90).

plan for each of the sites.[12] Each plan contains a "statement of significance", and assigns "relative values" to each of the elements that compose the huts, including the buildings' interior and exterior, artefacts, and environs (AHT-NZ 2004: 86-89).

Plainly tourism access influences how vulnerable a site might be, particularly with respect to cumulative impacts (Box 9.1). As noted previously (Chapter 3), a sensitive site that is logistically convenient to tour operators is potentially more vulnerable to cumulative impacts than an equally sensitive site that is less accessible to visitors.

Box 9.1 – Tourism levels and sensitivity of some of the sites examined in this research

	Sensitivity low	Sensitivity high
Tourism high	Whaling station and base ruins at Deception Is., Antarctica Airship mooring mast at Ny-Ålesund, Svalbard Abandoned quarry at Ny-London, Svalbard	"Heroic era" huts and associated features, Ross Is., Antarctica Whaling factory ruins at Smeerenburg, Amsterdam Is., Svalbard
Tourism low	Old mining area, Ny-Ålesund, Svalbard	Scattered sealing sites in King George Is., Antarctica

9.3.3. *Impact assessment and historic sites*

This research has demonstrated the diversity of behavioural traces associated with polar tourism. However, to what extent does the behavioural signature of tourism constitute an "impact" in a management sense? The International Association for Impact Assessment defines the impact of an actual or proposed action as "...the difference between what would happen with the action and what would happen without it" (Sénecal 2009: 1). Measured against this standard, even apparently unremarkable behaviour traces might be regarded as impacts. However, in management terms, the impact of tourism activities depends on the relative significance of the site (how important it is) weighted against the significance of the changes (how "bad" they are).

12 In legal terms, the huts are internationally protected as Historic Sites and Monuments and also have the status of protected areas (Antarctic Specially Protected Areas (ASPAs) in the terminology of the Protocol). The conservation plans do not have such a legal standing.

The values attributed to historic sites and areas vary according to different instruments (Table 9.3). Ashworth and Graham note that the value assigned to elements of the past does not lie in their intrinsic merits, but rather on a "...complex array of contemporary values, demands, and even moralities." (2005: 7). Similarly, for Pérez-Juez Gil "cultural heritage" is not the intrinsic quality of an artefact but rather an *extrinsic* quality that had been adjudicated *a posteriori* from its creation and, as such, is vulnerable to be changed by the adoption of different criteria in the future (Box 9.2). Such classifications separate items that are inherently the same, but classified in different ways (Pérez-Juez Gil 2006: 23) (see e.g. Fig. 5.9). In Antarctica and Svalbard the legal criteria for protection are not intangible values *per se* but tangible cultural remains as defined by the 1991 Protocol and the 2001 Svalbard Act, respectively, and related instruments.

Contemporary approaches to heritage management require the understanding of the cultural and social values that make a particular cultural item significant, and the identification of what attributes or elements of a cultural item embody and represent those values (Worthing and Bond, 2010: 1-4). The relative values of specific site components also constitute a factor in assessing the significance of impacts. For instance, the conservation plan for historic huts in the Ross Sea area categorises different objects and parts of the huts and their environs according to their contribution to the heritage value of the whole. For this, a four-tier scale has b9een used, from "iconic" (items that in themselves contribute to the value of the entire site), "original and highly significant", "original and of less significance", and "modern". This categorisation enables the planning of management actions and the definition 99of permissible levels of intervention for specific site components (AHT-NZ, 2004: 87-89). Potential visitation impacts would be more or less significant depending on the relative value of particular elements being impacted.

The basic management tool to assess the impacts of current or proposed activities is ex-ante Environmental Impact Assessment (EIA). EIA has been defined as "The process of identifying, predicting, evaluating and mitigating the biophysical, social, and other relevant effects of development proposals prior to major decisions being taken and commitments made" (IAIA, 1996). To determine the significance of impacts it is necessary to make judgments about what is important, desirable or acceptable (Lawrence, 2007a and refs). Protection criteria in Antarctica and Svalbard are described in legal instruments as lists of "negative" behaviour or consequences of such behaviour.[13]

EIA in Svalbard specifically applies to cultural heritage. The criteria of assessment vary according to zoning regulations, and depend on whether the activity has "more than an insignificant effect on the natural environment outside land-use planning areas" or "... a significant and long-term effect on the environment inside a land-use planning area" (Norway, 2002: § 3).

13 In Antarctica, "Listed Historic Sites and Monuments shall not be damaged, removed or destroyed." (Protocol Annex V, Art. 8 (4)). In Svalbard "No person may damage, dig up, move, remove, alter, cover up, conceal or disfigure protected structures and sites or movable historical objects, including any security zone, or initiate measures that may entail a risk of this happening." (Svalbard Act, § 42).

Cumulative impacts may be taken into consideration if the Governor of Svalbard deems it necessary, although the term "cumulative impacts" per se is not used in the legislation. In Antarctica the impact criteria revolve around the term "minor or transitory" impact (less, no more, or more than) although the thresholds between these categories are not defined in the Protocol (Hemmings and Roura, 2003; Bastmeijer and Roura, 2008). EIA does not normally apply to Antarctic historic values (Bastmeijer and Roura, 2008).

Monitoring is "the collection of activity and environmental data both before (baseline monitoring) and after activity implementation (compliance and impact monitoring)" (Morrison-Saunders et al. (2007). Monitoring can be part of EIA follow up, but can also be regarded as a stand-alone activity. In Antarctica and Svalbard monitoring, as EIA, is applied almost exclusively to the natural environment (e.g. Walton and Shears, 1994; Reid, 2007). There have been some systematic monitoring efforts at representative historic sites in Svalbard (Governor of Svalbard, 1999), more recently in the context of the Svalbard Environmental Protection Fund, but this is not a tool that is broadly applied at present. Several historic sites in Antarctica are actively managed, and to that extent there is an awareness of what happens there, but there is no formal monitoring.

Overall, there are no systematic mechanisms to judge what the impact of visitation at historic sites in Antarctica and Svalbard might be. The legal criteria in place are not nuanced to the type of changes usually brought about by tourism, but rather they describe substantive changes associated with, for instance, intentional damage. Value judgements on tourism-related impacts are made on case by case bases at sites that are actively managed. At this point, a key guiding principle for the work of the International Polar Heritage Committee, a scientific committee of ICOMOS, should be noted, which states:

> *.... that all evidence of previous / finished human activity in the polar areas has a potential significance for the documentation and the understanding of the history of these areas and should be expertly assessed with an eye to possible designation as a cultural heritage site before being altered or removed* (IPHC, n.d.).

This principle largely aims to ensure that the active management of historic sites does not result in the unnecessary loss of historic information. However, it is also applicable to the use of these sites as tourism destinations and attractions.

Box 9.2– Common categories of cultural significance values

After Worthing and Bond, 2001 – sorted alphabetically.

- Aesthetic
- Archaeological
- Architectural/technological
- Artistic
- Associational
- Commemorative values
- Ecological
- Economic
- Educational
- Environmental
- Historical
- Inspirational
- Recreational
- Scenic and panoramic
- Social value
- Spiritual and religious
- Symbolic/iconic

9.3.4 Managed and unmanaged sites, and the "time capsule" concept

The conduct of polar tourism depends, in part, on creating illusions about the places visited. The illusion of wilderness is important with respect to the tourism value of the natural world – cruise ships in the Antarctic carefully avoid meeting each other so that the passengers can maintain the illusion of wilderness (de Korte, 2005). The counterpart of the "wilderness experience", with respect to the cultural heritage, is the "time capsule experience". Here visitors not only walk into a space vastly different from that of their everyday lives – the polar wilderness – but also seemingly step into a different time. The 1989 Oxford English Dictionary defines a "time capsule" as "a container used to store for posterity a selection of objects thought to be representative of life at a particular time." By definition, a true "time capsule" is intentionally designed and deposited with a view to be found in the future (e.g. Ascher, 1979), whereas in the Polar Regions "time capsules" result from processes of deposition, such as abandonment, rather than the intentional placement of a container.

It is difficult to put one's finger on precisely what makes a "time capsule" experience in the Polar Regions, but the following two excerpts may help to illustrate it. Stehberg *et al.* (2010) describe a site that had long been abandoned:

> *On little visited islands (such as Rugged Island) caves have been located that have been completely undisturbed by humans from the 1820s until surveyed and excavated in recent years. One located by those authors in 2005 had timber artefacts*

> *and seal skin moccasins lying on the surface, and bottles scattered around the fireplaces.* (Stehberg *et al.*, 2010).

In contrast, some other polar sites – such as Scott's *Terra Nova* hut – are being carefully restored "…to undo inappropriate interventions and reflect more authentically its heroic era use" (AHT-NZ, 2004: 101) – in a way, as if time had stood still. A visitor to the *Terra Nova* hut, writing several decades after the hut had first been revisited since its abandonment in 1917, notes that:

> *There is something disingenuous about Scott's hut, however, just as there is about the myth [of RF Scott]. The mummified penguin lying open-beaked and akimbo next to a copy of the Illustrated London News had been placed there by the New Zealand Antarctic Heritage people, and Ponting's photograph show that Scott's desk is not the original (the replacement was brought over from the Cape Royds hut)…Still, I saw them everywhere. A gap in a row of cup hooks, the dented rubber of a tossed Wellington boot, a carefully re-rolled bandage, the whiff of Ponting's developing fluid in his tiny darkroom, a half-spent candle in a chipped candlestick – perhaps it was the whistling of the wind, but I could have sworn if I turned around I would find them tramping back, spent dogs at their heels* (Wheeler, 1999: 55).[14]

In the first example above the site was found as left by the original occupants more than 150 years earlier, while in second example the site was restored to look, to the extent possible, as it did when it was in use by its original occupants in the first two decades of the XX century. However, a site that is actively maintained is, by default, no longer a "time capsule". In their seminal 1996 book, J. Tunbridge and G. Ashworth discuss the concept of heritage dissonance, which they describe as "discordance or lack of agreement and consistency as to the meaning of heritage". Depending on how a polar historic site is represented in tourism and other narratives, there may be a dissonance between what the site is and what it is purported to be. This may lead to cases of "staged authenticity", that is, when tourists are not aware of the staging of the scene and take it as real (MacCannell, 1973). Of course, tourists may be aware of the lack of authenticity in an attraction, which may then be described as a "contrived" attraction (Cohen 1979:26-27). Tourism, heritage conservation, and territorial politics in the Polar Regions are all closely related (see e.g. Chapters 4 and 5; Timothy 2010), so there is a risk that polar "time capsules" are created or-recreated solely for tourist consumption, turning them into a polar version of "touristic spaces" (in the terminology of MacCannell, 1973) at the expense of the historic values assigned to past activity sites.

---

14 A fictional character of the novelist E. Arthur, who visited the *Terra Nova* hut while writing her novel, describes a similar feeling: "The objects in the hut were *belongings*, the only word for them; they belonged still to the men who had selected them and used them, and then left them here, but who still were touching them gently" (1994: 506; original emphasis).

Overall, the concept of "time capsule" is not a precise term, rather reflects the impression that a particular site may causes to a visitor. This impression results from a combination of interactions of various kinds, and the emotions that these generate in the visitor (Boivin, 2008; quoted in Schiffer, 2010: 14). Conceptually it could be said that a historic site in a polar setting can be regarded as a "time capsule" only when it remains unmanaged (*i.e.* in an archaeological context), not when it just looks so as a result of active conservation and management (*i.e.* in a systemic context). In practice, these distinctions have little bearing on how the sites are packaged for tourism consumption, and would only matter to discerning visitors.

Table 9.1 – Integrated assessment of case study sites: Effects of tourism

Site	Visitation level (2000-2010)	Assessment
Deception Island		
• Ruins of whaling station, British base at Whalers Bay (ca. 0.5 km^2)	High > 10,000 visitors annually	A large scale site. Ongoing decay by wind and corrosion. Seasonally high visitation pressure resulting in (possibly minor?) cumulative effects and temporary but recurrent effects (e.g. footprints). Smaller features may be affected by ongoing visitation. Site subject of "managed decay".
• Ruins of Chilean research station at Pendulum Cove (ca. 0.1 km^2)	High > 5,000 visitors annually	A relatively large site in ruins. High visitation may modify the ruins in non-obvious ways. Mostly unmanaged site.
Fildes Peninsula		
• Numerous remains of early 19th century sealing camps (< ca 100 m^2 each) along ca. 25 km of coastline	Assumed low	A scatter of small sites. None or little apparent visitation pressure, but sites could easily be destroyed by visitors in non obvious ways. The sites are sensitive as they are little known, have low visibility, and are not legally protected nor actively managed.
Ijsfjorden, Svalbard		
• Pyramiden: Abandoned Soviet/Russian mining complex (>1km^2)	Medium. 1,000 day visitors annually plus visits by Longyearbyen residents in winter.	A very large industrial and habitation complex that has been "mothballed" and is subject to ongoing decay. Present levels of tourism are bound to have minimal physical effects.
Kongsfjord		
• Ny-London: Abandoned marble quarry at (*ca.* 0.6 km^2)	Medium. 1,000 day visitors annually plus visits by Ny-Ålesund residents	A large industrial site ranging from ruins to usable buildings. Minor tourism pressure combined with pressure by natural processes. Tourism may affect some of the smaller items of the site. The site's management combines active conservation with managed decay.
• Ny-Ålesund: Airship mooring mast and associated features; old mining area (*ca.* 1 km^2).	High. An estimate 10,000 visitors for the mooring mast; few to associated monuments.	The airship mooring mast is a single large feature. Tourism pressure and effects on the access to site and around it rather than feature itself. Interactions of visitors with the feature are common, but have no apparent substantive effects.

Site	Visitation level (2000-2010)	Assessment
• Ny-Ålesund: Old mining area (*ca.* 1 km^2).	Low. Some residents may visit the area, which is currently off limits for organised tourism.	A control site. The site is largely unmanaged, albeit included in local land management plan. Some features are actively preserved.
NW Spitsbergen		
• Dutch whaling station remains at Smeerenburg (*ca.*0.4 km^2).	Medium. Some 1,000 annually	One large site with several ruins and graves. Contains sensitive features that could easily be destroyed by visitation. Comparatively high visitation pressure resulting in low possibility of damage, but damage could be severe. Possibly erosion of tundra vegetation and localized erosion focused around some key features.
• Trapper's huts (< 1000 m^2) e.g. Texas Bar	Generally low.	Small sites in various conditions and sensitivities. Usually closed to visitors and often maintained. Some erosion around the huts.
Ross Island		
• Remains from the Heroic era of exploration - huts, memorial crosses (ca 10 m^2 – 10,000 m^2) at Hut Point Peninsula; Cape Evans.	Medium. Some 1,000 annually; about 2/3 of the visitors are staff members from nearby bases; due to remote location tourist levels are likely to remain low.	Two "Heroic era" huts, several memorials and crosses. The sites have survived in (and despite) a harsh environment for some 70 years and are subject to active conservation. Most obvious changes related to conserving and displaying the site rather than to visitation itself. Paths and localized erosion around features. The huts contain valuable artefacts that could be damaged or removed. Less valuable items lying around the hut possibly being trampled on or moved around. Visitor numbers at the huts are capped to 2,000 annually.
• IGY building at Scott Base, Pram Point.	Low. Mostly personnel from Scott Base.	The building is managed as part of the rest of the base and while protected is still used for some purposes. Weathering. Minimal visitation pressure. Some items in display inside, some of which are protected to avoid being removed.
• Former site of Greenpeace's World Park Base (*ca.* 1,000 m^2)	Low. A handful of visitors but no organised tourism.	A small control site with no tourism. Limited visible evidence of earlier operations at the site. Minimal visitation pressure, although the site is used as a helicopter landing site and, recently, as a campsite.

10 General conclusions

10.1 INTRODUCTION

Antarctica and Svalbard lacked indigenous people and were regarded as *terra nullius*. Following discovery they became the subject of successive waves of activity centred on exploration, the exploitation of natural resources, and the assertion of territorial claims. The timing and tempo of these developments varied in each region on account of different historic and geographic contexts, but some parallels are apparent. Both regions are subject to international treaties from which substantially different regimes of governance have emerged: a national (Norwegian) regime for Svalbard, and an international regime for Antarctica. Under these regimes both regions have been designated as protected natural reserves; some natural resources are still exploited; and science has become a major activity as well as the focus of international relations.

Tourism became established in Antarctica and Svalbard at different times and in different manners, but in both regions it has consolidated into a major presence. Tourism in these regions is regarded as a legitimate activity and some locations have become regular tourism destinations. However, the recent trajectory of tourism has raised concerns about its environmental effects, particularly cumulative effects, and has led to discussions and initiatives about the regulation and management of this activity.

Contemporary polar tourism has a primary focus on nature and wildlife, and increasingly on the conduct of certain activities in a polar setting, rather than polar attractions *per se*. Polar tourism also gravitates towards sites that contain the material remains of earlier activities, or historic sites, many of which are protected by law. Historic sites in the Polar Regions usually have the essential elements – sights, stories, and access – to become "tourist attraction systems" (in the terminology of Leiper, 1990), and therefore are of interest to the tourism industry.

In this context, the central aim of this thesis is to examine the effects of tourism on historic sites in Antarctica and Svalbard, and the implications for management. This chapter will answer the central research question, linked as required with the various themes explored through this thesis.

10.2 CENTRAL RESEARCH QUESTION AND STRATEGY OF INQUIRY

The central research question of this thesis is formulated in Chapter 1:

> *What are the effects of tourism on polar historic sites, and what are the implications of this for the management of tourism and historic sites?*

Cultural heritage and tourism are both research and management issues. As a research subject, tourism at historic sites may be best approached as an interaction between people and material objects ("behaviour"). The concepts and methods of behavioural archaeology (Schiffer, 2010 and refs.) provide a useful framework to analyse this interaction. Archaeology, in this context, is used in the sense of Reid *et al.,* (1974; 1975), meaning that the boundaries of this discipline have been expanded from its traditional focus on the study of past cultural systems to include the study of present material culture in current behavioural systems.

Describing tourism as a behavioural system has several advantages over other tourism research approaches. First, it recognises that the major patterns of tourism are different to the patterns of other behavioural systems. This is reflected, for instance, in terms of the number of people involved, the frequency and duration of the visits, and the type of activities carried out. Secondly, it examines tourism as a site formation process in which tourism "impacts" are artefacts of tourism, that is, they are the effects (or changes) that result from the interaction between tourists and historic sites (including both cultural remains as well as the landscape itself) by adding to, removing from, or reworking natural or cultural materials. These changes take place in the context of other cultural and non-cultural (natural) site formation processes affecting the sites. Finally, this approach serves to describe and interprets behaviour and its consequences independently from *a priori* value judgements about impact significance.

Research findings may have management implications. Examining the impacts of tourism from this perspective requires a dual translation process: first from the language of management to that of behavioural archaeology, where the issues concerning the conduct of tourism are analysed from a new perspective; and from there back to a management language. The equivalence of terms from both perspectives was shown in Chapter 1 (Box 1.3).

10.2.1. Research themes and associated sub-questions

Tourism and historic sites are multifaceted concepts, both individually and in their interaction, with dimensions that implicate complex political matters concerning, among others, the assertion of alternative narratives of historic events, the defence of territorial claims and strategic interests, and the commercial use of resources in the Polar Regions. This research has explored several interconnected themes, while recognising that there are other relevant themes that have not been addressed here. Each theme has been explored through one or more case studies addressing specific sub-questions (Box 1.1). Chapters 3-8 answered those questions.

- **Tourism as a behavioural system and a site formation process, in the context of other (natural and cultural) formation processes:** Tourism is one of multiple processes affecting historic sites, both locally and regionally. Tourism may contribute to the modification of sites that are already undergoing change due to a range of natural and cultural processes. This topic has been discussed throughout this thesis, particularly in Chapters 4 and 7;
- **The legal designation of historic sites as heritage:** Assessing whether or not a change to something is an impact relies on a judgement as to the value of what is being impacted. In both Antarctica and Svalbard the criteria and processes used to categorise the sites and remains of past activity as "heritage" are defined by law. Since both regions are protected wilderness areas there is a basic tension between the protection of the natural and of the cultural heritage that is inherent to

this designation. Designation of past activity sites as heritage has also geopolitical implications. Protection and management are ritual activities of possession; management infrastructure serves as a symbol of administrative control (in the sense of Seed, 1995).This topic has been discussed primarily in Chapter 5;

- **The use of cultural heritage as tourist attractions.** The attractiveness of historic remains is not necessarily obvious. Tourism narratives play a central role in the process of constructing meaning and assigning value to historic remains, which helps to create consumable tourist products and tourism attractions. Information about the sites influences tourist behaviour and therefore the degree of exposure or protection of the sites. This topic has been discussed primarily in Chapter 6;
- **Researching the interaction of tourism and historic sites.** Given the multifaceted and dynamic nature of tourism and historic sites, finding methods to study their interaction required some research in itself. The use of rapid, low technology approaches is appropriate to the Polar Regions, where site visits may be infrequent, opportunistic, and brief. More comprehensive research methods might be required for different (or more specific) research or management needs, which would in turn require a greater investment of resources. This topic has been discussed in Chapter 2 and through this thesis.
- **The management of tourism and historic sites**. The effectiveness of current management approaches, and the assessment and monitoring of cumulative impacts at historic sites are central considerations in the development of management regimes for tourism. Tourism tends to focus on important site features so that it may have a greater impact on elements that have a greater relative value than the rest of a site. Cumulative impacts are potentially a cause of concern. This topic has been discussed through the thesis, in particular in Chapters 3, 7, and 8.

What emerges from these various examinations is that tourism is a complex, multi-scalar issue. Multiple perspectives are required to explain how tourism works; a management-focused framework alone is insufficient.

10.2.2. Behavioural archaeology perspectives on the effects of tourism

Tourism is not static but rather subject to synchronic variation and diachronic change on different spatial and temporal scales. Tourism operates in a hierarchy of behavioural scales – from tourism as a whole, to individual visitors. Key processes in these various scales include the recruitment and consolidation of particular sites as tourism destinations; the patterns of behaviour displayed by visitors, both in groups and individually, during shore visits; and discrete interactions between visitors and site features of various kinds. The overall footprint of tourism occurs on all these different scales (i.e. the historic site as a regular tourism destination; a tourism activity area imposed over a historic activity area; and effects on specific site components).

The behaviour of tourists during landings – the minimum unit of analysis – can be described on the basis of a basic repertoire of actions (or forms of behaviour). The resulting interactions between tourists and elements of a historic site are primarily visual (navigating, sightseeing, sight recording), acoustic (instructions, narratives), and mechanical (contact with materials of different kinds with feet, hands, or the body).

Some types of interactions – certain forms of behaviour and access to some areas – are regulated, which limits but does not entirely preclude that they take place. Tourism effects on historic features may be a consequence of behaviour directed towards a particular feature, or an ancillary consequence of behaviour directed to a different historic feature or to e.g. fauna, flora, landscapes, or people. Tourist behaviour may transform historic sites through wear and tear from normal behaviour, by accident, or intentionally.

In behavioural terms, a pattern of visitor activity may result in a corresponding "archaeology" of visitation traces. There is a pattern to individual visits and to the resulting direct and cumulative traces of visitation. The traces of visitation at historic sites may be of two kinds: additive and reductive (or accumulation traces and erosion traces). These define a *tourist activity area* superimposed over, but not necessarily identical to, a historic activity area.

Assessing the footprint of tourism on historic sites can be considered as an issue of conservation (maintenance) of the site's condition *i.e.* from a certain condition at the time of a visit or visits to a different condition afterwards. Behaviour can change the basic dimensions of artefacts (in the sense of Rathje and Schiffer, 1982; Schiffer, 1987) i.e. attributes of form, such as size and shape; spatial location; frequency; and relation to associated artefacts. In essence, these are changes to the condition and integrity of individual artefacts and assemblages of artefacts, and can be regarded as tourism "artefacts". The cumulative footprint of tourism can be characterised as mostly "nibbling" effects affecting discrete elements or parts of a site. Plainly greater tourism levels result in greater chances of accidental or intentional damage, and greater chances of cumulative effects.

Historic sites that are not actively preserved will eventually decay, whether or not they are used as tourism destinations, over a period that may span decades to possibly a few hundreds of years (e.g. Weiner, 2010: 62). However, tourism may help accelerate the process of decay (or increase the rate of change) of a site or its components. To an extent, the central research question above can be phrased differently as: *What is the effect of tourism on historic sites relative to the effect of natural processes on those sites?* Natural processes affecting historic sites have a greater transformation potential than organised tourism visits. Barring cultural activities that modify substantially the sites, such as demolition and clean ups, natural processes will be primary factors in the evolution of a historic site over time. However, tourism is more selective than natural processes as it "cherry picks" certain elements of a site that serve as tourism attractions and where impacts can concentrate and accumulate. Tourism may also disturb minor traces of past activity that are located in between more conspicuous remains. In addition, some historic sites may be substantially modified to accommodate tourism.

Historic sites are a palimpsest that reflects multiple activities and processes that take place after the abandonment of a site by the original occupants, each of which superimposes its own traces above previous cultural traces. In this context, tourism is one of several post-deposition processes affecting historic sites. An alternative question to the one in the precedent paragraph is: *What would happen if tourism at a particular historic site ceased?* Over a time span of a few hours to a few seasons many or most tourism traces would become invisible to the naked eye or disappear altogether. Traces such as graffiti would be longer lasting and would be identifiable as originating in an earlier visitation. Broken artefacts and features would blend with the "noise" of site decay caused by other site formation processes. Conceptually some traces may result from the removal of historic artefacts, which would simply disappear from the archaeological record. Ultimately, lasting traces constitute the lasting effects of tourism at historic sites. Some of these might not be readily identifiable because of "equifinality" – the process of reaching the same final state from different initial conditions and in different ways (Chapter 4).

Some historic sites are protected *for* tourism (and associated economic or territorial interests) rather than solely for their historic values. In this context, there is a potential for tourism pressures to transform authentic historic sites into "touristic spaces" (in the terminology of MacCannell, 1973). At some locations activities such as passport stamping, posting post cards and buying T-shirts have become the most significant component of a visit. Turning heritage into a commodity may help preserve the fabric of heritage but also contribute to the degradation of the value of a site as a repository of archaeological information.

So to summarise an answer to the first central question of this research:

What are the effects of tourism on historic sites in Antarctica and Svalbard?

1. Tourism behaviour at historic sites consists mainly of a basic repertoire of actions reflecting interactions of different kinds between people and the elements that compose historic sites (both natural and cultural). Some actions may involve mechanical interactions with these elements, and result in behavioural traces of different kinds.
2. The tourism "footprint" consists of an assemblage of behavioural traces that result from tourist behaviour. Visible tourism traces reflect a fraction of the type of actions and level of activities that take place during tourism visits at historic sites. Some behaviour may not result in visible traces – these may exist but be masked in the general "noise" of site decay. In some locations, some forms of behaviour may result in no effects at all.
3. The effect of organised tourism visits on historic sites is minor relative to the effect of natural processes on those sites, albeit focused on significant features..
4. Tourism is part of a sequence of formation processes that create and modify the archaeological record (i.e. material traces of past activities). Tourism traces, in turn, are modified by other natural and cultural processes. Other site formation processes may erase or obliterate the traces of tourism, or conversely contribute to preserve and enlarge them.
5. Information (instructions, narratives) influences tourism behaviour and resulting effects on historic sites.
6. The overall tourism footprint is multi-scalar and affects historic sites in at the level of regions (or cluster of sites); individual sites; and individual components of a site, both movable and unmovable.
7. Tourism is linked to other activities that take place at historic sites, and so are the combined effects of these various activities.

10.2.3 Management perspectives on the effects of tourism

The previous section has described a pattern of behavioural traces associated with tourism behaviour that can be identified at historic sites. To what extent does the assemblage of traces that result from tourism constitute "impacts" in a management sense? An answer to this question depends on value judgments on the significance of these traces relative to the value assigned to historic sites or its components.

Judging the significance of impacts is usually done through the environmental impact assessment process prior to the commencement of an activity, or monitoring the environment subsequently. Environmental management in Antarctica and Svalbard tends to prioritise the natural environment over the historic environment. This responds to the designation of these areas as natural reserves. In both regions there are instruments to establish the significance of impacts but these are not applied systematically to the historic environment. The application of preliminary EIA to the historic environment, and subsequent monitoring of activities is limited to a few sites, or non-existent. The selection of specific sites where tourism is now conducted on an annual basis has not been the subject of prior impact assessments to determine whether or not tourism was an appropriate use of the sites. Concerns have arisen after tourism was already established at certain places. The impact of tourism on the archaeological value of sites should be assessed with a view to minimising the effect of these activities, as required, through the use of management instruments as discussed above. Management tools and processes commonly used to protect the natural environment, including with respect to tourism, could be usefully applied to the historic environment.

From the point of view of site management, the problem with the tourism footprint lays not so much in *what* this footprint is but *where* it is located – that is, some parts of a site may be more sensitive or valuable than others. Tourism has a potential to modify the cultural remains and other traces of previous activities that exist at historic sites, which are valuable – among other reasons – for the information they hold about the past. Some tourism impacts may affect important parts of a historic site and degrade the archaeological record. However, the legal criteria that define damage to the cultural heritage, while strict, are not always sufficiently nuanced to describe the sort of changes caused by tourism. Changes that are minor in spatial extent, or ephemeral, are not usually regarded as impacts in practice, and result in no management actions.

To prevent these effects, some dedicated management tools may be required to regulate the movement of people within historic sites. Establishing minimal distances from sensitive historic features or parts of a site that contains the scattered remains of artefacts; and the use of established tracks may prevent impacts at historic sites that contain both resilient and sensitive features entwined with each other. Limits in numbers of visitors (e.g. at a site and through the year), as currently used in some Antarctic historic huts, may be used at particularly sensitive sites.

Information serves to both influence behaviour and to increase appreciation of the sites, and thus can minimise forms of behaviour that result in impacts. Tour guides are, in a way, at the forefront of the tourism "attack" on historic sites, but also constitute the first line of defence, since one of their responsibilities is to inform tourists and also have a degree of authority over them. An alternative is to give more responsibility for the enforcement of rules to tour guides. A guide certification scheme would ensure that consistent sets of rules are applied to a certain standard, and that a modicum of high quality information about the sites is provided. However, "management by information" by itself is not sufficient to prevent impacts, and notwithstanding the importance of enhancing the role of guides, some form of "external" monitoring by the authorities may also required.

Management actions need to cover different spatial and temporal scales – the establishment of tourism destinations; the choreography of group visits to historic sites; and the behaviour of individual tourists.

Different activities and disciplines have different and sometimes conflicting interests, and are interested in historic sites for different sets of values. To a degree, the contemporary management of historic sites and impacts on these sites has been shaped by the interests and perspectives of those who first reached a site after the original occupants, whether these were archaeologists, conservators, environmental managers, tour operators (and tourists), or visitors of other kinds. The management of historic sites should consider the competing interests of different disciplines and activities, and the potential impacts associated with these. These various factors should be considered so that historic sites are managed sustainably both as tourism sites and as archaeological resources.

Site specific guidelines, increasingly in use in Antarctica and Svalbard, are useful to micro-manage individual sites. However, human behaviour is both creative and flawed, and while it can be guided it cannot be regulated in minute detail. A broad brush management option is to restrict access to some historic sites on precautionary grounds, while other sites remain open to visitation. This would ensure that some significant sites are preserved and would serve (to a degree) as a control of historic sites used for tourism. The alternative is to allow access to all historic sites knowing that at the very least there will be some cumulative effects on or around significant elements of the site, and potentially more significant impacts.

This discussion highlights the need for more general strategic approaches to manage tourism. Strategic considerations are also required so that all historic sites in a certain region are managed as components of a coherent whole rather than as disconnected entities. The comment is largely applicable to various kinds of sites in the Polar Regions – there is generally no EIA or other formal process conducted prior to a site turning into a tourism destination; rather, tourism becomes established at certain sites, and years (or even decades) later actions are taken to manage it. Adopting a strategic approach to site management would seem to be more easily achievable in Svalbard than in Antarctica. This is a consequence of the single-state governance and management regime in place, in contrast to the international regime in place in Antarctica, and also of the strategic priority already given to Svalbard tourism by Norwegian authorities (e.g. Vikam 2011). Plainly tourism policies, or lack of them, have a downstream effect on how sites are managed, including historic sites.

So as to answer the second part of the central research questions:

What are the management implications of the effects of tourism on polar historic sites?

1. A range of management tools currently applied to manage the natural environment could be applied too to historic sites. This includes in particular impact assessment and monitoring, as well as various hierarchically defined area management regimes applicable from the regional to the site specific.
2. Dedicated management tools for historic sites may also be required, including measures to actively or passively preserve sites that have been afforded legal protection.
3. Tourism impacts on historic sites should be minimised and managed regardless of the ongoing decay of historic sites because tourism tends to focus on significant site features.
4. The impact of tourism on the archaeological value of sites should be assessed to prevent that tourism (or other activities) diminish the information value of historic sites.
5. The role of tour guides should be enhanced so that they have more responsibility for the enforcement of rules, and a minimum of basic qualifications that enable issuing consistent guidelines.
6. Management actions need to be based on strategic considerations and a precautionary approach rather than solely on site-specific considerations.
7. The management of historic polar sites should aim to prevent conflicts of use between different users, and preserve different sets of values.

10.3 REFLECTION ON RESEARCH METHODS AND FUTURE RESEARCH

Many historic sites, while protected, are not actively preserved but left to "weather gracefully". Assessing how tourism and tourists contribute to the transformation of historic sites is tantamount to assessing how these contribute to an ongoing process of decay. In this research, the effects of tourism at historic sites were assessed through an integrated approach that combined behavioural observations; site surveys of different kinds; and repeat photography as the main research methods, complemented by additional methods (Chapter 2). How effective were these various assessments?

As noted previously, the archaeological record is incomplete with respect to the record in the occupied site (i.e. the ethnographic record), and it is difficult to assess whether something is not there because it has been degraded or because it was never there. In addition, another factor influencing tourism effects is the issue of "equifinality". Historic features may decay in a number of ways, including through the effects of tourism. Visual examination alone – in the field or through photographic analysis –, though essential, may not be sufficient to interpret and explain particular changes. These factors emphasise the importance of observing tourism as it takes place in order to assess how it affects the sites visited, rather than relying solely on the examination of visitation traces. However, conducting tourism observations has its own difficulties that may affect the quality of observational data. These include reactivity by the subjects; "inattentional blindness" by the observer (Simon and Chabris, 1999); and observer consistency.

Repeat photography can be used to detect time serial changes to specific features. The fundamental requirement for this method requires an awareness of its limitations (Chapter 4), and a conceptual framework to interpret the changes such as, in this research, concepts of site formation processes. Furthermore, dedicated monitoring has substantive advantages, albeit less flexibility, than the opportunistic approach used in this research.

Overall, no single research method provides all the necessary answers. In this research, the integration of research methods enabled multiple complementary perspectives of the interaction of tourism and specific historic sites to be obtained, and reasonably informed assessments of the sites' status and change, and of the contribution of tourism change (Tables 2.3 and 9.1) to be made. The key word here is *integration.*

Potentially useful research methods not used in this research, or only as complementary research methods, might include, *inter alia*, the use of experimental methods; the use of micro-archaeology approaches and techniques; longer periods of observation at visited and control sites; more detailed site surveys and inventories; the use of time-serial photography or time-lapse video to monitor change; and dedicated research on blogs posted by polar tourists and other actors on the World Wide Web.

This research served to highlight future research needs and opportunities. Behavioural archaeology theory and methods can be applied to a broader range of activities and polar environments than used in this research. These may include the interaction of tourism with natural sites, and the effect of site formation processes on the cultural (but not necessarily historic) environment. The approaches developed in this research could be used to investigate the intersection of human behaviour (people and their artefacts or equipment) with the natural environment. Furthermore, behavioural perspectives could be used to assess the effectiveness of regulations and to monitor compliance. Finally, there is scope for the study of cultural heritage and tourism (individually and taken together) from perspectives that emphasise the geopolitical context in different parts of the Polar Regions, including aspects of contemporary uses of cultural heritage, and narratives about site significance and multiple versions of historic events.

10.4 CLOSING REMARKS

Some representations of polar historic sites in Antarctica and Svalbard as reflected in, for instance, tourism brochures, the popular media, and contemporary debates about tourism management or heritage conservation, describe these sites as veritable "time capsules". The polar environment protects many sites from the type of decay encountered in temperate regions, but nevertheless historic sites experience gradual but significant decay. In addition, despite often being in remote locations, polar historic sites are the focus of a range of contemporary activities that contribute to further transformation of the sites. The "time capsule" appearance is sometimes the result of careful work to prevent or to undo changes that occurred over time, so as to replicate the appearance of a site at a certain point in time in the past.

Tourism activities both in Antarctica and in the Arctic are broadly guided by the adage "Take only photographs, leave only footprints" – which bears some similarity to the findings of this research. A common argument is that responsible tourism has no discernible or at least no significant effects on its destinations and is therefore inherently sustainable. In effect, the adage above is sometimes used as a statement of fact rather than as a declaration of intent. An alternative representation of polar tourism depicts its recent trajectory as a cause of concern, and regards tourism as a resource activity with a somewhat intrusive presence in wilderness areas (e.g. Nuttall, 2010).

In fact, there are elements of truth in both perspectives: Some concerns about tourism might be exaggerated or at least not yet proven by the facts; and tourism facilitates public access to the polar cultural heritage, which may well be the visitors' own heritage. However, it is also the case that some narratives about tourism emphasise the positive aspects of this activity while dismissing the significance or even the possibility of tourism impacts. Tourism tends to take over the sites it uses, sometimes to the exclusion of other uses or values, which potentially causes impacts of different kinds as well as creates conflicts of use.

Overall, some tourism narratives deny the existence of change at historic sites that are used as tourism destinations, and the potential role of tourism as an agent of transformation. The tourism industry has traditionally rejected the need for regulation, or at least *restrictive* regulation, and has promoted instead the adoption of self-regulation. In this, the industry has been proactive and has adopted high (but non-mandatory) operational standards. Ensuring continued access to the resources (sites) has been the driver behind the adoption of this position. Self regulation tends to guide *how* activities should be conducted, but does not normally address the more strategic *if* question regarding the appropriateness of conducting tourism at certain sites.

Binding and non-binding guidelines of different kinds have now been adopted in both Antarctica and Svalbard, and these have contributed to minimise many direct impacts of tourism. In addition, tourist visits are seasonal and brief, and there are natural thresholds in the potential of individual visitors to transform the places they visit. All these factors limit the capacity of tourism to modify historic sites. However, sites that become tourism destinations will continue to be visited repeatedly for the foreseeable future, and will therefore be exposed, in a sense, to permanent human activity. Tourism tends to focus on the most important elements of a site, and cumulative impacts remain a cause of concern because they are likely to occur even when visits are conducted responsibly. In addition, while the vast majority of visitors are expected to behave appropriately, the probability of accidental or intentional damage increases with the increasing intensity of site use.

As demonstrated in this thesis, tourism "impacts" can be examined from a more detached perspective when they are regarded as tourism "artefacts". Many tourism "artefacts" are ephemeral and of no great significance for the long term preservation of historic sites. Other "artefacts", however, can be more lasting and/or affect important elements of a site. Overall, the potential for tourism visits to modify historic sites is comparatively minor relative to the potential effects of natural processes, but cannot be discounted entirely. At many historic sites the whole is more important than individual parts taken alone, and some tourism impacts on some historic features may well be significant.

Generically, whether tourists should or should not be allowed to visit historic sites is a moot discussion, because both alternatives are valid in different cases. Rather, the point is how to balance tourism and other activities with historic site protection and preservation. Some of these sites may need to be closed off to visitors in order to preserve the tangible heritage and the archaeological record; others – possibly a majority – would withstand a certain level of visitation. Here, site managers should accept that some effects will occur as a result of tourism, and that these effects need to be managed and monitored. Tourism can be both a cause and a consequence of change, and create its own dynamics so that some schemes to manage tourism may result in making sites more attractive to visitors. Overall, historic sites can be kept in their raw state as a decaying legacy of past events, including as repositories of archaeological information. Alternatively, historic sites can be altered to suit the expectations of polar travellers, the needs of the tourism and heritage industries, and the geopolitical interests of states.

The material remains to be found in the Polar Regions narrate the history of past events in these regions. When tourists set foot on these sites they help to maintain the memory of what happened there by looking at the site and listening to narratives about past events. However, tourist behaviour may erase some traces of past activities and replace them with new traces and a new narrative. Management decisions need to be made with the awareness that historic sites in the Polar Regions are not "frozen in time" but rather are used for a range of contemporary purposes and transformed as a result. Paraphrasing Borges, and remembering the earlier voyagers to the Polar Regions, one may think they would be pleased to know their stories are still told, in one version or another; time is both loss and memory, and these certainly coexist at historic polar sites.

Acknowledgments

Doctoral research can be sometimes a lonesome endeavour, but it requires the contribution and support of many people. The acknowledgement section in other theses I examined range from the comprehensive to the absurdly minimalistic. Here I have tried to be reasonably inclusive, while recognising that there are many people who helped directly or indirectly to complete this thesis that I may have failed to mention by name. My apologies and thanks to you.

My thanks for their help to produce this thesis go, without implicating them in my interpretations, to the people and institutions listed below.

The idea of starting a doctoral degree was implanted in my brain by Prof. Dr. Louwrens Hacquebord (University of Groningen) and Dr. Kees Bastmeijer[1] (University of Tilburg). I had many enjoyable conversations with Prof. dr. Hacquebord, my supervisor, on issues relevant to this research and to the broader geopolitical context of the Polar Regions. I benefited enormously from his experience and knowledge of these regions. Aside his contribution in several different ways, Dr. Bastmeijer is a contributing author of Chapter 3. This reflects quite a number of years of cooperation on Antarctic issues.

I also benefited from the sharp comments made by Dr. Peter Groote, the copromotor of this thesis, about various aspects of my research, which helped greatly to move things forward.

This research was funded by the University of Groningen, Groningen Institute of Archaeology, as part of the University's Ubbo Emmius Scholarship Programme. Funding included a stipendium, travel to several national and international conferences, and fieldwork in Svalbard. I enjoyed the independence and flexibility of this programme that enabled me to combine my research interests with research perspectives that were new to me and, I think, relevant to the line of research of the Arctic Centre. The Research Council of Norway provided partial funding to attend the Oslo Science Conference 2010, where some findings related to this research were presented.

I am grateful to the various institutions and individuals that invited, funded, transported, hosted or otherwise enabled me to travel to the Antarctic and Svalbard repeatedly and to conduct fieldwork there. Prof. Dr. Louwrens Hacquebord and Dr. Maarten Loonen facilitated my fieldwork in Spitsbergen. In particular, it was a pleasure to work with Dr. Loonen at the Netherlands Arctic Station in Ny-Ålesund. Opportunistic fieldwork in Antarctica was enabled by participation in projects with the Antarctic and Southern Ocean Coalition (ASOC), and Gateway Antarctica, University of Canterbury, New Zealand. Polar events do not happen solely in the Polar Regions, and I am thankful to ASOC for the possibilities of attending Antarctic Treaty Consultative Meetings and other meetings of the Antarctic Treaty System where issues relevant to this research are discussed.

1 Kees Bastmeijer, Professor of Nature Conservation and Water Law, Tilburg Law School, Tilburg University, The Netherlands. Email: c.j.bastmeijer@uvt.nl

I thank Argentina's National Directorate of the Antarctic, Antarctica New Zealand, the Netherlands Arctic Station (2007), Oceanwide Expeditions, and the University of Jena, Germany, for logistic support of various kinds that were essential for this research. Particular thanks to Gary Evans, Per Kyrre Reymert, Dr. Michelle Rogan-Finnemore, Dr. Maarten Loonen, and Dr. Hans-Peter Ulrich, Miguel Ponce and Base Decepción staff (2001–2002), and Oleg Sakharov and Bellingshausen Base staff (2005–2006). Reaching specific sites was made easier with the assistance or company of Wolfgang Dinter and Steffen Vogt (at Fildes Peninsula, KGI, Antarctica), Wojtek Moskal (around Ny-Ålesund, Svalbard), and Scott Base's Antarctic Field Team (2007–2008),

I am grateful to Professor Dr. Louwrens Hacquebord (University of Groningen, Arctic Centre), Dr. Susan Barr (Norway's Directorate of Cultural Heritage), and Per Kyrre Reymert (Government of Svalbard) for sharing their knowledge about Svalbard´s history and cultural heritage. On Antarctic matters I enjoyed, and benefited from, discussions with Jim Barnes (The Antarctic and Southern Ocean Coalition), Dr. Kees Bastmeijer, Ambassador Jorge Berguño (Instituto Antártico Chileno), Janet Dalziell (Greenpeace International), Dr. Alan Hemmings (University of Canterbury), Christian Lambrechts (United Nations Environment Programme), Dr. Ariel Mansi (Argentine Ministry of Foreign Affairs), and Dr. Tina Tin (ASOC). Dr. Ko de Korte, Frits Steinhausen, Monika Schillat, and Rolf Stange kindly shared their experiences on the inner workings of the tourism industry. Grete Marstein kindly acted as Norwegian interpreter. I was honoured to interview some veterans of Ny-Ålesund's mining times, Asbjørg Skogmo, Oskar Bye, and Lilly Kristensen. Dr. Alan Simpson (University of Waikato, New Zealand) and Dr. Enrique del Acebo (University of El Salvador, Argentina) provided useful advice on conducting doctoral research.

Colleagues from the Arctic Centre provided commentary and support of all kinds through the years and made my regular trips to Groningen enjoyable. I thank Ypie Aalders, Dr. Peter Abbink, Dr. Dag Avango, Nienke Boschman, Dr. K.I.M. van Dam, Prof. dr. Hacquebord, Ulf Gustafsson, Hidde de Haas, Frigga Kruse, Melissa van der Heide, Dr. Maarten Loonen, Cecilia Sandström, Annette Scheepstra, and Jan Wilhem Veluwenkamp. I remember in particular interesting exchanges about fieldwork and research in the Polar Regions (sometimes at the pub up the road) and about the Antarctic Treaty System. I thank Ulf in particular for introducing me to the Scandinavian sport of rolling skiing, which helped to keep both my fitness and my sanity.

Comments about specific chapters were provided variously by Dr. Dag Avango, Dr. Peter Capellotti, Janet Dalziell, Dr. Maarten Loonen, Dr. Erik Jaap Molenaar, Dr. Jessica O'Reilly, Dr. Jan Wilhem Veluwenkamp, and several anonymous reviewers. Prof. Dr. Haquebord and Dr. Peter de Groote reviewed the entire thesis more than once. I also thank the review panel for assessing my dissertation: Dr. Susan Barr, Prof. dr. Kees Bastmeijer, Prof. dr. Daan Raemaekers, and Prof. dr. Frank Vanclay.

Photographs supplementing my own collection were kindly provided by Dr. Susan Barr, Dr. Antonio Bendala, Dr. Alan Hemmings, Per Kyrre Reymert (on behalf of the Governor of Svalbard), Professor Dr. Hacquebord, Dr. Loonen, Dr. Rubén Stehberg, Christopher Weaver and Mike Salvarezza, and Dr. Rodolfo Werner. The Tromsø Polar Museum and the Norwegian Polar Institute kindly allowed the use of two of their photographs.

It was a pleasure to work with Roelf Barkhuis, Rosemary Harrison, Hannie Steegstra, and Nynke Tiekstra while producing this document.

Some thanks extend far away and long ago: I first travelled to the Antarctic in 1980-81; first attended a meeting of the Antarctic Treaty System in 1992; and first travelled to Svalbard in 1998, albeit all too briefly. Many colleagues along the way have helped in a myriad of ways that have contributed to increase my knowledge of the Polar Regions, and some of that is reflected in this thesis. Many Antarctic trips with Greenpeace International provided a unique view of this unique region, as did earlier trips with the Argentine Antarctic Program. Special thanks to Capt. Arne Sorensen and Capt. Robert Graham taking me to the Antarctic and back safely on several occasions.

I spent quite a few long periods away from home in the course of this thesis, and when at home I was often busy at night or on weekends. I thank dear Janet Dalziell, Isabel Roura and Matías Roura – i.e. my wife and children – for their love and support over these years as I travelled back and forth on polar-related matters; and my mother, Ethel Benvenuto, who often came to Amsterdam to lend a hand while I was away. I hope that I will now have more time for you.

I have been fortunate to have developed close ties with several different countries by birth, blood, or chance: Argentina, Italy, The Netherlands, and New Zealand. I am grateful for the various opportunities these countries gave me with regards to, *inter alia*, education, funding, and travel. These are the places where I have roots, family and friends, and with them I will always be linked in one way or another.

Ricardo M. Roura
Amsterdam, March 2011

References[1]

Aalders, Y., Avango, D., Gustafsson, U., de Haas, H., Hacquebord, L., Hartnell, C., Kruse, F. & de Pasqual S. (2008). *LASHIPA 5. Archaeological expedition on Spitsbergen 27 July - 17 August 2008.* Groningen: Arctic Centre, University of Groningen. Available from http://www.let.rug.nl/arctic/lashipa_web/LASHIPA-5_report-2008.pdf

Aas, S. (2005). New perspectives on the *Italia* tragedy and Umberto Nobile. *Polar Research, 24*(1-2), 5–15.

Acero, J.M. & Aguirre, C.A. (1994). A monitoring research plan for tourism in Antarctica. *Annals of Tourism Research, 21*(2), 295-302.

AECO (Association of Arctic Expedition Cruise Operators)(n.d.). http://www.aeco.no/objectives.htm. Accessed 15 November 2010.

AHDR (Arctic Human Development Report) (2004). Akureyri: Stefansson Arctic Institute

Ahlstrom, R.V.N. (1992). Casual repeat photography: An illustration from Hopi architectural history. *Journal of the Southwest, 34*(2), 166-186.

Ahmad, Y. (2006). The scope and definitions of heritage: from tangible to intangible. *International Journal of Heritage Studies 12*, 292–300.

AHT-NZ (Antarctic Heritage Trust – New Zealand) (2008). A walk up Ob Hill. Antarctic Conservation Blog 27 March 2008. Retrieved from http://www.nhm.ac.uk/nature-online/earth/antarctica/blog/?cat=11&paged=4. Accessed 8 July 2009.

AHT-NZ (Antarctic Heritage Trust – New Zealand) (2004). *Conservation Plan, Scott's Hut, Cape Evans.* Christchurch New Zealand: New Zealand Antarctic Heritage Trust.

AHT-NZ (Antarctic Heritage Trust – New Zealand) (n.d.). Scott's expedition base, Cape Evans. Retrieved from: http://www.heritage-antarctica.org/AHT/CapeEvans. Accessed 8 July 2009.

AHT-UK (Antarctic Heritage Trust – United Kingdom) (n.d.). *Port Lockroy.* Retrieved from http://www.ukaht.org/portlockroy.htm (Accessed 29 June 2009).

Amundsen, R. & Ellsworth, L. (1927), *First crossing of the Polar Sea.* New York: George H. Doran Company.

Andersen M. & Aars J. (2008). Short-term behavioural response of polar bears (*Ursus maritimus*) to snowmobile disturbance. *Polar Biology 31*, 501–507.

Antarctica 21 (n.d.). *The first air cruise to Antarctica.* Retrieved from http://www.antarcticaxxi.com.

Antarctica New Zealand (2000). *Proceedings of the Antarctic Tourism Workshop held by Antarctica New Zealand on June 23, 2000.* Christchurch, New Zealand: Antarctica New Zealand. (In file with authors).

Arenz, B. & Blanchette, R. (2008). East Base, SOS: Assessment of deterioration and recommendations for conserving this important Antarctic historic site. In: S. Barr & P. Chaplin (Eds.). *Historical polar bases: Preservation and management,* ICOMOS Monuments and Sites Series N° XVII (pp. 78-84). Oslo: ICOMOS International Polar Heritage Committee.

1 Antarctic Treaty Consultative Documents and other Antarctic publications are available from www.ats.aq.

Argentina (2006). *Tourism development in the Antarctic Peninsula: A regional approach.* Information Paper 31, Edinburgh: XXIX Antarctic Treaty Consultative Meeting.

Argentina (2010). *Proposal for the discussion of aspects related to the management of Historic Sites and Monuments.* Working Paper 047, Punta del Este, Uruguay: XXXIII Antarctic Treaty Consultative Meeting.

Argentina (2010). *Report of clean-up efforts by the Argentinean National Antarctic Program in the area of the Neko Harbour refuge (north-west coast of the Antarctic Peninsula).* Information paper 023, Punta del Este, Uruguay: XXXIII Antarctic Treaty Consultative Meeting.

Argentina & Sweden (2007). *Directrices para la visita del turismo a Cerro Nevado* [Visitor guidelines for tourism at Snow Hill Island]. Working Paper 40, New Delhi: XXX Antarctic Treaty Consultative Meeting

Argentina, Chile, Norway, Spain, the United Kingdom and the United States (2006). *Deception Island Antarctic Specially Managed Area (ASMA) Management Group.* Information Paper 19, Edinburgh: XXIX Antarctic Treaty Consultative Meeting.

Arlov, T. B. (1996). *Svalbards historie.* Oslo: Aschehoug.

Arlov, T.B. (2005). The discovery and early exploitation of Svalbard. Some historiographical notes. *Acta Borealia 22:1*, 3-19.

Ascher, R. (1961). Analogy in archaeological interpretation. *Southwestern Journal of Anthropology 17*, 317-325.

Ascher, R. (1974). How to build a time capsule. *Journal of Popular Culture, 8*:2, 241-253.

Ashworth, G.J. & Graham, B. (2005). *Senses of place: Senses of time,* Ashgate: Aldershot.

ASOC (Antarctic and Southern Ocean Coalition) (2000). *Antarctic Strategic Environmental Assessment: Application to the growing Antarctic tourism industry.* Information Paper 10. The Hague, The Netherlands: XII Special Antarctic Treaty Consultative Meeting.

ASOC (Antarctic and Southern Ocean Coalition) (2002a). *ATCM papers, discussions, and recommendations relating to tourism and non-governmental activities.* XXV Antarctic Treaty Consultative Meeting Information Paper 52 rev. 1, Warsaw: XXV Antarctic Treaty Consultative Meeting.

ASOC (Antarctic and Southern Ocean Coalition)(2002b). Regulating Antarctic tourism. Information Paper 83. XXV Antarctic Treaty Consultative Meeting/IP 83,

ASOC (Antarctic and Southern Ocean Coalition) (2004). Report on the XXVII Antarctic Treaty Consultative Meeting. Retrieved from http://www.asoc.org.

ASOC (Antarctic and Southern Ocean Coalition) (2008). A decade of Antarctic tourism: Status, change, and actions needed. Information Paper 41, Kyiv, Ukraine: XXXI Antarctic Treaty Consultative Meeting.

ASOC (Antarctic and Southern Ocean Coalition) & UNEP (the United Nations Environment Programme) (2003). *A review of inspections under Article 7 of the Antarctic Treaty and Article 14 of its Protocol on Environmental Protection, 1959- 2001.* Information Paper 118-Rev. 1. Saint Petersburg, Russia: XXVI Antarctic Treaty Consultative Meeting.

ASOC (Antarctic and Southern Ocean Coalition)(2002). *Regulating Antarctic tourism.* Information Paper 83. Warsaw, Poland: XXV Antarctic Treaty Consultative Meeting.

Asko, I , Kuitunen, M.T., Jalava, K. (2010). Developing the RIAM method (rapid impact assessment matrix) in the context of impact significance assessment. *Environmental Impact Assessment Review 30* (2): 82–89.

ATCPs (Antarctic Treaty Consultative Parties)(1994). *Final Report of the XXVII Antarctic Treaty Consultative Meeting.* Kyoto, Japan: XXVII Antarctic Treaty Consultative Meeting.

ATCPs (Antarctic Treaty Consultative Parties)(2001). *Final Report of the XXIV Antarctic Treaty Consultative Meeting.* St. Petersburg, Russia: XXIV Antarctic Treaty Consultative Meeting.

ATCPs (Antarctic Treaty Consultative Parties)(2003). *Final Report of the XXVI Antarctic Treaty Consultative Meeting.* Madrid, Spain: XXVI Antarctic Treaty Consultative Meeting.

ATCPs (Antarctic Treaty Consultative Parties)(2004). *Final Report of the XXVII Antarctic Treaty Consultative Meeting.* Cape Town, South Africa: XXVII Antarctic Treaty Consultative Meeting.

ATCPs (Antarctic Treaty Consultative Parties) (2005). *ATCM Guidelines for Environmental Impact Assessment in Antarctica.* Stockholm, Sweden: XXVIII Antarctic Treaty Consultative Meeting.

ATCPs (Antarctic Treaty Consultative Parties)(2009). *Final Report of the XXXIII Antarctic Treaty Consultative Meeting.* Punta del Este, Uruguay: XXXIII Antarctic Treaty Consultative Meeting.

Auburn, F.M. (1982). *Antarctic law and policy.* London: Hurst & Co Publishers.

Australia (2010). *Topic Summary: Footprint.* Working Paper 49, Punta del Este, Uruguay: XXXIII Antarctic Treaty Consultative Meeting.

Australia, Peru & United Kingdom (2005). *Report of Joint Inspections under Article VII of the Antarctic Treaty and Article 14 of the Environmental Protocol.* XXVIII Antarctic Treaty Consultative Meeting, Working Paper 032, Stockholm, Sweden: XXVIII Antarctic Treaty Consultative Meeting.

Avango, D., Hacquebord, L., Aalders, Y., De Haas, H., Gustafsson, G. & Kruse, F. (2011). Between markets and geo-politics: natural resource exploitation on Spitsbergen from 1600 to the present day. *Polar Record 47(1). 29-39.*

Bailey, G. (2007). Time perspectives, palimpsests and the archaeology of time. *Journal of Anthropological Archaeology 26*: 198–223

Barr, S., C. Lüdecke, & D. Olynyk. 2005. Polar Heritage. In: Truscott, M, M. Petzet & J. Ziesemer (Eds.). *Heritage at Risk: ICOMOS World Report 2004-2005 on Monuments and Sites in Danger (pp.* 265-267). Munich: K.G. Saur Verlag.

Barr, S. 2008. The effects of climate change on cultural heritage in the Polar Regions. In: M. Petzet & J. Ziesemer (Eds.). *Heritage at risk. ICOMOS world report 2006/2007 on monuments and sites in danger.* (pp. 203-205). Altenburg: E. Reinhold-Verlag.

Barr, S. & R. Downie. 2002. Report of a site survey of Whalers Bay. Appendix 4 to: Argentina, Chile, Norway, Spain, UK, USA, ASOC & IAATO. 2002. *An international expedition to Deception Island.* Information Paper 028. Warsaw, Poland: XXV Antarctic Treaty Consultative Meeting.

Barr, S., Downie, R. & Sánchez, R. (2004). Whaler's cross on Deception Island. *Polar Record,* 40 (01). 69-70.

Barr, S. & Chaplin, Paul (Eds.) (2004). Cultural heritage in the Arctic and Antarctic regions. Oslo, Norway ICOMOS International Polar Heritage Committee (IPHC)., Publication A

BAS (British Antarctic Survey). 2001. *Port Lockroy Base A. Visitors guide.* Cambridge: British Antarctic Survey.

BAS (British Antarctic Survey). (n.d.) *British Research Stations and Refuges – History.* http://www.antarctica.ac.uk/about_bas/our_history/stations_and_refuges/index.php. Accessed 10 June 2008.

Basberg, B (forthcoming). Antarctic tourism and the maritime heritage. *International Journal of Maritime History.*

Bastmeijer, C., Lamers, M. & Harcha, J. (2008). Permanent land based facilities for tourism in Antarctica: The need for regulation. *Review of European Community and International environmental Law* (RECIEL), *17 (1)*, 84-99.

Bastmeijer C.J. (2003), *The Antarctic environmental protocol and its domestic legal implementation*, dissertation, International Environmental Law and Policy series, Vol. 65, Kluwer Law International, The Hague, 2003.

Bastmeijer, K., & Roura, R. (2004). Regulating Antarctic tourism and the precautionary principle. *The American Journal of International Law 98* (4). 763-781.

Bastmeijer, K. & Koivurova T. (2008). In: K. Bastmeijer & T. Koivurova T (Eds.). Theory and practice of transboundary environmental impact assessment (pp. 175-219). Monographs Series on Legal Aspects of Sustainable Development. Leiden: Brill/Martinus Nijhof Publishers.

Bastmeijer, K. (2003). *The Antarctic environmental protocol and its domestic legal implementation*. The Hague: Kluwer Law International.

Bastmeijer, K. & Roura, R.M. (2008): Environmental Impact Assessment in Antarctica. In: K. Bastmeijer & T. Koivurova (Eds.): *Theory and Practice of Transboundary Environmental Impact Assessment*, pp. 175-219. Monographs Series on Legal Aspects of Sustainable Development. Leiden: Brill/Martinus Nijhof Publishers.

Bauer, T. G. (2001). *Tourism in the Antarctic – opportunities, constraints, and future prospects.* New York, NY, USA: Haworth Press,

Beck, P.J. (1986). *The international politics of Antarctica.* London: Croom Helm.

Beebe, J. (1995). Basic concepts and techniques of Rapid Appraisal. *Human Organization 54* (1), 42-51.

Berger, A.A. (2009). *What objects mean. An introduction to material culture.* Walnut Creek, CA, USA: Left Coast Press.

Bernard HR (2002). *Research methods in anthropology. Qualitative and quantitative approaches.* Oxford: Altamira Press.

Bizzarri, M. (2006). Il patrimonio culturale antartico. Breve excursus sulla normativa di tutela. [The Antarctic cultural heritage. A Brief on protection legislation.] *Studi Senesi 2*/2006, 345 – 368.

Blake, J. (2000). On defining the cultural heritage. *The International and Comparative Law Quarterly 49(1)*, 61-85.

Blanchette, R.A., Held, B.W. & Farrell R.L. (2002). Defibration of wood in the expedition huts of Antarctica: An unusual deterioration process occurring in the polar environment. *Polar Record 38 (207)*, 313-322.

Blanchette, R.A., Held, B.W., Jurgens, J.A. McNew, D.L., Harrington, T.C., Duncan, S.M. & Farrell. R.L. (2004). Wood-destroying soft rot fungi in the historic expedition huts. *Applied and Environmental Microbiology 70*(3), 1328–1335.

Blanchette, R.A.; Held, B.W.; Jurgens J.A.; Aislabie J; Duncan, S.M.; & Farrell, R.L. (2004). Environmental pollutants from the Scott and Shackleton expeditions during the 'Heroic Age' of Antarctic exploration. *Polar Record 40* (213). 143–151.

Boivin, N. (2008). *Material cultures, material minds.* Cambridge: Cambridge University Press.

Bogdewic, S.P. (1992). Participant observation. In: B.F. Crabtree & W.F. Miller (Eds.). *Doing qualitative research* (pp. 45-69). Newbury Park, California: Altamira Press.

Bore, R.R., Andreassen, I., Kristiansen, J.E. & Mellemstrand, C. (2009). *This is Svalbard. What the figures say.* Oslo, Kongsvinger: Statistics Norway (Statistisk sentralbyrå) (English version).

Brazil, China, Germany, Korea (ROK), Russian Federation (2006). *Possibilities for environmental management of Fildes Peninsula and Ardley Island. Proposal to establish an intersessional contact group.* XXIX Antarctic Treaty Consultative Meeting, Working Paper 022. Edinburgh, UK: XXIX Antarctic Treaty Consultative Meeting.

Bricher, P.K., Lucieer, A. & Woehler, E.J. (2008). Population trends of Adélie penguin (Pygoscelis adeliae) breeding colonies: a spatial analysis of the effects of snow accumulation and human activities. *Polar Biology 31*, 1397 – 1407.

Brown, R. N. R. (1915). Spitsbergen in 1914. *The Geographical Journal 46(1)*, 10-21

Brown, R. N. R. (1919). Spitsbergen, Terra Nullius. *The Geographical Review 7* (5), 311-321.

Bruner, E.M. (2005). *Culture on tour. Ethnographies of travel.* Chicago, IL, USA: The University of Chicago Press.

Bryan, G. W. & Bernard, H. R. (2003). Techniques to identify themes, *Field Methods 15*(1), 85-109.

Buchli, V. & Lucas, G. (Eds.)(2001). *Archaeologies of the contemporary past.* Routledge.

Buchli ,V. & Lucas, G. (2001). *The archaeology of alienation. A late 20th century British council house.* In: Buchli V & Lucas G (Eds.)(2001). *Archaeologies of the contemporary past* (pp. 158-167. London: Routledge,

Burger J. & Gochfeld, M. (2007). Responses of Emperor Penguins (*Aptenodytes forsteri*) to encounters with ecotourists while commuting to and from their breeding colony. *Polar Biology 30*, 1303-1313.

Bush, W. (2000a). Means and methods of implementation of Antarctic environmental regimes and national environmental instruments: An exercise in comparison. In: D. Vidas (Ed.), *Implementing the environmental protection regime for the Antarctic* (pp. 21-43). Dordrecht: Kluwer Academic Publishers.

Bush, W. (2000). Australian implementation of the environmental protocol. In: D. Vidas (Ed.), *Implementing the environmental protection regime for the Antarctic* (pp. 309-335). Dordrecht: Kluwer Academic Publishers.

Byrd, Richard E. (1968). Flight over the Pole. In: S. Manley & G. Lewis (Eds.), *Polar secrets. A treasury of the Arctic and the Antarctic,* Garden City, NY: Doubleday.

Cameron, C.M. (1999). *Hopi dwellings: architectural change at Orayvi.* Tucson: University of Arizona Press.

Cameron, J (1993). The precautionary principle: Core meaning, constitutional framework and procedures for implementation. Paper presented at the conference "The Precautionary Principle: A New Approach to Environmental Management," Institute of Environmental Studies, University of New South Wales, Sydney, Australia, September 20-21, 1993. (In file with author.)

Cameron, J. (2001). The precautionary principle in international law. In T. O'Riordan, J. Cameron & A. Jordan (Eds.). *Reinterpreting the precautionary principle.* London: Cameron May. (pp. 113-142). London: Cameron May.

Campbell, I.B., Balks, M.R. & Claridge G.G.C. (1993). A simple visual technique for estimating the impact of fieldwork on the terrestrial environment in ice-free areas of Antarctica. *Polar Record 29 (171).* 321-328.

Campbell, I.B. & Claridge, G.G.C. (1987). *Antarctica: Soils, weathering processes and environment.* Amsterdam: Elsevier.

Campbell, I.B., Claridge, G.G.C & Balks, M.R. (1998). Short and long-term impacts of human disturbance on snow-free surfaces in Antarctica, *Polar Record, 34* (188), 15-24.

Carman, J. (2002). *Archaeology and heritage. An introduction.* London: Continuum.

Ceballos-Lascuráin, H. (1991). *Tourism, ecotourism and protected areas.* The state of *nature-based tourism around the world and guidelines for its development.* Gland, Switzerland: International Union for the Conservation of Nature.

CEP (Committee for Environmental Protection)(2007). *Report of the Committee for Environmental Protection (CEP X)*, Annex E. New Delhi, 30 April 2007 – 11 May 2007: New Delhi, India: CEP X.

CEP (Committee for Environmental Protection) (2010). *Report of the Committee for Environmental Protection (CEP XIII) Punta del Este, May 3 – 7, 2010.* Punta del Este, Uruguay: CEP XIII.

CEP (Committee for Environmental Protection) (2004). *Interim Final Report of the VIIth CEP Meeting, Cape Town, South Africa, May 24-29, 2004.* Cape Town, South Africa: CEP VII.

Chaplin, P. (2008). Case study: Captain Robert Falcon Scott's 1910-13 British Antarctic expedition hut at Cape Evans. In: M. Petzet & J. Ziesemer (Eds.). *Heritage at risk. ICOMOS World Report 2006/2007 on monuments and sites in danger.* (pp. 206-208). Altenburg: E. Reinhold-Verlag.

Cherry-Garrard, A . (1922/1937). *The worst journey in the world.* London: Chatto and Windus.

Chile, France, Germany, Italy, Spain (1992). *Draft Annex to the Protocol, regulation concerning tourism and non-governmental activities*, XVII Antarctic Treaty Consultative Meeting Working Paper 1, Venice: XVII Antarctic Treaty Consultative Meeting.

Chile (2007a). *Antarctic Protected Area system: Revised list of Historic Sites and Monuments (Measure 3, 2003). Draft guidelines for its application.* XXX Antarctic Treaty Consultative Meeting, WP 38.

Chile (2007b). *Avances al plan de gestión territorial, manejo ambiental y conservación del patrimonio histórico de la base Gabriel González Videla, Territorio Antártico Chileno.* XXX Antarctic Treaty Consultative Meeting, IP 94.

Chile (2008). *Recuperación del Sitio Histórico Nº 56 Base Aérea Antártica Pdte. Gabriel González Videla*, Information Paper 12, Kiev: XXXI Antarctic Treaty Consultative Meeting.

Chile (2010). *Proposed Modification to Historic Site Nº 37.* Working paper 067. Punta del Este, Uruguay: XXXIII Antarctic Treaty Consultative Meeting.

Cobley, N.D. & Shears, J.R. (1999). Breeding Performance of Gentoo penguins (*Pygoscelis papua*) at a colony exposed to high levels of human disturbance, *Polar Biology 21*, 355-360.

Cochran, C & Collinge, I. (1994). *Conservation survey of abandoned British bases in the Antarctic Peninsula region.* Unpublished report. UKAHT (United Kingdom Antarctic Heritage Trust). (In file with author.)

Cohen, E. (1979). Rethinking the sociology of tourism. *Annals of Tourism Research Jan/Mar 1979*, 18-35.

Cohen, E. (1984). The sociology of tourism: Approaches, issues, and findings. *Annual Review of Sociology, 10*, 373-392.

Cole, D. (2004). Exploring the sustainability of mining heritage tourism. *Journal of Sustainable Tourism* 12 (6), 480-494.

Collier Jr., J., & Collier, M. (1967/1986), *Visual anthropology. Photography as a research method.* Albuquerque: University of New Mexico Press.

Conway, W.M. (1906). *No man's land: A history of Spitsbergen from its discovery in 1596 to the beginning of the scientific exploration of the country.* Cambridge [Cambrigeshire]: The University Press.

Conway, [W.] M. (1919). The political status of Spitsbergen. *The Geographical Journal 53* (2), 83-91.

Cooper, L.M. & Sheather, W.R. (2004). Integrating cumulative effects assessment into UK strategic planning: Implications for the European Union SEA Directive. *Impact Assessment and Project Appraisal 22* (1), 5-16.

Crosbie, K. (1999). Interactions between Skuas *catharacta sp.* and Gentoo penguins *Pygoscelis papua* in relation to tourist activities at Couverville Island, Antarctic Peninsula. *Marine Ornithology 27,* 195-197.

Culik, B., Adelung, D. & Woakes, A.J. (1991). The effect of disturbance on the heart rate and behaviour of adélie penguins (pygoscelis adeliae) during the breeding season. In K.R. Kerry & G. Hempel (Eds.) *Antarctic ecosystems. Ecological change and conservation* (pp. 177 – 182). Berlin: Springer-Verlag,.

Dastidar, P.G. & Persson, O. (2005). Mapping the global structure of Antarctic research vis-a- vis Antarctic Treaty System. *Current Science 89 (9)*, 1552-1554.

Davis PB (1998). Understanding visitor use in Antarctica: The need for site criteria. *Polar Record 34* (188), 45-52.

de Korte, K. (2005). A tourist in the Arctic. In: N. Boschman, L. Hacquebord & J.W Veluwenkamp (Eds.). *Het topje van de ijsberg. 35 jaar Arctish Centrum (1970-2005).* (pp. 79-82). Groningen: Barkhuis Publishing.

De Poorter, M. & Dalziell, J.C. (Eds.)(1996). *Cumulative environmental impacts in Antarctica: Minimization and management.* Proceedings of an IUCN Workshop on Cumulative Impacts in Antarctica, Washington DC, USA, September 18-21, 1996. Gland, Switzerland: IUCN.

Det Kongelige Miljøverndepartement [Royal Ministry of the Environment] (2001), *Svalbard-miljøloven. Lov 15 juni 2001 nr 79 om miljøvern på Svalbard* [*Svalbard Environmental Protection Act. Act of June 15, 2001, no. 79, relating to the protection of the environment in Svalbard*], Oslo, Det Kongelige Miljøverndepartement [Royal Ministry of the Environment].

Dewalt, K. M., Dewalt, B., & Wayward, C. B. (1998). Participant observation. In: H. R. Bernard (Ed.). *Handbook of methods in cultural anthropology* (pp. 259-300). Walnut Creek, California: Altamira Press.

Drewett, P. (1999): Field archaeology: An introduction. London: UCL Press.

Driveness, E.-A. & Jølle, H. D. (2006). *Into the ice. The history of Norway in the Polar Regions*, Oslo: Glydendal Akademisk.

Eagles, PFJ (2003). *International trends in park tourism: A macro view of park tourism finance 2*, paper presented at World Parks Congress, Durban, South Africa (September 8 - 19, 2003), available at http://www.conservationfinance.org/WPC/WPC_documents/Apps_12_Eagles_v1.pdf.

Ecuador (2007). *Turismo en Barrientos [Aitcho Islands, Antarctica].* Information Paper 105. New Delhi: XXX Antarctic Treaty Consultative Meeting.

EIS (Extreme Ice Survey). (n.d). URL: http://www.extremeicesurvey.org.

Emerson, R.M., Fretz, R.I., & Shaw, L.L. (1995). *Writing ethnographic notes.* Chicago, IL, USA: The University of Chicago Press.

Enzenbacher, D.J. (1992). Antarctic tourism and environmental concerns, *Marine Pollution Bulletin 25(9-12)*, 258-265.

Enzenbacher D.J. (1994). Antarctic Tourism: An overview of 1992/93 season activity, recent developments and emerging issues. *Polar Record 173* (30), 105-116.

Enzenbacher, D.J. (1995). The regulation of Antarctic tourism. In C.M. Hall & M.E. Johnston (Eds.) *Polar tourism: Tourism in the Arctic and Antarctic regions* (pp. 179-215), Chichester, UK: John Wiley & Sons.

European Commission (2002): *Using natural and cultural heritage to develop sustainable tourism in non-traditional tourist destinations.* European Commission, Directorate General Enterprise

(Tourism Unit). Retrieved from http://europa.eu.int/comm/enterprise/services/tourism/studies/ecosystems/heritage.htm.

European Environment Agency (2002). *The Arctic biogeographical region - warming up and changing?*. Retrieved from http://www.eea.europa.eu/publications/report_2002_0524_154909/biogeographical-regions-in-europe/page001.html

Evans, S.-L. (2007): *Heritage at risk: Cultural heritage management in the Antarctic.* Paper presented at ICOMOS 2007 – eXtreme heritage The Heritage of Polar Landscapes, 19-21 July 2007, James Cook University, Cairns, QLD, Australia. Available from http://www.aicomos.com/wp-content/uploads/sherrieleeevans.pdf.

Evenset, A. & Guttorm, C.N. (2011). *Environmental impacts of expedition cruise traffic around Svalbard.* Akvaplan-niva report no 4823 – 1. Tromsø, Norway: Akvaplan-niva.

Fogelson, N.(1992). *Arctic exploration and international relations 1900-1932. A period of expanding national interests.* Fairbanks: University of Alaska Press.

Fogg, G.E. (1992). *A history of Antarctic science.* Cambridge: Cambridge University Press.

Fox, A.J., & Cziferszky, A. 2008. Unlocking the time capsule of historic aerial photography to measure changes in Antarctic Peninsula glaciers. *Photogrammetric Record 23*(121), 51-68.

France, L. (1997). *Sustainable tourism.* London: Earthscan.

Fraser, W.R. & Patterson, D.L. (1997) Human disturbance and long-term changes in Adélie penguin populations: A natural experiment at Palmer Station, Antarctic Peninsula. In B. Battaglia, J. Valencia & D.W.H. Walton (Eds.) *Antarctic communities: Species, structure and survival* (pp. 445-452). Cambridge: Cambridge University Press.

Gamble, C. (2008). *Archaeology. The basics.* London: Routledge.

Gee, D. EIA and the Antarctic tourism industry. 24th Annual Conference of the International Association for Impact Assessment, 'Whose Business is it? Impact Assessment for Industrial Development,' April 24-30, 2004, Vancouver, Canada, Abstracts Volume, at 27-28.

Geitz, M. (Ed.) (2004). *Cruise tourism on Svalbard – A risky business?* Oslo: WWF International Arctic Programme.

Germany & Chile. 2007. *Progress report on the discussion of the international working group about possibilities for environmental management of Fildes Peninsula and Ardley Island.* Information Paper 022 rev.1. New Delhi, India: XXX Antarctic Treaty Consultative Meeting.

Giese, M. (1998). Guidelines for people approaching breeding groups of Adélie penguins (*Pygoscelis adeliae*), *Polar Record 34*: 287-292

Giese, Melissa (1996). Effects of human activity on Adélie penguin *Pygoscelis adeliae* breeding success. *Biological Conservation, 75(2).* 157-164.

Gifford-González D.P., Damrosch David B. Damrosch, Debra R. Damrosch, J. P., & Thunen, R.L. (1985). The third dimension in site structure: An experiment in trampling and vertical dispersal. *American Antiquity 50 (4).* 803-818.

Goldberg, P. & Macphail, R.I. (2006). *Practical and theoretical geoarchaeology.* Malder, MA, USA: Blackwell Publishing.

Gorecki, P. (1985). Ethnoarchaeology: The need for a post-mortem enquiry. *World Archaeology* 17 (2) (Ethnoarchaeology, Oct., 1985), 175-191.

Governor of Svalbard (1999a). *Overvåkning av kulturmjilo på Svalbard. Målsetting, metode, lokaliteter og overvåkning.* [Monitoring of the cultural environment on Svalbard. Objectives, methods, sites and monitoring]. *Sysselmannens rapportserie Nr.3/1999.* Longyearbyen: Governor of Svalbard.

Governor of Svalbard (1999b). Helhetlig plan for miljøtiltak i gruveområdet i Ny-Ålesund (Kings Bay). [Comprehensive plan for environmental initiatives in the mining area in Ny-Ålesund (Kings Bay)]. *Sysselmannens rapportserie* Nr.2. Longyearbyen: Governor of Svalbard.

Governor of Svalbard (2000). Kulturminneplan for Svalbard 2000-2010. [Cultural heritage plan for Svalbard 2000-2010] *Sysselmannens rapportserie.2/2000.* Longyearbyen: Governor of Svalbard.

Governor of Svalbard (2006a). *Reiselivsstatistikk for Svalbard 2006* [Travel Statistics for Svalbard 2006], Longyearbyen: Governor of Svalbard.

Governor of Svalbard (2006b) Turisme og friluftsliv på Svalbard. Utvikling, politiske føringer, rammebetingelser, utfordringer og strategier. [Tourism and outdoor activities on Svalbard. Development, political constraints, framework, challenges, and strategies.] *Sysselmannens rapportserie 1*/2006. Longyearbyen: Governor of Svalbard.

Governor of Svalbard. 2009. *Proposal for amendments to protection regulations.* Retrieved from http://www.sysselmannen.no/hoved.aspx?m=44365&amid=2487842. (Accessed 4 April 2009).

Governor of Svalbard (2009). New heavy oil and traffic prohibition http://www.sysselmannen.no/hoved.aspx?m=44365&amid=2836074 (article dated 21 September 2009). Accessed 30 November 2009.

Governor of Svalbard (n.d). *Ny-London. The saga of Mansfield and the marble quarry on Blomstrand Peninsula.* Miljøvernavdelingen N-9171. (Brochure). Longyearbyen: Governor of Svalbard.

Graham, B., Ashworth, G.J., & Tunbridge, J.E. (2000). *A geography of heritage. Power, culture and economy.* London: Arnold.

Greenpeace International (1991). *1990/91 Antarctic Expedition Report.* Amsterdam: Greenpeace International. (In file with author.)

Greenpeace International (1992). *1991/92 Antarctic Expedition Report.* Amsterdam: Greenpeace International. (In file with author.)

Greenpeace (1994). *1992/93 Antarctic Expedition Report.* Amsterdam: Greenpeace International. (In file with author.)

Greenpeace (2002). Arctic environment melts before our eyes. 7 August 2002. Retrieved from: http://www.greenpeace.org/international/news/glaciers-melt-before-our-eyes. Accessed 15 November 2008.

Green., K. & Moore, T. (1983/2010): Archaeology: An introduction. (5th edition 2010) London: Routledge.

Guerring, J. (2007). *Case study research: Principles and practices.* Cambridge: Cambridge University Press.

Gunn, C. (1972), *Vacationscape: Designing tourist regions.* Austin, TX, USA: University of Texas.

Guyomard, A.-I. (2008): La protection du patrimoine culturel en Antarctique [Cultural heritage protection in Antarctica] In: J.A.R. Nafziger & T. Scovazzi (Eds.). *Le patrimoine culturel de l'humanité / The Cultural Heritage of Mankind.* (pp. 687-727). Martinus Nijhoff Publishers.

Haase, D., Storey, B., McIntosh, A., Carr, A. Gilbert, N. (2007). Stakeholder perspectives on regulatory aspects of Antarctic Tourism. *Tourism in Marine Environments, 4 (2-3),* 167-183.

Hacquebord, L., (1992). Hector station on Deception Island (South Shetland Islands, Antarctica), an environmental assessment study of a whaling station. *Circumpolar Journal (1-2),* 72-97.

Hacquebord, L. (2004). Dutch Cultural Heritage in the Arctic. In: S. Barr & P. Chaplin (eds) *Cultural heritage in the Arctic and Antarctic regions,* ICOMOS Monuments and Sites Series N° VII (pp 73-78). Oslo: ICOMOS International Polar Heritage Committee.

Hacquebord, L., Steenhuisen, F. & Waterbolk, H. (2003). English and Dutch whaling trade and whaling stations in Spitsbergen (Svalbard) before 1660. *International Journal of Maritime History XV*(2), 117-134.

Hall, C.M. & Johnston, M.E. (Eds.) (1995). *Tourism in the Arctic and Antarctic regions.* John Wiley and Sons.

Hall, C.M. & Saarinen, J. (2010). Tourism in the Arctic and Antarctic Regions. Introduction - Definitions, locations, places, and dimensions. In: C.M. Hall & J. Saarinen (Eds.). *Tourism and change in the Polar Regions: Climate, environment and experiences* (pp.1-43). London: Routledge.

Hall, F.C. 2001. *Ground-based photographic monitoring.* United States Department of Agriculture Forest Service, Pacific Northwest Research Station General Technical Report PNW-GTR-503.

Hall, F.C. 2002. *Photo point monitoring handbook.* United States Department of Agriculture Forest Service Pacific Northwest Research Station General Technical Report PNW-GTR-526.

Hanoa, R. (1993). *Kings Bay Kull Kompani A/S 1917-1992. Fra gruvedrift til forskningsservice på Svalbard.* Oslo: Schibsted.

Hansom, J.D. & Gordon, J.E. (1998). *Antarctic environments and resources. A geographical perspective.* Essex, NY, USA: Longman.

Harrowfield, D.L. (2005). Archaeology on ice. A review of historical archaeology in Antarctica. *New Zealand Journal of Archaeology (26).* 5-28.

Headland, R. K. 2002. *Whalers' cemetery; Deception Island, South Shetland Islands. In: Valencia, J. & R. Downie.* Workshop on a Management Plan for Deception Island. *Santiago de Chile: Instituto Antártico Chileno: 93-95.*

Headland, R.K. (1994). *Historical development of Antarctic tourism. Annals of Tourism Research* 21(2). 269–280.

Headland, R. K (1989/2009). *Chronological list of Antarctic expeditions and related historical events.* Cambridge: Cambridge University Press.

Heilen M.P., Schiffer M.B. & Reid J.J. (2008). Landscape formation processes. In B. David & J. Thomas (Eds.): *Handbook of landscape archaeology* (pp. 601-608). Walnut Creek, CA, USA: Left Coast Press.

Hemmings A.D. (Fall 2000). Icewatch. *Living Planet*, Fall 2000, pp.31–43.

Hemmings, A.D. & Roura, R. (2003). A square peg in a round hole: Fitting impact assessment under the Antarctic environmental protocol to Antarctic tourism. *Impact Assessment and Project Appraisal. 21*(1). 13-24.

Hemmings AD, Cuthbert A & Dalziell J (1991). Non-governmental activities and the protection of the Antarctic environment: A paper for the government of New Zealand. Wellington, New Zealand: The Antarctic and Southern Ocean Coalition. [In file with author].

Hemmings A. D. & Stephens, T. (2010). The extended continental shelves of sub-Antarctic Islands: Implications for Antarctic governance. *Polar Record 46 (4).* 312-327.

Hemmings, A.D. 2007. Globalisation's cold genius and the ending of Antarctic isolation. Chapter 13 in L.K. Kriwoken, J. Jabour & A.D. Hemmings (Eds). *Looking South: Australia's Antarctic Agenda* pp. 176-190. Annadale, NSW, Australia: The Federation Press.

Henare, A., Holbraad, M. & Wastell, S. (2007). Introduction: Thinking through things. In: Henare A, Holbraad M & Wastell S (Eds.). *Thinking through things: Theorising artefacts ethnographically.* P.1-32. London: Routledge.

Herz, M. & Davis, J. (2002). *Cruise control. A report on how cruise ships affect the marine environment.* The Ocean Conservancy, May 2002. Retrieved from: http://www.oceanconservancy.org/site/DocServer/cruisecontrol.pdf?docID=141.

Hind, John (2008). Art attack. *The Observer* Sunday January 27, 2008. http://observer.guardian.co.uk/magazine/story/0,,2245950,00.html.

Hjelle, A. (1979): Some aspects of the geology of northwest Spitsbergen. *Norsk Polarinstitutt skrifter 167*: 37-62.

Hofman, R.& Jatko, J. (2000) *Assessment of the possible cumulative environmental impacts of commercial ship-based tourism in the Antarctic Peninsula area*, Proceedings of a Workshop held in La Jolla, California, June 2000. Washington DC: National Science Foundation.

Höver, J. (1927). The hangar and the mooring masts. In: R. Amundsen & L. Ellsworth (Eds.) (1927), *First crossing of the Polar Sea* (pp. 25-44), New York: George H. Doran Company.

Howard, P. (2003). *Heritage. Management,interpretation, identity.* London: Continuum.

Howe, D. (1999).Ground handling and mooring. In G. A. Khoury & D.J. Gillet (Eds.), *Airship technology* (pp. 297-320). Cambridge Aerospace Series 10.

Hughes, J. (1994). Antarctic historic sites: The tourism implications. *Annals of Tourism Research* (21). 281-294.

Hughes, J. & Davis, B. (1995). Tourism at Historic Sites and Monuments. In: C.M. Hall & M.E. Johnston: *Polar tourism: Tourism in the Arctic and Antarctic regions* (pp. 235-255). John Wiley and Sons.

Hunter, C. & Green, H. (1995). *Tourism and the environment: A sustainable relationship?* London and New York: Routledge.

IAATO (International Association of Antarctica Tour Operators) (2001). Report of IAATO under Art. III(2) of the Antarctic Treaty subsection 2.2.4, July 2001. In: ATCPs (Antarctic Treaty Consultative Parties) (2001) Final Report of the XXIV Antarctic Treaty Consultative Meeting (170-195). Saint Petersburg, Russia: XXIV Antarctic Treaty Consultative Meeting.

IAATO (International Association of Antarctica Tour Operators) (2002). Chairman's Report from the Aspen Meeting on Antarctic Tourism, Information Paper 30. Warsaw, Poland: XXV Antarctic Treaty Consultative Meeting.

IAATO (International Association of Antarctica Tour Operators) (2003). Overview of Antarctic Tourism 2002–2003 Antarctic Season, Information Paper 71. Madrid, Spain: XXVI Antarctic Treaty Consultative Meeting.

IAATO (International Association of Antarctica Tour Operators) (2004a) : Overview of Antarctic tourism 2003-2004 Antarctic season. Information Paper 63. Cape Town, South Africa: XXVII Antarctic Treaty Consultative Meeting.

IAATO (International Association of Antarctica Tour Operators) (2004b). Six-year survey of the dominant tourist activities and trends since the ratification of the environmental protocol and a five-year estimated forecast of upcoming activities. Antarctic Treaty Meeting of Experts, Tromsø-Trondheim 22-25 March 2004. ATME Paper 11. Tromsø-Trondheim, Norway: Antarctic Treaty Meeting of Experts.

IAATO (International Association of Antarctica Tour Operators) (2004c). *Report of IAATO 2003-2004.* XXVII Antarctic Treaty Consultative Meeting Information Paper 68. Cape Town, South Africa: XXVII Antarctic Treaty Consultative Meeting.

IAATO (International Association for Impact Assessment) (2004d*). Bylaws of May 2, 2004.* Retrieved from http://www.iaato.org/bylaws.html.

IAATO (International Association of Antarctica Tour Operators) (2007). *Tourism statistics.* Retrieved from www.iaato.org/stats/.

IAATO (International Association of Antarctica Tour Operators) (2010): *About IAATO: Bylaws. June 24, 2010.* Retrieved from http://www.iaato.org/bylaws.html. Accessed 15 November 2010.

IAATO (International Association of Antarctica Tour Operators) (n.d./3). *Tourism statistics.* Retrieved from: http://www.iaato.org/tourism_stats.

IAATO (International Association for Impact Assessment)(n.d./2). *About IAATO. Objectives.* Retrieved from: http://www.iaato.org/objectives.html/. Accessed 24 September 2004.

IAATO (n.d./1): Marine Wildlife Watching Guidelines, retrieved from http://www.iaato.org/wildlife.html. Accessed 24 September 2004.

IAIA (International Association for Impact Assessment) (2009). *What is impact assessment?* Fargo, ND, USA: International Association for Impact Assessment. http://www.iaia.org/publicdocuments/special-publications/What%20is%20IA_web.pdf. Accessed 15 January 2010.

ICIMOD (International Centre for Integrated Mountain Development). 2008. Himalaya: Changing landscapes. Retrieved from http://www.changing-landscapes.com. Accessed 10 November 2008.

ICOMOS (International Council on Monuments and Sites). 2008. *Recommendations from the Scientific Council Symposium: Cultural heritage and global climate change. Adopted in Pretoria, South Africa, 7 October 2007. Final Draft: 21 March 2008.* Retrieved from http://www.international.icomos.org/climatechange/index.htm. Accessed 15 November 2010.

Ingold, T. (2007). *Lines. A brief history.* London: Routledge.

IPHC (International Polar Heritage Committee) (n.d.). *About the IPHC. Origins and objectives.* Retrieved from http://www.polarheritage.com/index.cfm/AboutIPHC. Accessed 22 October 2010.

Jabour, J. (2009). National Antarctic Programs and their impact on the environment. In K.R. Kerry & M.J. Riddle (Eds.). *Health of Antarctic wildlife: A challenge for science and policy* (pp. 211-230). Berlin Heidelberg: Springer.

Jabour, J. & Weber, M. (2008). Is it time to cut the Gordian knot of polar sovereignty? *Review of European Community & International Environmental Law 17 (1).* 27-40.

Jackson A.(Ed.) (1995). *On the Antarctic horizon: Proceedings of the International Symposium on the Future of the Antarctic Treaty System*, Ushuaia, Argentina, March 20-24, 1995. (In file with author.)

Jacobsson, M. 2007. The Antarctic Treaty system: legal and environmental issues – future challenges for the Antarctic Treaty system. In: G. Triggs. & A. Riddell (editors): Antarctica. Legal and environmental challenges for the future. London: British Institute of International and Comparative Law.

Jørgensen, R. 2005. Archaeology on Svalbard: Past, present and future. *Acta Borealia 22 (1).* 49-61.

Kaltenborn, B.P., & Emmelin, L. (1993). Tourism in the high north: Management challenges and recreation opportunity spectrum planning in Svalbard, Norway. *Environmental Management Vol. 17 (1).* 41-50.

Killick, David (7 January 2009). Mawson's Hut a tourist magnet. *The Mercury.* http://www.themercury.com.au/article/2009/01/07/48365_tasmania-news.html

Kings Bay AS (2007), *Ship calls summer 2007. Ny-Ålesund Harbour.* Unpublished document, updated 26 June 2008. (In file with author.)

Kings Bay (2009). *Land use plan.* Ny-Ålesund, Svalbard: Kings Bay A.S. Available from http://www.kingsbay.no/index.php?option=com_content&view=article&id=300&Itemid=187

Kipfer, B.A. (2007). *The archaeologist's fieldwork companion.* Malden, MA, USA: Blackwell Publishing.

Koivurova,T. (2005). Environmental protection in the Arctic and Antarctic: Can the polar regimes learn from each other?, *International Journal of Legal Information 33(2),* 204-218.

Kolltveit, B. (2006). Deckchair explorers: The origin and development of organised tourist voyages to north and south Polar Regions. *International Journal of Maritime History* 18 (2), 351-369.

Kriwoken, L.K. & Rootes, D. (2000). Antarctic tourism, tourism on ice: Environmental Impact Assessment of Antarctic tourism. *Impact Assessment and Project Appraisal 18(2)*, 138-150.

Kumar, K. (1993). *Rapid appraisal methods.* Washington D.C., USA: The World Bank.

Lamers, M. (2009). The Future of Tourism in Antarctica: Challenges for sustainability. PhD Dissertation. University of Maastricht, The Netherlands.

Lamers, M., Amelung, B. & Stel, J. (2010). Business as (un)usual. An integrated scenario analysis of Antarctic tourism". In: Hall, M. & J. Saarinen (Eds.). *Tourism and change in Polar Regions: Climate, environment and experiences* (pp. 247-262). London: Routledge.

LaMotta, V.M. & Schiffer, M.B. (2001). Behavioural archaeology: Toward a new synthesis. In: I. Hodder (Ed.), *Archaeological theory today* (pp. 14-64). Cambridge, UK: Polity Press,

Lawrence, D.P. (2007). Impact significance determination - Back to basics. *Environmental Impact Assessment Review 27*, 755–769.

Laws, R. (1991). Unacceptable threats to Antarctic science. *New Scientist*, March 30.

Leary, D. (2008). *Looking beyond the International Polar Year: Emerging and re-emerging issues in international law and policy in the Polar Regions report and recommendations from an international experts symposium held at the University of Akureyri, Iceland, 7-10 September 2008.* United Nations University Institute of Advanced Studies, Japan and the University of Akureyri, Iceland.

Lee, R. (2000). *Unobtrusive methods in social research.* Buckingham: Open University Press.

Leiper, N. (2004). *Tourism Management.* London: Arnold.

Leiper, N. (1990). Tourist attraction systems. *Annals of Tourism Research 17*, 367-384.

Leiper, N. (1979). The framework of tourism. Towards a definition of tourism, tourist, and the tourist industry, *Annals of Tourism Research Oct / Dec 1979*, 390-407.

Lewis Smith R.I. & Simpson H.W. (1987). Early nineteenth century sealers' refuges on Livingston Island, South Shetland Islands, *British Antarctic Survey Bulletin 74*: 49-72.

Ligtendag W (2002). Dirck Gerritsz Pomp & Antarctica. In: Dirck Gerritsz Pomp alias Dirck China, Symposiumbundel, Vereniging Oud Enkhuizen, 29-44.

Litchfield, P.W. (1927). Lighter-than-air craft. *Annals of the American Academy of Political and Social Science (131),* 79-85.

Lowenthal, D. (2005). Natural and cultural heritage. *International Journal of Heritage Studies 11(1)*, 81-92.

Lyman, R. L. (2004). The concept of equifinality in taphonomy. In: N.D. Munro & G. Bar-Oz (Eds.). Special volume on debating issues of equifinality in ungulate skeletal part studies. *Journal of Taphonomy 2,* 1-4.

Lynch H.J., Crosbie K., Fagan, W.F. & Naveen, R. (2009). Spatial patterns of tour ship traffic in the Antarctic Peninsula region. *Antarctic Science 22,* 123-130.

MacCannell, D. (1973). Staged authenticity: Arrangements of social space in tourist settings. *American Journal of Sociology* 79(3), 589-603.

MacCannell, D. (1976). *The tourist. A new theory of the leisure class.* New York, NY, USA: Schocken Books.

Mackay, R (2005). Ice, icon and identity: The meaning of Mawson's huts. In J. Lydon & T. Ireland (Eds.). *Object lessons: Archaeology & heritage in Australia* (Ch. 6). Melbourne, Vic.Australia : Australian Scholarly Publishing.

Madsen, J. Tombre, I. & Elde N.E. (2009). Effects of disturbance on geese in Svalbard: implications for regulating increasing tourism. *Polar Research 28*, 376–389.

Maher P. (2007). Advocating for Antarctic wilderness: Short-term visits and human values. *USDA Forest Service Proceedings* RMRS-P-49, 170-177.

Marstrander, L. (1999). Svalbard cultural heritage management. In: U. Wråkberg (Ed.). *The centennial of S. A. Andrée´s North Pole expedition* (pp. 119 – 135). Stockholm: Royal Swedish Academy of Sciences, Center for History of Science

Martin P. & Bateson (1986/1993). *Measuring behaviour. An introductory guide.* 2nd edn. 1993. Cambridge, UK: Cambridge University Press.

Masiokas, M.H., Villalba, E., Luckman, B.H., Lascano, M.E., Delgado, S. & Stepanek, P. (2008). 20th-century glacier recession and regional hydroclimatic changes in northwestern Patagonia. *Global & Planetary Change 60* (1/2), 85-100.

Mason, P. (1994). A visitor code for the Arctic? *Tourism Management 15 (2)*, 93–7.

Mathisen, T. 1954. *Svalbard in international politics 1871-1925: The solution of a unique international problem.* Norsk Polarinstitutt Skrifter 101. Oslo: Brøggers Forslag.

McKercher, B. & du Cros, H. (2002). *Cultural tourism. The partnership between tourism and cultural heritage management.* New York: The Haworth Hospitality Press.

Molenaar, E.J. (2001). CCAMLR and Southern Ocean Fisheries, *The International Journal of Marine and Coastal Law L. 16*(3), 465-499.

Molenaar, Erik Jaap (2005). Sea-borne tourism in Antarctica: Avenues for further intergovernmental regulation. *The International Journal of Marine and Coastal Law 20(2),* 247-295 (49).

Morrison-Saunders A, R Marshall & J Arts (2007). EIA Follow-up International Best Practice Principles, *Special Publication Series No. 6*, International, Fargo, ND, USA: Association for Impact Assessment. Available:http://www.iaia.org/publicdocuments/pdf/special-publications/SP6.pdf

Mulvaney, K (2003). *The whaling season: An inside account of the struggle to stop commercial whaling.* Washington D.C.: Island Press.

Nash, K. & Løkken, G. (1990): Oppmåling Sommeren 1990, 1:1000. Longyearbyen: Governor of Svalbard (Map – In file with author).

Naveen, R., Forrest, S.C., Dagit, R.G., Blight, L.K., Trivelpiece, W.Z., Trivelpiece, S.G. (2001). Zodiac landings by tourist ships in the Antarctic Peninsula region, 1989-99. *Polar Record 37*(201), 121-132.

Nelson, S.B. (1993). Airships in the Arctic. *Arctic 46(3)*, 278-283.

New Zealand (2001). Report to CEP IV on the question of collection of antarctic meteorites by private expeditions. Working Paper 09. Saint Petersburg, Russia: XXIV Antarctic Treaty Consultative Meeting/WP 9 (2001);

New Zealand (2010). The concept of human footprint in the Antarctic. Working Paper 49, XXXIII Antarctic Treaty Consultative Meeting. Punta del Este, Uruguay: XXXIII Antarctic Treaty Consultative Meeting.

New Zealand & Australia (2006). *Regulation of land-based infrastructure to support tourism in Antarctica.* XXIX Antarctic Treaty Consultative Meeting/WP045 rev.1

New Zealand & United States of America (2006). *Clean up of abandoned Cape Hallett Station.* XXIX Antarctic Treaty Consultative Meeting, IP115.

New Zealand, United Kingdom & United States. 2006. *Ross Sea protected area inspections 2006.* Working Paper 034. Edinburgh: XXIX Antarctic Treaty Consultative Meeting.

New Zealand (2006). *An update on the Antarctic Visitor Site Assessment Scheme: VISTA.* Information Paper 011. Edinburgh: XXIX Antarctic Treaty Consultative Meeting.

NHM ([United Kingdom] National History Museum) (n.d.) *Antarctic heritage and conservation. Dilemmas.* Retrieved from http://www.nhm.ac.uk/nature-online/earth/antarctica/dilemmas/index.html. Accessed 5 November 2008.

Nobile, U. (1961). *My Polar flights.* London: Frederick Muller.

Norway (2001). *Report of the 2001 Norwegian Antarctic inspection under Article VII of the Antarctic Treaty and Article 14 of the Protocol on Environmental Protection to the Antarctic Treaty.* XXIV Antarctic Treaty Consultative Meeting, Working Paper 25. Saint Petersburg, Russia: XXIV Antarctic Treaty Consultative Meeting.

Norway (2002): *Regulations relating to environmental impact assessment and delimitation of the land-use planning areas in Svalbard Laid down by Royal Decree of 28 June 2002 pursuant to section 47 and section 59, last paragraph, of the Act of 15 June 2001 No. 79 relating to protection of the environment in Svalbard (the Svalbard Environmental Protection Act).* Available from: http://www.regjeringen.no/mobil/nb/dep/md/dok/lover_regler/forskrifter.html?id=705

Norway (2004). Chairman's report from Antarctic Treaty Meeting of Experts on tourism and non-governmental activities in Antarctica. XVII Antarctic Treaty Consultative Meeting Working Paper 4. Cape Town, South Africa: XVII Antarctic Treaty Consultative Meeting.

Norway (2004). *Polar tourism: Experience gained and lessons learned from Svalbard.* Antarctic Treaty Meeting of Experts, Tromsø-Trondheim 22-25 March 2004. ATME Paper 24. Tromsø-Trondheim, Norway: Antarctic Treaty Meeting of Experts.

Nuttall, M. 2010. Narratives of history, environment and global change. Expeditioner-tourists in Antarctica. In C.M. Hall & J. Saarinen (Eds.): *Tourism and change in the Polar Regions: Climate, environment and experiences.* 204-214 London: Routledge.

Nyssen, J., Haile, M., Naudts, J., Munro, N., Poesen, J. Moeyersons, J., Frankl, A., Deckers, J. & Pankhurst, R. (2009). Desertification? Northern Ethiopia re-photographed after 140 years. *Science of the Total Environment 407*, 2749-2755.

O'Connor, T. & Evans, J.G. (1999/2005). *Environmental archaeology: Principles and methods.* Phoenix Mill, UK: Sutton Publishing.

O'Riordan T., J. Cameron, & A. Jordan (Eds.), (2001). *Reinterpreting the precautionary principle*, London: Cameron May.

Orrego Vicuña, E. (1948). *Terra Australis. Diario de la primera expedición antártica chilena.* Santiago de Chile: Empresa Editora Zig Zag.

Palazzi, R.O. (1993). Antártida y archipiélagos subantárticos: Factores para su análisis. Segunda Parte: Los factores variables. [Antarctica and Subantarctic islands: Factors for analysis. Part Two:The variable factors.] *Revista de la Escuela Superior de Guerra Aérea de la Fuerza Aérea Argentina*, (Publicación N° 182) Buenos Aires, Argentina: Escuela Superior de Guerra Aérea de la Fuerza Aérea Argentina.

Patterson, D. L.; Easter-Pilcher, A. & Fraser, W.R. (2003). The effects of human activity and environ mental variability on long-term changes in Adélie penguin populations at Palmer Station,

Antarctica. In: A.H.L. Huiskes, W.W.C. Gieskes, J. Rozema, R.M.L. Schorno, S.M. Van der Vies & W.J. Wolf, (Eds.), *Antarctic biology in a global context* (pp. 301-307). Leiden: Backhuys Publishers.

Paulsen, B. (2006). *Rapport fra veteransbesøk 2006* [Report from the veterans' visit 2006]. Unpublished report. Ny Ålesund: Kings Bay AS. (In file with author.)

Pearce, P.L. (2005). *Tourist behaviour: Themes and conceptual schemes.* Clevedon, UK: Channel View Publications.

Pearson M., & Stehberg, R. (2006). Nineteenth century sealing sites on Rugged Island South Shetland Islands. *Polar Record 42 (04).* 335-347.

Pearson, M.; Stehberg, R., Zarankín, A.; Senatore, M.X.; & Gatica, C. (2010). Conserving the oldest historic sites in the Antarctic: The challenges in managing the sealing sites in the South Shetland Islands. *Polar Record 46 (236).* 57–64 (2010).

Pearson, M. & B. McGowan. 2000. *Mining heritage places assessment manual.* Canberra: Australian Council of National Trusts and Australian Heritage Commission. Available from http://www.environment.gov.au/heritage/ahc/publications/commission/books/mining-heritage-places.html

Pedersen, T. (2009). Norway's rule on Svalbard: Tightening the grip on the Arctic islands. *Polar Record 45 (2).* 147–152.

Pérez-Juez Gil, A. (2006). *Gestión del patrimonio arqueológico* [Archaeological heritage management]. Barcelona: Ariel.

Pfeiffer, S. & Peter, H.-U. (2004). Ecological studies toward the management of an Antarctic tourist landing site (Penguin Island, South Shetland Islands). *Polar Record 40 (215)*, 345-353.

Pfeiffer, S., Buesser, C., Mustafa, O. & Peter, H.-U. (2007). Tourism growth and proposed management solutions in the Fildes Peninsula region (King George Island, Antarctica). *Tourism in Marine Environments 4* (2-3). 151-168.

Poland, J. S., Riddle, M.J. & Zeeb, B.A. (2003). Contaminants in the Arctic and the Antarctic: A comparison of sources, impacts, and remediation options. *Polar Record 39* (211). 369–383.

Polunin, N. (1945). Plant life in Kongsfjord, West Spitsbergen, *The Journal of Ecology 33(1),* 82-108.

Powell, P.B., Kellert, S.R. & Ham, S.H. (2008). Antarctic tourists: Ambassadors or consumers? *Polar Record 44 (230),* 233–241.

Prestvold, K. (2003). *Ijsfjorden. A journey through the nature and cultural history of Svalbard.* Longyearbyen: Governor of Svalbard.

Pretes, M. (2002). Tourism mines and mining tourists. *Annals of Tourism Research, 29(2)*, 439-456.

Racing the Planet (n.d.). The last desert – Antarctica. http://www.4deserts.come/thelastdesert (Accessed September 2008).

Rapp, G. (Rip) & Hill, C.L. (2006). *Geoarchaeology. The earth-science approach to archaeological interpretation.* New Haven, CT, USA: Yale University Press.

Rathje, W. (2001). Integrated archaeology. A garbage paradigm. In: Buchli V & Lucas G (2001). *Archaeologies of the contemporary past* (pp. 63-76). London, UK: Routledge.

Rathje, W.L. & Schiffer, M.B. (1982). *Archaeology.* Fort Worth, TX, USA: Harcourt Brace Jovanovich Colleague Publishers.

Rawlins, D. (2000). Byrd's heroic 1926 North Pole failure. *Polar Record 36(196)*, 25-50.

Rees, W.G. (2005). A new bipolar map projection. *Polar Record 41 (218).* 215-222.

Reid, J.J., Schiffer, M.B. & Neff, J.M. (1974). Expanding archaeology. *American Antiquity 39*,125-126.

Reid, J.J., Schiffer, M.B. & Neff, J.M. (1975). Behavioural archaeology: Four strategies. *American Anthropologist (77)*.836-48.

Richardson M.G. (2000). Regulating tourism in the Antarctic: Issues of environment and jurisdiction. In: D. Vidas (Ed.), *Implementing the environmental protection regime for the Antarctic* (pp. 71-90). Kluwer Academic Publishers: Dordrecht.

Riffenburgh, B. (1998). Impacts on the Antarctic environment: Tourism vs. government programmes, *Polar Record 34(190),* 193-196.

Rogers, G.F., Malde H.E., & Turner R.M. (1984). *Bibliography of repeat photography for evaluating landscape changes.* Salt Lake City: University of Utah Press.

Roura, R. (2004). Monitoring and remediation of hydrocarbon contamination at the former site of Greenpeace's World Park Base, Cape Evans, Ross Island, Antarctica. *Polar Record 40 (212),* 51-67.

Roura, R (2007). Greenpeace. In: B. Riffenburgh (Ed.). *The Encyclopedia of the Antarctic* (pp. 472-474). New York: Routledge.

Roura, R. 2008a. Antarctic scientific bases: Cultural heritage and environmental perspectives 1983-2008. In S. Barr & P. Chaplin (Eds.). *Historical polar bases: Preservation and management,* ICOMOS Monuments and Sites Series N° XVII (pp 38-50). Oslo: ICOMOS International Polar Heritage Committee.

Roura, R. (2008b). Vignetten van het Hoge Noorden. Toerisme in het Arctisch gebied [Vignettes from the High North. Tourism in the Arctic region.] In D. Veerman (Ed.). *Tegenpolen. Op expeditie naar contrastrijke poolgebieden* [Poles apart. In expedition to contrasting Polar Regions] (pp. 114-122). Zutphen, The Netherlands: Walburg Pers.

Roura, R. (2008c). *Spot the diffeЯence: Monitoring the transformation of Ny-London, a cultural heritage site in Spitsbergen.* Polar Symposium 14 May 2008 (p. 36), Nederlandse SCAR Commissie (KNAW) and The Nederlands Polair Programma (NWO).

Roura, R. (2009). Cultural heritage, tourism narratives, and tourism impacts: A case study of the airship mooring mast at Ny-Ålesund, Svalbard. May 2009. Special Issue on polar tourism. *Téoros 28(1),* 29-38.

Roura, R.M. (2010a). Monitoring the transformation of historic features in Antarctica and Svalbard: Local processes and regional contexts. Polar Record 46(04), 289-311.

Roura R (2010b). Cultural heritage tourism in Antarctica and Svalbard: Patterns, impacts, and policies. In: Michael Hall & Jarkko Saarinen (Eds.). *Tourism and change in the Polar Regions: Climate, environment and experiences* (pp.180-203). London: Routledge.

Roura, RM (2010c). *Being there: Tourist behaviour, impacts, and compliance as assessed through Antarctic tourism Web logs.* [Abstract] Oslo Science Conference 2010. Oslo, 8-12 June 2010. Available from: http://ipy-osc.no/abstract/382612

Roura, R. & Pérez Muñoz, C. (2002). *Evaluation of sites with human activities at Deception Island. Recommendations for a management plan. A field report of the Antarctic and Southern Ocean Coalition submitted to the Deception Island Management Group.* Washington DC: The Antarctica Project. (In file with author.)

Roura, R.M., dos Santos Afonso, M., Pérez Muñoz, C. & Tin, T. (2008). *Tourism and the human footprint at Deception Island, South Shetland Islands, Antarctica.* Abstract Volume, SCAR/IASC Open Science Conference, St. Petersburg, Russia, July 8-11 2008 (p. 452) (Poster). St. Petersburg, Russia: Arctic and Antarctic Research Institute.

Roura, R. & Loonen, M.J.J.E. (2009). Status and change of a historic mining site at Ny-Ålesund, Spitsbergen (Svalbard). Symposium *Sustainable use of resources in Polar Regions*, Groningen Barnouilliborg/Academie Gebouw 22 October 2009 (p. 49)(Poster). Groningen, The Netherlands: Willem Barentsz Polar Institute.

Roush, W., J.S. Munroe, & D.B. Fagre (2007). Development of a spatial analysis method using ground-based repeat photography to detect changes in the Alpine treeline ecotone, Glacier National Park, Montana, U.S.A. *Arctic, Antarctic and Alpine Research 39*(2): 297–308.

Royal Ministry of the Environment (2001). *Svalbard Environmental Protection Act. Act of 15 June 2001 No. 79 relating to the protection of the environment in Svalbard.* Oslo: Royal Ministry of the Environment.

Russian Federation (2008). *Proposals for regulating the [sic] adventure tourism and non-governmental activity in the Antarctic.* Working Paper 50, Kiev: XXXI Antarctic Treaty Consultative Meeting.

Ryan, G.W. & Bernard, H. R. 2003. Techniques to identify themes. *Field Methods* 15:1, 85-109.

Sander, G., Holst, A. & Shears, J. (2006). *Environmental impact assesment of the research activities in Ny-Ålesund 2006.* (Kortrapport nr. 4/Brief Report Series no 4). Tromsø, Norway: Norwegian Polar Institute.

Sanger, J. (1996). *The compleat observer? A field guide to observation.* Qualitative Studies Series 2. London: Routledge Falmer.

Save the Huts (n.d.). *The use of huts is being changed.* Retrieved from http://www.savethehuts.com/think_again_open_letter_those_who_love_our_polar_heritage/change_use. (Accessed 18 May 2009).

Serrano, E. (2007). Tourism in the South Shetland Islands. Recent changes in local destinations and activities. *Tourism in Marine Environments 4* (2-3). 221-236.

New Zealand (2000). *Report to CEP IV on the question of collection of Antarctic meteorites by private expeditions.* Working Paper 9. The Hague,The Netherlands: XII Special Antarctic Treaty Consultative Meeting.

SCAR (Scientific Committee for Antarctic Research)(2000). *Antarctic meteorites.* Working Paper 19. The Hague, The Netherlands: XII Special ATCM.

Schiffer, M. B. (1972). Archaeological context and systemic context. *American Antiquity 37* (2), 156-165.

Schiffer, M.B. (1983). Toward the identification of formation processes. *American Antiquity, 48* (4), 675-706.

Schiffer, M.B. (1987), *Formation processes of the archaeological record*, Salt Lake City: University of Utah Press.

Schiffer, M.B. (1995), *Behavioural archaeology. First principles.* Salt Lake City: University of Utah Press.

Schiffer, M.B. (2010), Behavioural archaeology: Principles and practice. London: Equinox.

Schiffer, M.B. & Miller, A.R. (1999): *The material life of human beings: Artifacts, behavior, and communication.* London: Routledge.

Scott, K. (2003). Institutional developments within the Antarctic Treaty System, *International and Comparative Law Quarterly 473* (52). 473-487.

Scott, S.V. (2001). How cautious is precautious? Antarctic tourism and the precautionary principle. *International and Comparative Law Quarterly*, 4 (50), 963-971.

Scott, R.F. (1913): *Scott's last expedition.* (L, Huxley - Ed.). London: McMillan.

Seed, P. (1995). *Ceremonies of possession in Europe's conquest of the New World, 1492-1640.* New York: Cambridge University Press.

Senécal, P., Goldsmith, B., Conover, S., Sadler, B. & Brown, K. (1999). *Principles of Environmental Impact Assessment best practice.* Fargo, ND, USA: International Association for Impact Assessment, retrieved from http://www.iaia.org/publicdocuments/special-publications/Principles%20of%20IA_web.pdf. Accessed 15 January 2010.

Smith, R.I.L. (1995). Colonization by lichens and the development of lichen-dominated communities in the maritime Antarctic. *Lichenologist 27* (6). 473-483.

Smith, T. (2007). Repeat photography as a method in visual anthropology. *Visual Anthropology 20* (2). 179 - 200.

Snyder, J. (2007). Tourism in the Polar Regions. The sustainability challenge. Paris, France: United Nations Environmental Programme and Washington DC, USA: The International Ecotourism Society

Sobo, E.J. & de Munck, V.C. (1998). The forest of methods. In: V.C. de Munck, & E.J. Sobo (Eds.). *Using methods in the field. A practical introduction and casebook* (pp. 13-37). Walnut Creek, CA, USA: Altamira Press.

Spenneman, D.H.R. (2007). Extreme cultural tourism. From Antarctica to the Moon. *Annals of Tourism Research 34*(4), 989-918.

Splettstoesser, J. (2000). IAATO's stewardship of the Antarctic environment: A history of tour operator's concern for a vulnerable part of the world. *International Journal of Tourism Research 2* (1), 47-55.

Splettstoesser, J. (1999). Antarctica tourism: successful management of a vulnerable environment. In T.V. Singh & S., Singh (Eds.), *Tourism development in critical environments* (pp. 137-148). New York: Cognizant Communication Corporation (CCC) Press.

Splettstoesser, J. (2001). Markers and signs for Historic Sites and Monuments in Antarctica. In: Valencia, J. & R. Downie (Eds.). *Workshop on a Management Plan for Deception Island* (pp. 73-92). Santiago de Chile: Instituto Antártico Chileno.

Splettstoesser, J. & Folks, M. (1994), Environmental guidelines for tourism in Antarctica. *Annals of Tourism Research 21* (2), 231-244.

Splettstoesser, J., Landau, D. & Headland, R.K. (2004). Tourism in the forbidden lands: The Antarctica experience. In: T.V. Singh (Ed.). *New horizons in tourism. Strange experiences and stranger practices* (pp. 27-36). Wallingford, Oxfordshire, UK: CABI Publishing.

Splettstoesser, J. & Landau, D. (2001). Tourism at Deception Island. In: Valencia, J. & R. Downie (Eds.). *Workshop on a Management Plan for Deception Island* (pp. 44-52). Santiago de Chile: Instituto Antártico Chileno.

Statistics Norway (2005). *Svalbard Statistics 2005. Cultural heritage in Svalbard, by type and function.* Retrieved from http://www.ssb.no/english/subjects/00/00/20/nos_svalbard_en/nos_d330_en/tab/017.html

Statistics Norway (2005). *Svalbard Statistics 2005. Geographical survey. Islands and glaciers.* Retrieved from http://www.ssb.no/english/subjects/00/00/20/nos_svalbard_en/nos_d330_en/tab/012.html.

Stehberg, R. (2004). Archaeologists document historical heritage in the South Shetland Islands. In S. Barr & P. Chaplin (Eds.). *Cultural heritage in the Arctic and Antarctic regions,* ICOMOS Monuments and Sites Series N° VII (pp 69-72). Oslo: ICOMOS International Polar Heritage Committee.

Stehberg, R., M. Pearson, A. Zarankin, X. Senatore, & C. Gatica. 2008. Protection and preservation of the oldest sites of the Antarctic: The case of Fildes Peninsula and Byers Peninsula in the South Shetland Islands. In S. Barr & P. Chaplin (Eds.). *Historical polar bases: Preservation and management (pp.* 85-93). ICOMOS Monuments and Sites Series N° XVII. Oslo: ICOMOS International Polar Heritage Committee.

Stewart, E., Draper, D., & Johnston, M. (2005). A review of tourism research in the Polar Regions. *Arctic 58*(4), 383-394.

Stonehouse, B. (1990). A traveller's code for Antarctic visitors, *Polar Record 26*(156), 56-58.

Stonehouse, B. (1992). Monitoring shipborne visitors in Antarctica: A preliminary field study. *Polar Record 28* (166). 213-218.

Strijbos, Jan P (1957): *Svalbard. Zwerftocht langs de koele stranden van Spitsbergen.* [Svalbard. A ramble along the cold beaches of Spitsbergen]. Amsterdam: L.J. Veen.

Sugden, D. (1982). *Arctic and Antarctic. A modern geographical synthesis.* Oxford: Basil Blackwell.

Tape, K., Sturm, M. & Racine, C. (2006). The evidence for shrub expansion in Northern Alaska and the Pan-Arctic. *Global Change Biology 12*, 686-702.

Thomson, R. B. (1977): Effects of human disturbance on an Adélie penguin rookery and measures of control. In: G.A. Llano (Ed.) *Adaptations within Antarctic Ecosystems. Proceedings of the Third SCAR Symposium on Antarctic Biology.* Smithsonian Institute, Washington DC, 1177 – 1180.

Thompson, M. (1979). *Rubbish theory. The creation and destruction of value.* Oxford: Oxford University Press.

Timothy, D.J. & Boyd, S.W. (2003). *Heritage and tourism.* Harlow, Essex, UK: Pearson Education - Prentice Hall.

Timothy, D.J. (2010). Contested place and the legitimisation of sovereignty through tourism in Polar regions. In: C.M. Hall & J. Saarinen (Eds.). *Tourism and change in the Polar Regions: Climate, environment and experiences* (pp. 288-301). London: Routledge.

Tin, T., Hemmings, A.D. & Roura. R. (2008). Pressures on the wilderness values of the Antarctic continent. *International Journal of Wilderness 14*(3). 7-12.

Tracey, P.J. (2001). *Managing Antarctic tourism.* PhD dissertation. University of Tasmania, Hobart, Australia.

Triggs, G.D. (Ed.). (1987). *The Antarctic legal regime: Law, environment and resources.* Cambridge: Cambridge University Press.

Triggs, G. & Riddell, A. (2007). *Antarctica. Legal and environmental challenges for the future.* London: British Institute of International and Comparative Law.

Tunbridge, J & Ashworth, G (1996). *Dissonant heritage: The management of the past as a resource in conflict.* Chichester: Wiley.

Turner, J. & Overland, J. (2009). Contrasting climate change in the two polar regions. *Polar Research 28*, 146–164

Trathan, P.N., Forcada, J., Atkinson, R., Downie, R.H. & Shears, J. (2008). Population assessments of gentoo penguins (*Pygoscelis papua*) breeding at an important Antarctic tourist site, Goudier Island, Port Lockroy, Palmer Archipelago, Antarctica. *Biological Conservation 141*: 3019–3028.

Ulfstein, G. (1995). *The Svalbard Treaty. From terra nullius to Norwegian sovereignty.* Oslo: Scandinavian University Press.

United Kingdom (2004a). *Proposed amendment of Recommendation XVIII-1 (1994). Site guidelines for sites visited by tourists.* XXVII Antarctic Treaty Consultative Meeting, Working Paper 026, Cape Town: XXVII Antarctic Treaty Consultative Meeting.

United Kingdom (2004b). *Area protection and management. Historic Site No. 71 - Whalers Bay, Deception Island. Salvage of the de Havilland Single Otter from Whalers Bay, Deception Island by the British Antarctic Survey.* Working Paper 45, Cape Town: XXVII Antarctic Treaty Consultative Meeting.

United Kingdom (2006a), *Report of the CEP Intersessional Contact Group on site guidelines for visitors to Antarctica.* Working Paper 1, Edinburgh, UK: XXIX Antarctic Treaty Consultative Meeting.

United Kingdom (2006b). *Site guidelines for Goudier Island, Port Lockroy.* Working Paper 40, Edinburgh, UK: XXIX Antarctic Treaty Consultative Meeting.

United Kingdom & Germany (1999b): *Antarctic Treaty report of a joint inspection under Article VII of the Antarctic Treaty by United Kingdom and German observers.* XXIII Antarctic Treaty Consultative Meeting/WP23.

United Kingdom & Germany (1999a): Antarctic Treaty report of a joint inspection under Article VII of the Antarctic Treaty by United Kingdom and German observers. London: Foreign and Commonwealth Office, and Bonn: Auswartiges Amt. Available from http://www.ats.aq/e/ats_governance_listinspections.htm.

United Kingdom, France, & Ukraine (2010). *Report of an incident at Wordie House (HSM No. 62).* Working Paper 025. Punta del Este, Uruguay: XXXIII Antarctic Treaty Consultative Meeting.

United States of America (2004). Monitoring and assessment of activities: Approaches taken by the Antarctic Site Inventory. XXVII Antarctic Treaty Consultative Meeting Working Paper 11. Cape Town, South Africa: XXVII Antarctic Treaty Consultative Meeting.

United States of America (2006). *Report of inspections under Article VII of the Antarctic Treaty and Article 14 of the Protocol on Environmental Protection. November 12 – December 1, 2006.* Washington DC, USA: United States Department of State. (Subsequently submitted as *United States report of inspections.* Information Paper 10. New Delhi, India: XXX Antarctic Treaty Consultative Meeting).

United States of America (2008a). *Revised Management Plan for Antarctic Specially Protected Area No. 106 Cape Hallett, Northern Victoria Land, Ross Sea.* Antarctic Treaty Consultative Meeting, Working Paper 013.

United States of America (2008b). *Site guidelines for Cape Hallett, Northern Victoria Land, Ross Sea.* Antarctic Treaty Consultative Meeting, Working Paper 045.

Urry, J. (1992). The tourist gaze revisited, *American Behavioural Scientist 36*(2), 172-186.

Urry, J. (1990). *The tourist gaze.* London: Sage Publications.

Uruguay (2005). Visitors programme to the “Artigas” Antarctic Scientific Base (BCAA). XXVIII Antarctic Treaty Consultative Meeting Information Paper 056. Stockholm, Sweden: XXVIII Antarctic Treaty Consultative Meeting.

van Polanen Petel, T., Giese, M. & Hindell, M. (2008). A preliminary investigation of the effect of repeated pedestrian approaches to Weddell seals (*Leptonychotes weddellii*). *Applied Animal Behaviour Science 112* (1-2), 205-211.

Vanclay, F. & Bronstein, D.A. (Eds.) (1995). *Environmental and social impact assessment.* New York: John Wiley and Sons.

Veal, A.J. (1992/2006). *Research methods for leisure and tourism.* 3rd edn. 2006. Harlow, England: Prentice Hall.

Verschuuren, J.M. (2003). *Principles of environmental law. The ideal of sustainable development and the role of principles of international, European, and national environmental law*. Studies on Environmental Law, Baden-Baden: Nomos Verlag.

Viken, A. & Jørgensen, F. (1998). Tourism on Svalbard. *Polar Record 34* (189), 123-128.

Viken, A. (forthcoming). Tourism, research, and governance on Svalbard: A symbiotic relationship. *Polar Record*. doi:10.1017/S003224741000604 1.

Vistad, O.I., Eide, N.E., Hagen, D., Erikstad, L. & Landa, A. (2008). *Miljøeffekter av ferdsel og turism i Arktis. En litteratur- og forstudie med vekt på Svalbard.* [Environmental impacts from traffic and tourism in the Arctic. A review of the literature and state-of-knowledge with a focus on Svalbard.] (NINA Report 316). Trondheim: Norsk institutt for naturforskning.

Vogel, C. (24 January 2010). Woman Collides With a Picasso. *The New York Times*, 24 January 2010 http://www.nytimes.com/2010/01/25/arts/design/25arts-WOMANCOLLIDE_BRF.html.

Wainwright, M. (20 August 2010). Woman who urinated on war memorial flees court. *The Guardian* 20 August 2010. http://www.guardian.co.uk/uk/2010/aug/20/urinate-sex-war-memorial-blackpool

Walton, D.W. & Shears, J. (1993). The need for environmental monitoring in Antarctica: Baselines, environmental impact assessments, accidents and footprints. *International Journal of Analytical Chemistry 55*, 77-90.

Walton, D.W.H. (1990). Waste disposal – expectations and realities. Guest Editorial. *Antarctic Science 2*(2), 103-103.

Webb, E. J., Campbell D.T., Schwartz R.D. & Sechrest, L. (1966). *Unobtrusive measures: Non reactive research in the social sciences*. Chicago: Rand Mc Nally Company.

Weiner, S. (2010): *Microarchaeology. Beyond the visible archaeological record.* Cambridge: Cambridge University Press.

Wheeler, S. (1999). *Terra Incognita: Travels in Antarctica*. New York: The Modern Library.

White, K.J. (1994). Tourism and the Antarctic economy. *Annals of Tourism Research 21(*2), 245-268.

Wilson, R.P., Culik, B., Danfeld, R. & Adelung, D. (1991). People in Antarctica – how much do Adélie Penguins *Pygoscelis adeliae* care? *Polar Biology 11*, 363 – 370.

World Tourism Organisation (2002): *Québec Declaration on Ecotourism.* World Ecotourism Summit, Québec Canada, May 19-22, 2002. Madrid, Spain: World Tourism Organization and the United Nations Environment Programme. Retrieved from http://www.gdrc.org/uem/eco-tour/quebec-declaration.pdf.

World Wildlife Fund (WWF) (2004). *Cruise tourism in the Arctic.* Retrieved from http://www.panda.org/about_wwf/where_we_work/arctic/what_we_do/tourism/index.cfm (last modified May 28, 2004).

Worthing, D. & Bond, S. (2010). *Managing built heritage*. Oxford: Blackwell Publishing.

Yin, R.K. (1984/2003). *Case study research. Design and methods.* 3rd edition. Thousand Oaks: Sage Publications.

Yin, R.K. (Ed.) (2004). *The case study anthology*. Thousand Oaks: Sage Publications.

Epilogue

This research began formally in March 2005. Since then several events relevant to tourism and historic sites, individually and in interaction with each other, have taken place both in Antarctica and in Svalbard. These events highlight both the dynamic nature of polar tourism and the active role of polar historic sites in contemporary social systems, or systemic context.

In 2004 IAATO had predicted 32,500 ship borne Antarctic tourists for the 2008-09 season (Fig. 3.1). The actual number of tourists that season was over 37,000 ship-borne, airborne and land-based tourists.[1] However, this was substantially down from the peak-to-date of over 46,000 tourists in the previous season (2007-08). The reduction in tourist numbers demonstrated the effect of the global economic downturn of 2008, which continued into the 2009-2010 season with over 36,000 tourists. It is likely that the slowdown in the growth of Antarctic tourism represents a temporary effect of economic malaise and that tourism will continue to grow and diversify in the years to come.

The number of sites used for tourism landings in Antarctica has stabilised in the past two decades, but tourism has grown disproportionally at certain sites.

On November 23, 2007 the MV *Explorer* sank off the South Shetland Islands with no loss of life. She was the first ship purpose built for Antarctic tourism, and the first to sink. This was the most serious of several incidents involving cruise ships both in Antarctica and Svalbard in recent years. In February 2011 the Norwegian sailing vessel *Berserk* with a crew of three disappeared in the Ross Sea while supporting a two-person expedition to the South Pole. The vessel and its crew were never found.

Since the XXVII ATCM in Cape Town in 2004 (Chapter 3), Antarctic Treaty Parties have met at six ATCMs to discuss, *inter alia*, several issues relevant to this research. Antarctic tourism has also been discussed at several dedicated meetings outside ATCMs and at online fora of both the ATCM and the CEP. Since the Cape Town meeting the ATCM have agreed on mandatory instruments adding requirements concerning insurance and contingency planning,[2] and preventing landings from vessels carrying more than 500 passengers.[3] None of these instruments had become effective at the

1 The Antarctic tourism statistics quoted in this paragraph have been retrieved from http://www.iaato.org/tourism_stats.html, files "Tourism summary" for the corresponding season, accessed 9 August 2010.

2 Measure 4 (2004) *Insurance and contingency planning for tourism and non-governmental activities in the Antarctic Treaty Area.*

3 (ATCM) Measure 15 (2009) *Landing of Persons from Passenger Vessels in the Antarctic Treaty Area.*

time of writing. The Protocol's Annex VI on liability for environmental damage was signed in 2005,[4] and has not yet entered into force.

Non-mandatory instruments agreed in this period included site specific guidelines for a number of sites frequently used for the landing of tourists; and a Resolution outlining general principles of Antarctic tourism,[5] which is meant to inform and guide further work in managing this activity. One of these principles is that:

> *Tourism should not be allowed to contribute to the long-term degradation of the Antarctic environment and its dependent and associated ecosystems, or the intrinsic natural wilderness and historical values of Antarctica. In the absence of adequate information about potential impacts, decisions on tourism should be based on a pragmatic and precautionary approach that also incorporates an evaluation of risk. (ATCM Resolution 7, 2009; operative para. 2)*

New guidelines for the protection and management of HSMs have been adopted.[6] There have been proposals to apply the notion of "enhancement" to the Antarctic cultural heritage, a holistic concept that promotes the current and future use of cultural heritage, and which encompasses passive preservation, active conservation to extend the life of heritage, and dissemination through education (e.g. Argentina, 2010). If this approach is adopted tourism will become an integral part of the management of Antarctic historic sites.

Several new sites and monuments have been designated as Antarctic HSMs since 2005: a German hut from the 1970s; the tent erected at 90° S by Roald Amundsen and his colleagues on their arrival at the South Pole on 14 December 1911, whose precise location is unknown; a small island where Admiral Dumont D'Urville and his crew reportedly landed on 21 January 1840 (and where nothing remains of that event); a Chilean monument to the Antarctic Treaty dedicated in 2001 and designated for protection six years later; a British base and hut from the 1950s and 1970s, respectively; and a plaque commemorating the PM-3A Nuclear Power Plant, which operated at the USA's McMurdo Station in the 1950s. The number and diversity of these additions to the HSM list underscores a renewed interest in the symbolic role of this instrument in projecting a Party's historic and ongoing presence in Antarctica and in the Antarctic Treaty System.

In the meantime, the centenaries of significant events in Antarctic exploration are taking place, including the arrival at the South Pole by Roald Amundsen and his team in 1911. These anniversaries are resulting in tourism activities, both adventure and mainstream, focused on retracing the steps of those earlier events.

---

4 (ATCM) Measure 1 (2005) - *Annex VI to the Protocol on Environmental Protection to the Antarctic Treaty Liability Arising From Environmental Emergencies.*

5 (ATCM) Resolution 7 (2009) - *General Principles of Antarctic Tourism/.*

6 (ATCM) Resolution 3 (2009) *Guidelines for the designation and protection of Historic Sites and Monuments*

Svalbard has recorded a large increase in passenger traffic, including air traffic, private yachts, and commercial cruises of different kinds. The number of cruise liners visiting Svalbard briefly as part of a longer cruise has been stable at around 30 in recent years, but the number of passengers on these vessels has risen from around 20,000 in 2000 to just under 30,000 in 2008. Many of the tourists who arrive in Svalbard by plane join 4-7 day expedition cruises around the archipelago. The number of tourists participating in week-long coastal cruises has increased from around 5,000 in 2001 to more than 14,000 in 2008. Cruise tourists landed at around 100 different locations in 2000, while eight years later they landed at 150 different locations (Bore *et al.*, 2009, 16-17). The government of Svalbard has introduced new regulations that prohibit the use of heavy oils on vessels sailing within the three largest national parks of Svalbard. The objective of these restrictions is to avoid major pollution from heavy bunker oil in the event of an accident at sea and to limit the environmental damage caused if it should occur. Added to existing regulations, this new regulation will mean that heavy bunker oil will be prohibited within most of Svalbard's territorial waters. Heavy bunker oil is the type of fuel normally used by large capacity vessels.

Access restrictions – described as a "general traffic prohibition" - has also been introduced at eight protected cultural heritage sites "...in order to protect certain unspoiled and extremely valuable cultural heritage sites against wear and undesirable influence from human traffic." These include early Dutch and English whaling sites; a Russian hunting station, blubber ovens and cultural remains from the hunting of walrus; and a German weather station established during the Second World War. These cultural heritage sites were already automatically protected by the 2001 Svalbard Act, and the traffic restriction strengthened protection. The prohibition was adopted even though in the consultation process that preceded the adoption of guidelines the Svalbard Tourism Council dismissed precautionary action as "speculations and assumptions" (Governor of Svalbard 2008: 16).

The Swedish-Finnish-Swiss International Geophysical Year station Kinnvika (1957-1958) was given legal protection under a special order by Norway's Riksantikvaren (Directorate for Cultural Heritage). This was only the second protection order according to the 2001 Svalbard law the 2001 Svalbard Act for cultural heritage originating from after 1946. The order was the result of an upsurge in interest in using Kinnvika as a base for research during the IPY 2007-2009, which led to the decision of the cultural heritage authorities to more strictly control the use of the station (IPHC, 2009; Riksantikvaren, 2009).

In Antarctica, as in Svalbard, tourism has consolidated into a major presence. The number of designated historic sites has continued to increase. Precautionary action has been used in both regions, in different ways and at different scales, to manage the interaction between tourism and historic sites.

Appendix 1 – Acronyms and abbreviations

AECO	Association of Arctic Expedition Cruise Operators (www.aeco.no).
AHT	Antarctic Heritage Trust
ASMA	Antarctic Specially Managed Area
ASOC	The Antarctic and Southern Ocean Coalition (www.asoc.org)
ASPA	Antarctic Specially Protected Area
ATCM	Antarctic Treaty Consultative Meeting
ATCP	Antarctic Treaty Consultative Party
ATME	Antarctic Treaty Meeting of Experts
ATS	Antarctic Treaty System
CCAMLR	Commission for the Conservation of Antarctic Marine Living Resources (www.ccamlr.aq)
CCAS	Convention for the Conservation of Antarctic Seals
CEE	[Antarctic] Comprehensive Environmental Evaluation
CEP	Committee for Environmental Protection (www.cep.aq)
COMNAP	Council of Managers of National Antarctic Programs (www.comnap.org)
CRAMRA	Convention on the Regulation of Antarctic Mineral Resource Activities
CRM	Cultural Resource Management
DSPIR	Driving forces, State, Pressures, Impacts, Responses
EIA	Environmental Impact Assessment
GIS	Geographical Information System
GPS	Global Positioning System
HSM	[Antarctic] Historic Sites and Monuments
IAATO	International Association of Antarctica Tour Operators (www.iaato.org)
ICOMOS	International Council on Monuments and Sites
IEE	[Antarctic] Initial Environmental Evaluation
IGY	International Geophysical Year (1957-1959)
IPHC	International Polar Heritage Committee [of ICOMOS] (www.polarheritage.com)
IPY	International Polar Year
RIAM	Rapid Impact Assessment Matrix
SATCM	Special Antarctic Treaty Consultative Meeting
SCAR	Scientific Committee of Antarctic Research (www.scar.org)

Appendix 2 – Glossary

A necessarily idiosyncratic selection of terms relevant to this thesis collected from the disciplines of archaeology, cultural resource management, impact assessment, environmental law, geomorphology (physical geography), and tourism.

Act: The smallest unit of recurrent behaviour involving an artefact (Rathje and Schiffer, 1982).

Activity: (1) A set of related acts (Rathje and Schiffer, 1982). (2) (Antarctica): An event or process resulting from, or associated with, the presence of humans in Antarctica, and/or which may lead to the presence of humans in Antarctica. (ATCPs, 2005). (3) (Svalbard): Single-instance, recurring or continuous undertaking of a commercial or non-commercial nature (Svalbard Environmental Act 2001).

Activity area: The place an activity occurs, whether it is repeated many times or takes place once (Rathje and Schiffer, 1982).

Antarctic Circle: Parallel of latitude at 66° 30' S (Strahler, 1996, [2011]).

Antarctic Treaty Area: The area subject to the 159 Antarctic Treaty, located south of 60° S.

Antarctic Treaty System: The [1959] Antarctic Treaty, the measures in effect under that Treaty, its associated separate international instruments in force and the measures in effect under those instruments. (1991 Protocol definition).

Archaeological context: The state of artefacts in the archaeological record (Rathje and Schiffer, 1982).

Arctic Circle: Parallel of latitude at 66° 30' N (Strahler, 1996, [2011]).

Arctic tundra: A plant formation class within the tundra biome, consisting of low, mostly herbaceous plant, but with very small stunted trees, associated with the tundra climate (Strahler, 1996, [2011]).

Artefact: (1) Any object affected by human behaviour (Rathje and Schiffer, 1982). (2) An object made or modified for use by humans (Green and Moore, 1983, [2010]). (3) Any physical entity that is formed by human beings ("Object" is often used as synonym). (Caple, 2006:1).

Artefact scatter: A distribution of artefacts indicating the presence or former presence of a monument. thesaurus.english-heritage.org.uk/.

Assemblage: (1) Groups of artefacts that occur together (Caple, 2006:49). (2) Associated set of contemporary artefacts (Clarke, 1968: 188, quoted in Gamble, 2008: 56].

Association: Occurrence of two or more artefacts together (Rathje and Schiffer, 1982).

Behaviour: The interaction of one or more living individuals with elements of the material world (Reid *et al.*, 1974, 1975).

Behavioural system: A set of patterned behaviours that articulate a human group with the physical world and with other behavioural systems" (La Motta and Schiffer, 2001: 27).

Commodification [of heritage]: The creation of tradable commodities from resources, in this case selected elements from the past, which were previously not traded (Ashworth, 1991b, quoted in Graham *et al.*, 2000:143). "Tourism often involves the process known as commodification, the packaging of the heritage item to make it more saleable, which in turn makes it self-conscious." (Howard, 2003:50).

Condition: Refers to the physical condition of the elements that make the place important. A related concept is that of "integrity", which refers to the completeness of these elements (Pearson and McGowan, 2000:49).

Cultural formation processes: Human activities responsible for forming and modifying the archaeological record (Rathje and Schiffer, 1982).

Cultural heritage: Traces of past human activity, including structures and sites of all kinds, on or below the surface of the land, in the sea, on the seabed and in watercourses (Svalbard Environmental Act 2001).

Cumulative impacts: (1) "the results of additive and aggregative actions producing impacts that accumulate incrementally or synergistically over time and space."[1] (2) The combined impact of past, present, and reasonably foreseeable activities. These activities may occur over time and space and can be additive or interactive/synergistic (adapted from De Poorter and Dalziell, 1996) (ATCPs, 2005).

"Die in beauty": Passive preservation, letting nature take its course without interfering in order to prevent decay. (Prestvold, 2003:11). *Managed decay* - This involves extensive documentation, removal of hazardous material, like poisonous chemicals, followed by minimal intervention (Laser, 2006:21). *Negative management* – A form of management that implies consciously doing nothing (Howard, 2003:17).

Dissonance/dissonant heritage: Discordance or lack of agreement and consistency as to the meaning of heritage (Tunbridge and Ashworth, 1996, quoted in Graham *et* al., 2005:23).

Disturbance: Movement and damage of artefacts in archaeological context as the by-product of other activities (Rathje and Schiffer, 1982).

Ecofact: Any flora or fauna material found at an archaeological site; nonartifactual evidence that has not been technologically altered but that has cultural relevance, such as a shell carried from the ocean to an inland settlement (Kipfer, 2001).

Environmental formation processes: Biological, chemical, and physical aspects of the environment responsible for forming and modifying the archaeological record (Rathje and Schiffer, 1982).

1 This is the definition used by the International Association for Impact Assessment, <http://www.iaia.org>, which quotes as a source Vanclay and Bronstein (1995).

Equifinality: In taphonomy, the process of reaching the same final state from different initial conditions and in different ways (Lyman, 2004).

Exposure: The process of interaction between an identified potential output and an environmental element or value. (ATCPs, 2005)

Footprint: In Antarctica, spatial extent of disturbance (Jabour, 2009). The overall concept of human footprint in the Antarctic can be characterised by a "core" area of human presence and a surrounding "buffer" zone extending out to the distance at which the visually dominant core area or activity is no longer visible (New Zealand, 2010).

Historic: Beyond a certain age (Carman, 2002:45).

Historical: Associated with a particular historic event or person (Carman, 2002:45).

Integrity: See *condition*.

Impact/s: (1) (IA) The difference between what would happen with the action and what would happen without it. (IAIA, 2009: 1). (2) The changes that archaeological resources undergo as the result of some action (Rathje and Schiffer, 1982).

Impact, direct: A change in environmental components that results from direct cause-effect consequences of interaction between the exposed environment and outputs (ATCPs, 2005).

Impact, indirect: A change in environmental components that results from interactions between the environment and other impacts – direct or indirect. (Impact assessment) (ATCPs, 2005).

Impact, unavoidable: An impact for which no further mitigation is possible. (ATCPs, 2005)

Marker: Information about a [tourist] sight (MacCannell, 1976: 110). Conventionally this is only information attached or posted alongside the sight. However, MacCannell uses it to refer to any information about the sight; distinguishes off-site and on-site markers, whether information is found at the sight or separate from the sight; and separates the information itself from the vehicle in which it is contained, such as a plaque.

Mitigation: The use of practice, procedure or technology to minimise or to prevent impacts associated with proposed activities. (ATCPs, 2005).

Monitoring: Consists of standardised measurements or observations of key parameters (outputs and environmental variables) over time, their statistical evaluation and reporting on the state of the environment in order to define quality and trends. (adapted from SCAR/COMNAP Monitoring Workshop) (ATCPs, 2005).

Monument: "A large fixed structure...fixed landscape features, where the emphasis is placed on the external surface...whereas buildings have interior and exteriors and the focus is upon their function and sites have depth in the form of stratigraphy, monuments have solidity and shape" (Carman, 2002: 46).

Natural environment: Climate, terrain, geological resources, and flora and fauna (Rathje and Schiffer, 1982).

Output: A physical change or an entity imposed on or released to the environment as a result of an action or an activity. (ATCPs, 2005)

Palimpsest: A superimposition of successive activities, the material traces of which are partially destroyed or reworked because of the process of superimposition (Bailey, 2007: 203).

Periglacial: Processes, climates, and topographic features that are either in the immediate margins of former or existing glaciers or are influenced by cold temperatures (Rapp and Hill, 2006: 389).

Permafrost: Soil, regolith, and bedrock at a temperature below 0°C (32°F) found in cold climates (Strahler, 1996, [2011]).

Pole: Point at which the end of the Earth's axis of rotation intersects the Earth's surface (Strahler, 1996, [2011]).

Preservation "...implies actions and perhaps even interventions to maintain the object as it was found. Indeed, "preserve as found" is a key phrase in many areas of cultural heritage." (Howard, 2003:199).

Protection "...is an all-encompassing and largely passive term. All sorts of actions and interventions in the area, building or object might take place within the broad term "protection". A protected area may exclude people, or protect from building development without any active management or intervention." (Howard, 2003:199).

Reclamation: Activities that move objects from archaeological context back into systemic context (Rathje and Schiffer, 1982).

Reuse: Change in the user or use of an object (Rathje and Schiffer, 1982).

Site: A place that has material remains of human activities (Rathje and Schiffer, 1982).

Spitsbergen (Spitzbergen, Svalbard): A high Arctic archipelago, which according to the 1920 Treaty Concerning the Archipelago of Spitsbergen comprises all the islands situated between 74° to 81° N and 10° to 35° E. In Norwegian terminology Spitsbergen is the name of West Spitsbergen, the largest island, and the archipelago itself is referred to by the name of Svalbard.

Stabilization: The arrest of the processes of decay (ICOMOS NZ, 1993).

Svalbard: See Spitsbergen.

Systemic context: The state of artefacts as part of an adaptive system or society (Rathje and Schiffer, 1982).

Time capsule: Buried in foundations of new buildings etc. and containing objects typical of the present time, for discovery in the future. (The Concise Oxford Dictionary New Edition 1976 – print 1979).

Use-wear: The gradual attrition or accumulation of materials that occurs on an artifact during use (Rathje and Schiffer, 1982).

Weathering: The alteration of materials by environmental processes (Rathje and Schiffer, 1982).

Wear and tear: Damage sustained as result of ordinary use. (The Concise Oxford Dictionary New Edition, 1976, print 1979).

Appendix 3 – Cultural heritage in the 2001 Svalbard Act

ACT OF 15 JUNE 2001 NO.79 RELATING TO THE PROTECTION OF THE ENVIRONMENT IN SVALBARD

CHAPTER V. THE CULTURAL HERITAGE

§ 38 (fundamental principle)

Structures and sites and movable historical objects in Svalbard shall be protected and safeguarded as a part of Svalbard›s cultural heritage and identity and as an element of a coherent system of environmental management.

§ 39 (protected elements of the cultural heritage)

The following are protected:

a. structures and sites dating from 1945 or earlier;
b. movable historical objects dating from 1945 or earlier that come to light by chance or through investigations, excavation or in any other way;
c. elements of the cultural heritage dating from after 1945 that are of particular historical or cultural value and that are protected by a decision of the directorate.

Evidence of human graves of all kinds, including crosses and other grave markings as well as bones and bone fragments found on or below the surface of the ground and skeletal remains at slaughtering sites for walruses and whales, and associated with spring-guns for polar bears, are considered to be structures and sites and are protected irrespective of their age.

Around automatically protected structures and sites such as are mentioned in the first paragraph, litra a, or in the second paragraph, a security zone shall extend for 100 metres in all directions from the visible or known perimeter, unless the Governor prescribes that it shall be delimited differently.

In cases of doubt, the directorate will determine what is a protected structure or site or movable historical object, and it may rescind the protection of specified elements of the cultural heritage.

§ 40 (decisions on temporary protection)

When special reasons so indicate, the Governor may decide to grant temporary protection to structures and sites or movable historical objects such as are mentioned in section 39, first paragraph, litra c, until such time as the question of protection has been finally decided.

§ 41 (administrative procedure for protection decisions)

Decisions pursuant to sections 39 and 40 are individual decisions under the Public Administration Act. Such decisions shall be registered judicially and announced as the Governor considers appropriate.

§ 42 (the substance of protection)

No person may damage, dig up, move, remove, alter, cover up, conceal or disfigure protected structures and sites or movable historical objects, including any security zone, or initiate measures that may entail a risk of this happening.

In the security zone it is not permitted to erect tents, light fires or undertake similar activities.

The Governor may regulate or prohibit access and passage within the security zone if this in itself or over a period of time could undermine the purposes of protection.

§ 43 (ownership rights to protected movable historical objects)

Protected movable historical objects are State property when it is clear that there is no reasonable possibility of establishing ownership or tracing the owner. The directorate may hand over all or part of a find to the finder.

§ 44 (special duty of a head of undertaking to protect the cultural heritage)

When an undertaking is being planned, the head of undertaking shall at his own expense investigate whether the undertaking will affect structures and sites in one of the ways mentioned in section 42. If it is found that the undertaking will affect structures and sites, sections 57 and 58 will apply. The directorate will decide as soon as possible whether the undertaking may be carried out, and if so in what way.

If it only becomes apparent after work has begun that it may affect protected structures and sites in contravention of section 42, the Governor shall be notified pursuant to section 62, second paragraph, and the work halted. The directorate will decide as soon as possible whether the work may proceed and if so under what conditions.

Finds that are made during investigations or excavation or during an undertaking shall be documented and conserved at the expense of the head of undertaking unless otherwise decided pursuant to section 98. Finds shall be delivered within one year unless this time limit is extended by the Governor.

§ 45 (investigation, management and maintenance)

The directorate has the right to search for, excavate, examine and document protected elements of the cultural heritage. After such measures have been completed, structures and sites shall be restored to their previous condition unless otherwise determined by the directorate. The Governor may raise movable historical objects and take the necessary steps to preserve them.

The Governor may put in order or enclose protected structures and sites or take steps to look after and display them, including clearing the surrounding area, for example to avoid injury to humans and animals. Such measures may also be implemented in the security zone. Measures over and above ordinary maintenance, such as restoration, rebuilding or removal, require approval by the directorate.

Before measures are taken in respect of structures and sites, the owner or holder of rights shall be notified and given the opportunity to express an opinion pursuant to the provisions of section 16 of the Public Administration Act. Measures relating to structures and sites that are in use may only be implemented with the consent of the owner or user unless they are necessary in order to prevent decay.
If there is a danger of decay, the directorate may order the owner or user to carry out measures to counteract this.

§ 46 (export and return of movable historical objects)

No person may export from Svalbard protected movable historical objects that were found in or originate in Svalbard.

The prohibition of the first paragraph does not apply when it has been decided in or pursuant to regulations that finds shall be permanently or temporarily stored outside Svalbard.
In special cases, the directorate may issue permits for export.

The provisions of sections 23a - 23f of the Act of 9 June 1978 No. 50 concerning the Cultural Heritage, in so far as they relate to the return of unlawfully exported cultural objects, shall apply to Svalbard to the extent that they are appropriate.

Appendix 4 – Cultural heritage in the 1991 Antarctic Protocol

Annex V to the Protocol on Environmental Protection to the Antarctic Treaty: Protected Areas, Article 8 – Historic Sites and Monuments

1. Sites or monuments of recognized historic value which have been designated as Antarctic Specially Protected Areas or Antarctic Specially Managed Areas, or which are located within such Areas, shall be listed as Historic Sites and Monuments.
2. Any Party may propose a site or monument of recognized historic value which has not been designated as an Antarctic Specially Protected Area or an Antarctic Specially Managed Area, or which is not located within such an Area, for listing as a Historic Site or Monument. The proposal for listing may be approved by the Antarctic Treaty Consultative Parties by a measure adopted at an Antarctic Treaty Consultative Meeting in accordance with Article IX(1) of the Antarctic Treaty. Unless the measure specifies otherwise, the proposal shall be deemed to have been approved 90 days after the close of the Antarctic Treaty Consultative Meeting at which it was adopted, unless one or more of the Consultative Parties notifies the Depositary, within that time period, that it wishes an extension of that period or is unable to approve the measure.
3. Existing Historic Sites and Monuments which have been listed as such by previous Antarctic Treaty Consultative Meetings shall be included in the list of Historic Sites and Monuments under this Article.
4. Listed Historic Sites and Monuments shall not be damaged, removed or destroyed.
5. The list of Historic Sites and Monuments may be amended in accordance with paragraph 2 above. The Depositary shall maintain a list of current Historic Sites and Monuments.

Appendix 5 – List of interviewees

Date	Location	Description
21 March 2005	Ushuaia	Andrea Barros, Infuetur
21 March 2005	Ushuaia	Eric Bardie, yacht tour operator (SV *Phillis*)
22 March 2005	Ushuaia	Daniel Martinelli, tourism agent/IAATO
22 & 23 March 2005	Ushuaia	María Elena Daverio and Marie Jensen, tourism researchers, Universidad Nacional de la Patagonia
23 March 2005	Ushuaia	Gustavo Martínez - Tour operator (Antarpply)
23 March 2005	Ushuaia	Marisol Vereda, tourism researcher, Universidad Nacional de la Patagonia
28 & 30 March 2005	Buenos Aires	Rodolfo Sánchez, Patricia Ortúzar, environmental managers, Instituto Antártico Argentino
29 March 2005	Buenos Aires	Daniel Leguizamón, Secretaría de Turismo (Ministerial adviser; Director of Infuetur 1992-2000)
29 March 2005	Buenos Aires	Ariel Mansi, Ministry of Foreign Affairs
30 March 2005	Buenos Aires	Mariano Memolli - Director, Dirección Nacional del Antártico
31 March 2005	Buenos Aires	María dos Santos Afonso, Universidad de Buenos Aires
01 April 2005	Buenos Aires	Hugo Casela - Argentine Army's Antarctic OIC
05 April 2005	Buenos Aires	Javier García Cano, Universidad de Buenos Aires, Argentina's IPHC representative
06 April 2005	Buenos Aires	Pablo Wainschenker, documentary maker, tour guide
22 June 2005	Oslo	Susan Barr, Director IPHC
12 December 2005	Cambridge	John Shears and colleagues, BAS
07 July 2006	Christchurch, NZ	Neil Gilbert, Antarctica NZ
07 July 2006	Christchurch, NZ	Nigel Watson, NZAHT
18 September 2006	Longyearbyen Svalbard	Rune Bergstrom, Head of Environmental Department, Sysselmannen på Svalbard

Date	Location	Description
21 September 2006	Longyearbyen Svalbard	Karl Våtvik, Wilderness Centre, Svalbard
21 September 2006	Longyearbyen Svalbard	Sander Solnes, Conservator Svalbard Museum
22 September 2006	Longyearbyen Svalbard	Frigg Jørgensen, Executive Secretary, AECO
22 September 2006	Longyearbyen Svalbard	Per Kyrre Reymert, Cultural Heritage Adviser, Sysselmannen på Svalbard
22 September 2006	Longyearbyen Svalbard	Stefano Poli, tour operator (Poli Artici)
22 September 2006	Longyearbyen Svalbard	Stein Tore Pedersen, Executive Officer Tourism, Sysselmannen på Svalbard
22 September 2006	Longyearbyen Svalbard	Tove Eliassen, Daglig Leder, Svalbard Reiseliv
10 May 2007	New Delhi	Jorge Berguño, Instituto Antártico Chileno & Chile's Head of Delegation to the ATCM
19 June 2007	NW Spitsbergen	Rolf Stanger, Expedition Leader, Oceanwide Expeditions
30 June & 1 July 2007	Ny-Ålesund	Asbjørg Skogmo, Oskar Bye, Lilly Kristensen, mining veterans & leading members of the Kings Bay Veterans Association; Kjell Petter Stene, journalist
11 July 2007	Ny-Ålesund	Neil Cox, officer in charge, BAS station
11 July 2007	Ny-Ålesund	Bodil Paulsen, Kings Bay AS
12 July 2007	Longyearbyen Svalbard	Per Kyrre Reymert, Cultural Heritage Adviser, Sysselmannen på Svalbard
10 July 2008	St. Petersburg	Maria Gavrilo, Russian Antarctic Expedition

De "voetafdruk" van het pooltoerisme: het gedrag van toeristen op cultuurhistorische plaatsen op Antarctica en Spitsbergen

R. Roura

SAMENVATTING

1 INLEIDING

Dit onderzoek beschrijft en interpreteert de invloed van het pooltoerisme op historische plaatsen in Antarctica en Svalbard (ook bekend onder de naam Spitsbergen) en beoordeelt de implicaties hiervan voor het beleid. Ontdekkingsreizigers, walvisvaarders, zeehondenjagers, wetenschappers en anderen hebben veel materiële overblijfselen achtergelaten in de poolgebieden die van veel betekenis zijn omdat ze de geschiedenis van de exploratie en exploitatie van deze gebieden vertellen. Het hedendaagse pooltoerisme vertegenwoordigt een nieuwe fase in deze "exploratie" en exploitatie van de poolgebieden. Het potentieel voor de transformatie van historische plaatsen is toegenomen als gevolg van de aanzienlijke groei van dit pooltoerisme in de afgelopen decennia. Belangrijke cultuurhistorische plaatsen worden regelmatig opgenomen in de gebruikelijke toeristische routes en zijn ook onderwerp van gespecialiseerd toerisme. In deze context is de centrale onderzoeksvraag van dit proefschrift:

> *Wat zijn de effecten van het toerisme op historische plaatsen in de poolgebieden en wat zijn de implicaties hiervan voor het toeristisch beleid en voor het beheer van deze historische plaatsen?*

2 CONCEPTUEEL KADER

In deze studiele wordt het toerisme onderzocht vanuit het perspectief van de gedragsarcheologie zoals die door Schiffer (2010) is omschreven. Het woord "archeologie" doet misschien denken aan het opgraven van de overblijfselen van vroegere culturen. De gedragsarcheologie werd echter bedacht als een instrument om de interacties tussen mensen en materialen te kunnen bestuderen in alle tijden en plaatsen, inclusief aspecten van processen van landschapsformatie. Door menselijk gedrag ontstaat er interactie (akoustisch, chemisch, mechanisch, thermisch en visueel) tussen de mensen (toeristen) en hun artefacten (bijvoorbeeld laarzen, wandelstokken en fototoestellen) aan de ene kant en de externe omgeving (zowel natuurlijke als culturele) aan de andere. De veranderingen van de omgeving die het gevolg zijn van deze interacties kunnen ook worden beschouwd als "artefacten" van het toerisme. In termen van beleid zouden, deze toeristische "artefacten" beschouwd kunnen worden als "impacts", afhankelijk van een waardeoordeel over hun betekenis.

Dit conceptuele hoofdkader wordt aangevuld met twee theorieën. Het scheppen van toeristische attracties wordt geanalyseerd op basis van Leiper (1990). De benoeming van materiële overblijfselen van vroegere activiteiten als "erfgoed" wordt besproken vanuit het perspectief van de Rubbish Theory van Thompson (1979).

3 ONDERZOEKSMETHODEN

Dit proefschrift onderzoekt *case studies* van of Antarctica of Svalbard, of van beide regio's. Tussen 2005 en 2008 werden vier veldwerkonderzoeken uitgevoerd in Svalbard (twee onderzoeksperiodes) en Antarctica (twee onderzoeksperiodes). Vroeger veldwerk in Antarctica is ook opgenomen als onderdeel van dit onderzoek. De toegepaste onderzoeksmethoden zijn: site verkenning, etnografische waarnemingen en fotografische registratie. Aanvullende onderzoeksmethoden waren semi-gestructureerde interviews; observatie van deelnemers; de studie van toeristische weblogs; en analyse van toeristische statistieken. De gegevens die met behulp van de verschillende methodes werden verkregen werden geïntegreerd om tot beoordelingen van de diverse case studies te komen.

4 HOOFD THEMA'S

Dit onderzoek heeft verscheidene samenhangende thema's betreffende toerisme en cultuurhistorische plaatsen in de poolgebieden onderzocht. Tegelijkertijd wordt erkend dat er andere relevante thema's zijn die hier niet behandeld worden.De thema's die in dit proefschrift behandeld worden zijn:

Toerisme als gedrag: Toerisme speelt zich af in een hiërarchie van gedrags-schalen – van toerisme als geheel tot individuele bezoekers. De voornaamste processen in deze verschillende schalen omvatten de werving en consolidatie van bepaalde plaatsen als toeristische bestemmingen; de gedragspatronen die bezoekers vertonen, zowel groepsgewijs als individueel, tijdens bezoeken aan land; en afzonderlijke interacties tussen bezoekers en verschillende soorten plaatsgebonden elementen. De effecten van toerisme op historische locaties zijn het gevolg van interacties op al deze niveaus.Het toeristisch gedrag omvat een scala aan handelingen die kunnen leiden tot interacties tussen toeristen, hun artefacten, en de historische plaatsen en resulteren in de creatie van toeristische "artefacten". Visuele en gehoorinteracties (instructies, verhalen) kunnen toeristen aantrekken of juist afstoten van de elementen die deel uitmaken van een historische plaats. In termen van gedrag kan een patroon van bezoekersactiviteit resulteren in een overeenkomstige "archeologie" van onmiddellijke en cumulatieve sporen van bezoek die een toeristisch activiteitsgebied bepalen dat opgelegd wordt op een gebied van historische activiteit. Deze twee gebieden kunnen elkaar overlappen maar zijn niet noodzakelijk dezelfde. De toestand en de samenhang van de historische plaatsen kunnen veranderen door vertrapping en andere handelingen. Hierdoor kan schade aan het archeologische bodemarchief ontstaan.

Toerisme als proces van plaatsvorming, in de context van andere (natuurlijke en culturele) vormingsprocessen: Historische plaatsen in de poolgebieden zijn niet statisch maar onderhevig aan verandering en verval als gevolg van zowel culturele en non-culturele (natuurlijke) processen. Toerisme kan bijdragen aan dit proces. Recente veranderingen opeenvolgend in de tijd werden waargenomen in een aantal case studies van plaats elementen. Geen enkel element ging aanzienlijk achteruit tijdens de periode van controle; integendeel, verscheidene elementen werden op verschillende manieren hersteld. Veranderingen werden uitgelegd als een gevolg van een scala van culturele processen (inclusief beheer, onderzoek en toerisme) en natuurlijke processen (voornamelijk de activiteit van de wind). Lokale veranderingen vinden plaats in de context van bredere regionale ontwikkelingen in Antarctica en Svalbord.

Het aanduiden van plaatsen van vroegere activiteit als "cultureel erfgoed": De vraag of een verandering aan iets als impact beschouwd moet worden berust op een oordeel over de waarde van hetgeen wordt beïnvloed. In zowel Antarctica als Svalbord worden de criteria en processen die gebruikt worden om de plaatsen en overblijfselen van activiteiten uit het verleden ("historic sites") te classificeren als "erfgoed" door de wet bepaald. De aanwijzing van historische plaatsen als erfgoed heeft zowel milieutechnische als geopolitieke implicaties.

Het gebruik van cultureel erfgoed als toeristische attractie: Verhalen die verteld worden aan toeristen spelen een centrale rol in het proces van betekenisgeving en het toekennen van waarde aan historische overblijfselen. Dit helpt bij het creëren van aantrekkelijke toeristische producten en toeristische attracties. Informatie omtrent de plaatsen beïnvloedt het gedrag van de toerist en dus ook de mate van blootstelling of bescherming van de plaatsen. Toeristische verhalen maken het mogelijk voor de verschillende elementen van de toeristische attractie om samen te vallen en zo een geheel te vormen, met name op plaatsen die omwille hun oorspronkelijke functie of huidige toestand weinig van hun geschiedenis prijsgeven.

Onderzoeksmethoden om de wisselwerking van het toerisme met historische sites te bestuderen: De methoden die voor dit onderzoek geselecteerd werden maakten de tot standkoming van een snelle en redelijk geïnformeerde beoordeling van de status, recente wijzigingen, toeristische modaliteiten en eisen van beheer van de onderzochte plaatsen mogelijk.Een snelle, low-tech benadering is geschikt voor de poolgebieden waar bezoeken aan de sites zeldzaam, opportunistisch en kort kunnen zijn.Natuurlijk is het mogelijk dat uitgebreidere evaluatie methoden vereist zullen zijn voor andere (of meer specifieke) onderzoeks- of beheersbehoeften.

Het beheer van toeristische en historische plaatsen: Toerisme heeft de neiging om zich te concentreren op belangrijke elementen van een plaats met als gevolg dat het een grotere impact kan hebben op elementen die een grotere relatieve waarde hebben dan de rest van de site. De invloeden van toerisme zijn over het algemeen gelokaliseerd binnen sites, wat weerspiegeld kan worden in lokale vormen van beheer. Cumulatieve impacts zijn een mogelijke reden voor bezorgdheid. Enkele toegewijde beheersinstrumenten kunnen nodig zijn om o.a. het volgende te reguleren: de bewegingen van het verkeer van personen binnen historische plaatsen; minimum afstanden van kwetsbare elementen ; het gebruik van vaste paden en het bepalen van grenzen aan bezoekers aantallen.

5 CONCLUSIES

De toeristische “voetafdruk” bestaat uit een verzameling van gedrags-sporen die zich ontwikkelen op verschillende tijdelijke, ruimtelijke en gedragsmatige niveaus.

Toerisme vindt niet plaats op onveranderlijke plaatsen die “bevroren in de tijd” zijn, maar veeleer op plaatsen die onderhevig zijn aan verval. Het is er onderdeel van en draagt bij aan dit proces. Het toerisme kan de sporen van vroegere activiteiten uitwissen, zonder dat ze totaal verdwijnen. Het heeft minder effect op de historische plaatsen dan natuurlijke processen, maar het heeft de neiging zich te richten op de kwetsbare waardebepalende elementen van die plaats.

Toerisme kan effect hebben op de cultuurhistorische waarde van historische plaatsen. Dit effect moet worden beperkt ondanks het feit dat het een onderdeel is van het proces van verval van die historische plaatsen. Dit kan door bewaking van het verval en het opstellen van een milieueffect rapportage; door de inzet van gediplomeerde gidsen met verantwoordelijkheid voor regelgeving en door speciale voorzorgsmaatregelen.

Referentie:

Schiffer, M.B. (2010): *Behavioral archaeology: Principles and practice*. London: Equinox.

9 789077 922873
13W31285/ T3/ 9789077922873